Wood–polymer composites

Related titles:

Polymer nanocomposites
(ISBN 978-1-85573-969-7)
Polymer nanocomposites are a class of reinforced polymers with low quantities of nanometric-sized clay particles which give them improved barrier properties, fire resistance and strength. Such properties have made them valuable in components such as panels and as barrier and coating materials in automobile, civil and electrical engineering as well as packaging. *Polymer nanocomposites* provides a comprehensive review of the main types of polymer nanocomposite and their properties.

Design and manufacture of textile composites
(ISBN 978-1-85573-744-0)
This book brings together the design, manufacture and applications of textile composites. The term 'textile composites' is often used to describe a rather narrow range of materials, based on three-dimensional reinforcements produced using specialist equipment. The intention here though is to describe the broad range of polymer composite materials with textile reinforcements, from woven and non-crimp commodity fabrics to 3D textiles. Whilst attention is given to modelling of textile structures, composites manufacturing methods and subsequent component performance, it is substantially a practical book intended to help all those developing new products with textile composites.

Green composites: Polymer composites and the environment
(ISBN 978-1-85573-739-6)
There is an increasing movement of scientists and engineers dedicated to minimising the environmental impact of polymer composite production. Life-cycle assessment is of paramount importance at every stage of a product's life, from initial synthesis through to final disposal and a sustainable society needs environmentally safe materials and processing methods. With an internationally recognised team of authors, *Green composites* examines polymer composite production and explains how environmental footprints can be diminished at every stage of the life cycle. This book is an essential guide for agricultural crop producers, governmental agricultural departments, automotive companies, composite producers and material scientists dedicated to the promotion and practice of eco-friendly materials and production methods.

Details of these and other Woodhead Publishing books, as well as books from Maney Publishing, can be obtained by:

- visiting our web site at www.woodheadpublishing.com
- contacting Customer Services (e-mail: sales@woodhead-publishing.com; fax: +44 (0) 1223 893694; tel.: +44 (0) 1223 891358 ext. 130; address: Woodhead Publishing Limited, Abington Hall, Granta Park, Great Abington, Cambridge CB21 6AH, England)

If you would like to receive information on forthcoming titles, please send your address details to: Francis Dodds (address, tel. and fax as above; e-mail: francisd@woodhead-publishing.com). Please confirm which subject areas you are interested in.

Maney currently publishes 16 peer-reviewed materials science and engineering journals. For further information visit www.maney.co.uk/journals.

Wood–polymer composites

Edited by
Kristiina Oksman Niska and Mohini Sain

Woodhead Publishing and Maney Publishing
on behalf of
The Institute of Materials, Minerals & Mining

CRC Press
Boca Raton Boston New York Washington, DC

WOODHEAD PUBLISHING LIMITED
Cambridge England

Woodhead Publishing Limited and Maney Publishing Limited on behalf of
The Institute of Materials, Minerals & Mining

Published by Woodhead Publishing Limited, Abington Hall, Granta Park,
Great Abington, Cambridge CB21 6AH, England
www.woodheadpublishing.com

Published in North America by CRC Press LLC, 6000 Broken Sound Parkway, NW,
Suite 300, Boca Raton, FL 33487, USA

First published 2008, Woodhead Publishing Limited and CRC Press LLC
© 2008, Woodhead Publishing Limited
The authors have asserted their moral rights.

This book contains information obtained from authentic and highly regarded sources. Reprinted material is quoted with permission, and sources are indicated. Reasonable efforts have been made to publish reliable data and information, but the authors and the publishers cannot assume responsibility for the validity of all materials. Neither the authors nor the publishers, nor anyone else associated with this publication, shall be liable for any loss, damage or liability directly or indirectly caused or alleged to be caused by this book.

Neither this book nor any part may be reproduced or transmitted in any form or by any means, electronic or mechanical, including photocopying, microfilming and recording, or by any information storage or retrieval system, without permission in writing from Woodhead Publishing Limited.

The consent of Woodhead Publishing Limited does not extend to copying for general distribution, for promotion, for creating new works, or for resale. Specific permission must be obtained in writing from Woodhead Publishing Limited for such copying.

Trademark notice: Product or corporate names may be trademarks or registered trademarks, and are used only for identification and explanation, without intent to infringe.

British Library Cataloguing in Publication Data
A catalogue record for this book is available from the British Library.

Library of Congress Cataloging in Publication Data
A catalog record for this book is available from the Library of Congress.

Woodhead Publishing Limited ISBN 978-1-84569-272-8 (book)
Woodhead Publishing Limited ISBN 978-1-84569-457-9 (e-book)
CRC Press ISBN 978-1-4200-7611-0
CRC Press order number WP7611

The publishers' policy is to use permanent paper from mills that operate a sustainable forestry policy, and which has been manufactured from pulp which is processed using acid-free and elementary chlorine-free practices. Furthermore, the publishers ensure that the text paper and cover board used have met acceptable environmental accreditation standards.

Project managed by Macfarlane Book Production Services, Dunstable, Bedfordshire, England (e-mail: macfarl@aol.com)
Typeset by Godiva Publishing Services Limited, Coventry, West Midlands, England
Printed by TJ International Limited, Padstow, Cornwall, England

Contents

	Contributor contact details	xi
	Introduction K OKSMAN NISKA, Luleå University of Technology, Sweden and M SAIN, University of Toronto, Canada	**xv**
1	**Raw materials for wood–polymer composites** C CLEMONS, USDA Forest Service, USA	**1**
1.1	Introduction	1
1.2	Polymers: structure and properties	2
1.3	Wood: structure and properties	10
1.4	Sources of further information and advice	20
1.5	References and further reading	20
2	**Additives for wood–polymer composites** D V SATOV, Canada Colors and Chemicals Limited, Canada	**23**
2.1	Introduction	23
2.2	Lubricants and rheology control additives for thermoplastic composites	26
2.3	Coupling agents	29
2.4	Stabilizers	31
2.5	Fillers	33
2.6	Density reduction additives	36
2.7	Biocides	36
2.8	Product aesthetics additives	37
2.9	Flame retardants and smoke suppressants	38
2.10	Future trends	40
2.11	Conclusion	40

3	Interactions between wood and synthetic polymers	41

K OKSMAN NISKA, Luleå University of Technology, Sweden and A R SANADI, University of Copenhagen, Denmark

3.1	Introduction	41
3.2	The interface and interphase in composites	42
3.3	Wetting, adhesion and dispersion	43
3.4	Techniques to evaluate interfacial interactions and adhesion	48
3.5	Improving interface interactions in wood–polymer composites	60
3.6	Interphase effects on other properties	66
3.7	Conclusions	68
3.8	References and further reading	69

4	Manufacturing technologies for wood–polymer composites	72

D SCHWENDEMANN, Coperion Werner & Pfleiderer GmbH & Co. KG, Germany

4.1	Introduction	72
4.2	Raw material handling	72
4.3	Compounding technologies	79
4.4	Pelletising systems	90
4.5	Profile extrusion	95
4.6	Injection moulding	95
4.7	Sheet extrusion	98
4.8	Future trends	100
4.9	References	100

5	Mechanical properties of wood–polymer composites	101

M SAIN and M PERVAIZ, University of Toronto, Canada

5.1	Introduction	101
5.2	Mechanical performance of wood–polymer composites	101
5.3	General mechanical properties of wood–polymer composites and test methods	104
5.4	Critical parameters affecting mechanical properties of wood–polymer composites	109
5.5	Conclusions	116
5.6	References	116

6	Micromechanical modelling of wood–polymer composites	118

R C NEAGU, Ecole Polytechnique Fédérale de Lausanne (EPFL), Switzerland and E K GAMSTEDT, Kungliga Tekniska Högskolan (KTH), Sweden

6.1	Introduction	118
6.2	Elastic properties	119
6.3	Hygroexpansion	131
6.4	Strength	134
6.5	Conclusions	138
6.6	References	138

7	Outdoor durability of wood–polymer composites	142

N M STARK, USDA Forest Service, USA and D J GARDNER, University of Maine, USA

7.1	Introduction	142
7.2	Characteristics of raw materials	142
7.3	Changes in composite properties with exposure	145
7.4	Methods for protection	155
7.5	Future trends	161
7.6	Sources of further information and advice	162
7.7	References and further reading	162

8	Creep behavior and damage of wood–polymer composites	166

N E MARCOVICH and M I ARANGUREN, Universidad Nacional de Mar del Planta, Argentina

8.1	Introduction	166
8.2	Viscoelasticity and creep	167
8.3	Creep in wood–plastic composites	176
8.4	Creep failure and material damage	183
8.5	Conclusions and future trends	185
8.6	References	186

9	Processing performance of extruded wood–polymer composites	190

K ENGLUND and M WOLCOTT, Washington State University, USA

9.1	Introduction	190
9.2	Current extrusion processing methods for natural fiber–thermoplastic composites	191
9.3	Rheology of a wood fiber-filled thermoplastic	193

viii   Contents

9.4	Commercial wood–polymer composites	197
9.5	References	207

10   Oriented wood–polymer composites and related materials   208
F W MAINE, Frank Maine Consulting Ltd., Canada

10.1	Introduction	208
10.2	Orientation of polymers	208
10.3	Applications	212
10.4	Current developments	219
10.5	Future trends	225
10.6	References	225

11   Wood–polymer composite foams   227
G GUO, University of Southern California, USA, G M RIZVI, University of Ontario Institute of Technology, Canada and C B PARK, University of Toronto, Canada

11.1	Introduction	227
11.2	Structure and characterization of wood–polymer composite foams	229
11.3	Critical issues in production of wood–polymer composite foams	231
11.4	Fundamental mechanisms in blowing agent-based foaming of wood–polymer composites	235
11.5	Foaming of wood–polymer composites with chemical blowing agents	239
11.6	Foaming of wood–polymer composites with physical blowing agents	244
11.7	Foaming of wood–polymer composites with heat expandable microspheres	249
11.8	Void formation in wood–polymer composites using stretching technology	250
11.9	Effects of additives on wood–polymer composite foams	250
11.10	Summary and future trends	252
11.11	References	253

12   Performance measurement and construction applications of wood–polymer composites   257
R J TICHY, Washington State University, USA

12.1	Introduction	257
12.2	Performance measures and building codes	259
12.3	Wood–polymer composite properties	260
12.4	Building construction applications	265

| 12.5 | Conclusions | 270 |
| 12.6 | References | 271 |

13 Life-cycle assessment (LCA) of wood–polymer composites: a case study   273
T THAMAE and C BAILLIE, Queens University, Canada

13.1	Introduction: comparing wood–polymer and glass-fiber reinforced polypropylene car door panels	273
13.2	The life-cycle assessment process	274
13.3	Goal and scope definition	276
13.4	Inventory	282
13.5	Impact assessment	285
13.6	Interpretation	291
13.7	The possible effect of European Union legislation on end-of-life vehicles	295
13.8	Conclusions	296
13.9	Acknowledgements	297
13.10	References	297

14 Market and future trends for wood–polymer composites in Europe: the example of Germany   300
M CARUS and C GAHLE, nova-Institut, Germany and H KORTE, Innovationsberatung Holz & Fasern, Germany

14.1	Introduction	300
14.2	The development of the European market: the example of Germany	301
14.3	The most significant wood–polymer composite products in the European market	304
14.4	Future trends: markets	309
14.5	Future trends: processing and materials	311
14.6	Conclusions	316
14.7	Wood–polymer composite codes, standards, research and manufacturing in Europe	317
14.8	The nova-Institut and Innovationsberatung Holz und Fasern	322
14.9	Examples of wood–polymer composite products	325
14.10	References	329

15 Improving wood–polymer composite products: a case study   331
A A KLYOSOV, MIR International Inc., USA

| 15.1 | Introduction: wood–polymer composite decking | 331 |
| 15.2 | Brands and manufacturers | 332 |

15.3	Improving the performance of wood–polymer composite decking	333
15.4	Conclusions	352
15.5	References	353
	Index	354

Contributor contact details

(* = main contact)

Introduction

K. Oksman Niska*
Division of Manufacturing and
  Design of Wood and
  Bionanocomposites
Luleå University of Technology
Forskargatan 1
931 87 Skellefteå
Sweden
E-mail: kristiina.oksman@ltu.se

M. Sain
Faculty of Forestry
University of Toronto
33 Willcocks Street
Toronto
Ontario M5S 3B3
Canada
E-mail: m.sain@utoronto.ca

Chapter 1

C. Clemons
USDA Forest Service
Forest Products Laboratory
One Gifford Pinchot Drive
Madison, WI 53705-2398
USA
E-mail: cclemons@fs.fed.us

Chapter 2

D. Victor Satov
Canada Colors and Chemicals Ltd
Toronto
Canada
E-mail: vsatov@canadacolors.com

Chapter 3

K. Oksman Niska*
Division of Manufacturing and Design
  of Wood and Bionanocomposites
Luleå University of Technology
Forskargatan 1
931 87 Skellefteå
Sweden
E-mail: kristiina.oksman@ltu.se

A. R. Sanadi
Forest and Landscape
University of Copenhagen
Denmark

Chapter 4

D. Schwendemann
Coperion Werner & Pfleiderer GmbH
  & Co. KG
BU 2.1 LFT
Theodorstrasse 10
70469 Stuttgart
Germany
E-mail: Daniel.Schwendemann@
  coperion.com

Contributor contact details

Chapter 5
M. Sain and M. Pervaiz
Faculty of Forestry
University of Toronto
33 Willcocks Street
Toronto
Ontario M5S 3B3
Canada
E-mail: m.sain@utoronto.ca

Chapter 6
R. C. Neagu*
Laboratoire de Technologie des
  Composites et Polymères (LTC)
Ecole Polytechnique Fédérale de
  Lausanne (EPFL)
EPFL-STI-IMX-LTC
MXG 137 (Bâtiment MXG)
Station 12
CH-1015 Lausanne
Switzerland
E-mail: cristian.neagu@epfl.ch

E. K. Gamstedt
Department of Fibre and Polymer
  Technology
Kungliga Tekniska Högskolan (KTH)
Teknikringen 56
SE-100 44 Stockholm
Sweden
E-mail: gamstedt@kth.se

Chapter 7
Dr Nicole M. Stark*
Engineered Composites Science
USDA Forest Products Laboratory
One Gifford Pinchot Drive
Madison, WI 53726-2398
USA
E-mail: nstark@fs.fed.us

Douglas J. Gardner
Advanced Engineered Wood
  Composites Center
5793 AEWC Building
University of Maine
Orono, ME 04469
USA
E-mail:
  doug_gardner@umenfa.maine.edu

Chapter 8
N. E. Marcovich and M. I.
  Aranguren*
Institute for Research in Materials
  Science and Technology
  (INTEMA)
Facultad de Ingeniería – Universidad
  Nacional de Mar del Plata
Juan B. Justo 4302
(7600) Mar del Plata
Argentina
E-mail: marangur@fi.mdp.edu.ar
  aranguren@intema.gov.ar

Chapter 9
K. Englund* and M. Wolcott
Wood Materials and Engineering
  Laboratory
Washington State University
Pullman, WA 99164-1806
USA
E-mail: englund@wsu.edu
  Wolcott@wsu.edu

Chapter 10
F. W. Maine
Frank Maine Consulting Ltd
71 Sherwood Drive
Guelph
Ontario N1E 6E6
Canada
E-mail: maine.f@sympatico.ca

Chapter 11
G. Guo*
M. C. Gill Foundation Composites
  Center
Mork Family Department of
  Chemical Engineering and
  Materials Science
University of Southern California
3651 Watt Way
Los Angeles, CA 90089
USA
E-mail: gguo@usc.edu
  gguo@uiuc.edu

C. B. Park
Department of Mechanical and
  Industrial Engineering
University of Toronto
5 King's College Road
Toronto
Ontario M5S 3G8
Canada
E-mail: park@mie.utoronto.ca

G. Rizvi
Faculty of Engineering and Applied
  Science
University of Ontario Institute of
  Technology
2000 Simcoe Street North
Oshawa
Ontario L1H 7K4
Canada
E-mail: ghaus.rizvi@uoit.ca

Chapter 12
R. J. Tichy
Wood Materials and Engineering
  Laboratory
Washington State University
Pullman, WA 99164-1806
USA
E-mail: tichy@wsu.edu

Chapter 13
T. Thamae* and C. Baillie
Department of Chemical Engineering
Queens University
Kingston
Ontario K7L 3N6
Canada
E-mail:
  thimothy.thamae@chee.queensu.ca
  cbaillie@post.queensu.ca

Chapter 14
M. Carus* and C. Gahle
nova-Institut GmbH
Chemiepark Knapsack
50351 Hürth
Germany
E-mail: contact@nova-institut.de

H. Korte
Innovationsberatung Holz & Fasern
Lübsche Str. 77
23966 Wismar
Germany
E-mail: info@hanskorte.de

Chapter 15
A. A. Klyosov
MIR International Inc.
36 Walsh Road
Newton, MA 02459
USA
E-mail: aklyosov@comcast.net
  anatoleklyosov@ldicomposites.com

Introduction

K OKSMAN NISKA, Luleå University of Technology, Sweden
and M SAIN, University of Toronto, Canada

In the past ten years wood–polymer composite (WPC) has become a state-of-the-art commercial product with a growing market potential in the area of building, construction, and furniture. The market share of WPC in the area of automotives is also increasing in Europe and Asia. Although WPC has a long history in Europe from the beginning of the twenty-first century, the main commercialization has happened only since the early 1990s. A major manufacturing initiative was undertaken by small and medium enterprises (SMEs) in North America in the mid-1990s that resulted in a fully commercialized decking product for the building industry. Since then, many other innovative products have been commercialized in the United States and Canada. A major market trend is now to expand the product range in construction with enhanced mechanical performance and durability. In recent years we have seen that WPC products are slowly penetrating the European market, in automotive applications, furniture, and in building products.

On the research and development front WPC has gained significant popularity as evidenced by a threefold to fourfold increase in international symposia and workshops in the past five years. Major growth in the technology is coming from equipment design, process formulation and product design.

The importance of promoting new and improved knowledge in the field of WPC has prompted us to develop this book, the content of which provides a comprehensive insight into the commercial development, technological innovation and market avenues of WPC materials and products. The book contains 15 chapters, relating to raw materials used, fundamental developments and future trends in technology and industrial products, technical challenges, standardization, and market opportunities for WPC.

Each chapter of this book has been written by authors who have a long experience of WPC and we believe that this book is an excellent handbook for industrial and academic readers to get the most recent insight on the state-of-the-art of used materials, technologies and products of WPCs. Both editors of this book are active in promoting WPCs in Europe and North America, and we think that this book will be a very useful textbook for classroom teaching.

1
Raw materials for wood–polymer composites

C CLEMONS, USDA Forest Service, USA

1.1 Introduction

To understand wood–plastic composites (WPCs) adequately, we must first understand the two main constituents. Though both are polymer based, they are very different in origin, structure, and performance. Polymers are high molecular weight materials whose performance is largely determined by its molecular architecture. In WPCs, a polymer matrix forms the continuous phase surrounding the wood component. These matrix polymers are typically low-cost commodity polymers that flow easily when heated, allowing for considerable processing flexibility when wood is combined with them. These polymers tend to shrink and swell with temperature but absorb little moisture and can be effective barriers to moisture intrusion in a well-designed composite.

Wood itself contains polymers such as lignin, cellulose, and various hemicelluloses but has very different properties from the synthetic polymers with which it is most often combined. Wood is less expensive, stiffer, and stronger than these synthetic polymers, making it a useful filler or reinforcement. Though wood does not shrink and swell much with temperature, it readily absorbs moisture, which alters its properties and dimensions and can lead to biodegradation if not protected.

In this chapter, we explore the basic structure and properties of polymers and wood individually to lay a foundation for a greater understanding of the composites made from them. Basic concepts and properties are briefly summarized with emphasis on materials common to current commercial technology. Sources of further information are listed at the end of the chapter.

The Forest Products Laboratory is maintained in cooperation with the University of Wisconsin. This article was written and prepared by US Government employees on official time, and it is therefore in the public domain and not subject to copyright.

1.2 Polymers: structure and properties

Polymers are high molecular weight substances consisting of molecules that are, at least approximately, multiples of simple units (Carley, 1993). The word polymer comes from the Greek *poli*, which means many, and *meros*, which means parts (Osswald and Menges, 1996). Polymers can be natural (e.g. cellulose, collagen, keratin) or synthetic (e.g. polypropylene, polyethylene) in origin. A polymer is called a plastic when it has other materials such as stabilizers, plasticizers, or other additives in it.

Owing to the low thermal stability of wood flour, plastics that can be processed at temperatures lower than about 200 °C are usually used in WPCs. In North America, the great majority of WPCs use polyethylene as the matrix, though polypropylene, polyvinyl chloride, and others are also used (Morton *et al.*, 2003). The large use of polyethylene is due, in part, to that fact that much of the early WPCs were developed as an outlet for recycled film as well as the low cost and availability of recycled sources of polyethylene. Polypropylene is widely used in Europe.

1.2.1 Structure and organization

Molecular structure

Much of how a polymer performs is determined by its molecular structure. This structure is developed during the polymerization process where low molecular weight monomers are reacted to form long polymer chains. Table 1.1 shows the basic chemical structural units of several common polymers as well as their common abbreviations.

Table 1.1 Structural units for selected polymers with approximate glass transition (T_g) and melting (T_m) temperatures. Condensed from Osswald and Menges (1996)

Structural unit	Polymer	T_g (°C)	T_m (°C)
$-CH_2-CH_2-$	Polyethylene (PE)	−125	135
$-CH_2-CH(CH_3)-$	Polypropylene (PP)	−20	170
$-CH_2-CH(C_6H_5)-$	Polystyrene (PS)	100	—
$-CH_2-CH(Cl)-$	Polyvinyl chloride (PVC)	80	—
$-CO-C_6H_4-CO-O-CH_2-CH_2-O-$	Polyethylene-terephthalate (PET)	75	280

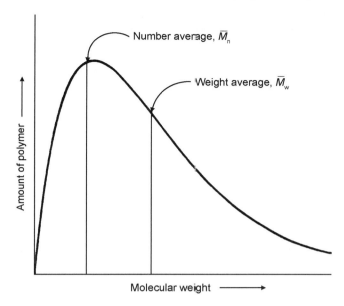

1.1 Typical molecular weight distribution of a polymer.

The number of repeat units in the polymer is the degree of polymerization. The molecular weight of the polymer is then the molecular weight of the repeat unit times the degree of polymerization. However, not all polymer chains have the same molecular weight and a distribution such as that shown in Fig. 1.1 is typical. The two most common measures of molecular weight are the number-average molecular weight (M_n) and the weight-average molecular weight (M_w), given by the following equations (Osswald and Menges, 1996):

$$M_n = \frac{\Sigma m_i}{\Sigma n_i} = \frac{\Sigma n_i M_i}{\Sigma n_i} \qquad 1.1$$

$$M_w = \frac{\Sigma m_i M_i}{\Sigma m_i} = \frac{\Sigma n_i M_i^2}{\Sigma n_i M_i} \qquad 1.2$$

where m_i is the weight, M_i is the molecular weight, and n_i is the number of molecules with i repeat units. The number-average molecular weights of many commercial polymers are typically about 10 000–100 000 (Billmeyer, 1984). A useful measure of the breadth of the molecular weight distribution curve is the polydispersity index, defined as the ratio of weight-average to number-average molecular weight ratio (M_w/M_n) (Osswald and Menges, 1996).

Polymers can contain one type of monomer (as with the homopolymers in Table 1.1) or multiple monomers (copolymers, terpolymers, etc.). In the latter, the arrangement of the repeat units can be controlled during polymerization to influence the performance of the polymer (see Fig. 1.2). One commercial

Random copolymer:	AABABBAABABBBABAAB
Alternating copolymer:	ABABABABABABABABAB
Block copolymer:	AAABBBAAABBBAAABBB
Graft copolymer:	AAAAAAAAAAAAAAAAAA
	B B B B
	B B

1.2 Copolymer types. A, B represent different repeat units.

example is random polypropylene copolymer where 1–8 wt% of ethylene is added during polymerization of the polypropylene to form a copolymer with improved clarity, slightly better impact properties, and enhanced flexibility (Kissel *et al.*, 2003). Another example is maleated polypropylene, a graft copolymer that is commonly used as a coupling agent to improve the adhesion between wood and polypropylene (see Chapter 3).

Tacticity is important in the arrangement of repeat units in polymers with asymmetrical repeat units. For example, polypropylene contains a methyl group (CH_3) attached to a carbon chain (see Table 1.1). The methyl groups can be attached to one side of the chain (isotactic), alternating sides of the chain (syndiotactic), or lack a consistent arrangement (atactic). Whereas atactic polypropylene is a soft material that is mainly used in applications such as sealants and caulks, highly isotactic polypropylene has more desirable properties in the solid state and is a large-volume, commercial polymer used in a wide variety of applications (Kissel *et al.*, 2003).

Some polymers have considerable branching of molecular chains, often as a result of side reactions during polymerization. The type and amount of branching can influence the structure and properties of the polymer. For example, Fig. 1.3 shows the branching in different polyethylenes. Branching can inhibit the ability of the molecules to pack together to form highly ordered (i.e. crystalline) regions. As a result, polyethylenes that have long branches (LDPE) or many branches (LLDPE) have lower density than those that do not (HDPE). In contrast to branching, where a molecule is still considered discrete, some polymers can form chemical links (called crosslinks) between molecular chains that can increase rigidity but also can restrict or prevent flow.

Molecular organization

Polymers are often categorized by their behavior, which is influenced by their molecular organization. *Thermosets* are polymers that crosslink extensively and, as a result, are substantially infusible and insoluble in their final state (Carley, 1993). Once cured, increasing temperature eventually leads to degradation rather than melting. Thermosets include epoxies, phenols, and isocyanates. Thermosets have been used with wood and other natural fibers since the early 1900s. One of the earliest wood–thermoset composites was in a phenol-formaldehyde and wood

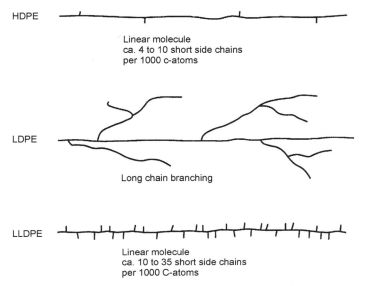

1.3 Branching in high-density polyethylene (HDPE), low-density polyethylene (LDPE) and linear low-density polyethylene (LLDPE). Reprinted with permission from Osswald and Menges (1996).

flour composite called Bakelite. Its first commercial product was reportedly a gearshift knob for Rolls-Royce in 1916 (Gordon, 1988). However, in recent years, the growth of thermosets as matrices in WPCs has not matched that of thermoplastics.

In contrast to its use in thermosets, large-scale use of wood flour in thermoplastics did not occur until the past several decades but recent growth has been great. Most of this is due to the rapid growth of exterior building products such as railings, window and door profiles, and especially decking. Unlike thermosets, thermoplastic polymers can be repeatedly softened by heating. When cooled, they harden as motion of the long molecules is restricted. If the polymer molecules remain disordered as they are cooled from the melt, they are considered amorphous thermoplastics and the temperature at which they solidify is known as the glass transition temperature (T_g). Polystyrene is an amorphous thermoplastic, for example.

Some thermoplastics form regions of highly ordered and repetitive molecular arrangements on cooling. These are called semicrystalline polymers since much, though not all, of the molecular structure is in an ordered state. A crystallinity of 40–80% is typical for common semicrystalline polymers such as polypropylene and polyethylene and depends on molecular architecture as well as processing history. In addition to a glass transition (T_g), semicrystalline polymers have crystalline melting points (T_m), above which temperature the crystal order disappears and flow is greatly enhanced.

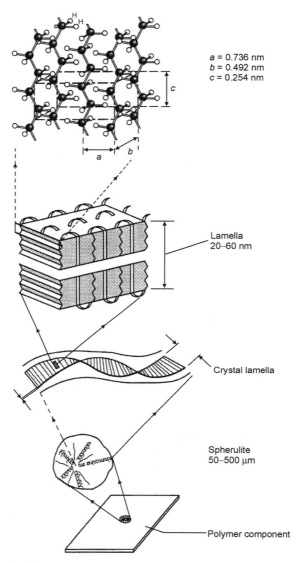

1.4 General molecular structure and arrangement of typical semicrystalline materials. Reprinted with permission from Osswald and Menges (1996).

Though the molecular and crystalline arrangement of polymers is complex, Fig. 1.4 shows a general hierarchy of ordering in a typical semicrystalline polymer. The crystalline regions form lamellae which, in turn, form spherulites. Spherulites are the largest domain with a specific order and their size is typically much larger than the wavelength of light, making semicrystalline polymers translucent and not transparent (Osswald and Menges, 1996). Though molecular architecture such as tacticity, polymer branching, and molecular weight are important parameters affecting crystalline structure, processing also plays a large role. For example, cooling rates can affect the types of crystals formed and the size and number of spherulites formed. Also, processing-induced molecular orientation can increase the final crystallinity of the solid polymer.

1.2.2 Properties

The properties of thermoplastic polymers are often highly dependent on the temperature at which they are measured and the speed at which they are tested. Generally speaking, when the temperature of a polymer melt is reduced below the melt temperature the material behaves as a leathery solid. If a polymer is semicrystalline, a crystal structure develops. As the temperature is further reduced below its glass transition, the amorphous portions solidify and form a glassy, stiff and, in some cases, brittle material (Osswald and Menges, 1996). Figure 1.5 shows how different polymer properties change with temperature.

While polymers have solid-like properties such as elasticity and dimensional stability, they also have liquid-like characteristics such as flow over time that depend on temperature, stress, and pressure. This tendency of a polymer to behave as it were a combination of a viscous liquid and an elastic solid is generally referred to as viscoelasticity (Carley, 1993). For example, most polymers have higher moduli when stress is rapidly applied versus when it is applied slowly. Also, some polymers tend to sag over time (i.e. creep) when bearing sustained loads, an important consideration in structural applications.

Typical room temperature properties of commonly used polymers in WPCs are summarized in Table 1.2. These values are provided to give a general indication of the polymer properties. However, the exact performance of these polymers is difficult to summarize since various grades of plastics are produced, whose performance has been tailored by controlling polymerization and additive content. Polyvinyl chloride in particular often contains a considerable amount of additives such as heat stabilizers and plasticizers, resulting in a wide range of processability and performance. However, some general comments can be made.

Generally speaking, though these polymers tend to have considerably lower mechanical performance than the so-called engineering plastics, the commodity plastics listed have reasonably good mechanical performance for many applications and low price. The polyethylenes are by far the most common polymers used in WPCs in North America. Polypropylene is common in other parts of the world. They absorb little moisture and act as effective moisture barriers. This is important since moisture sorption in WPCs can negatively affect the performance of the composite.

Though polyethylene and polypropylene are largely impervious to moisture, they are susceptible to degradation by UV radiation, and the use of light-stabilizing additives is common. The thermal expansion and contraction of

8 Wood–polymer composites

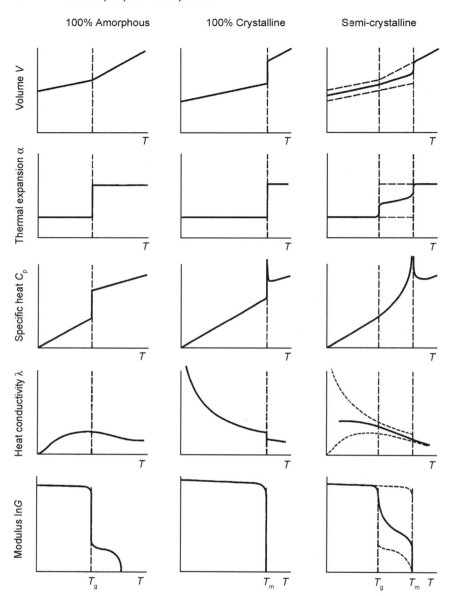

1.5 Trends in the polymer properties of thermoplastics as a function of temperature, T. Reprinted with permission from Osswald and Menges (1996).

polyethylene and polypropylene are significant and they tend to creep (or sag over time), especially under load or at high temperatures, limiting their structural performance. However thermal expansion and creep can be reduced with fillers and reinforcements.

Table 1.2 Typical room temperature properties of common polymers. Condensed from Osswald and Menges (1996)

Polymer	Density (g/cm³)	Tensile strength (MPa)	Tensile modulus (GPa)	Elongation at break (%)	Water absorption in 24 h (%)	Coefficient of thermal expansion ($K^{-1} \times 10^6$)	Thermal conductivity (W/m K)
Low-density polyethylene (LDPE)	0.91–0.93	8–23	0.2–0.5	300–1000	<0.01	250	0.32–0.40
High-density polyethylene (HDPE)	0.94–0.96	18–35	0.7–1.4	100–1000	<0.01	200	0.38–0.51
Polypropylene (PP)	0.90–0.92	21–37	1.1–1.3	20–800	0.01–0.03	150	0.17–0.22
Rigid polyvinyl chloride (PVC)	1.4–1.6	50–75	1.0–3.5	10–50	3–18	70–80	0.14–0.17

Polyvinyl chloride is also used in WPCs, although not nearly as much as polyethylene and polypropylene. Despite its good mechanical performance, complexities in formulating and processing as well as patent issues have limited broader use in WPCs.

More detailed information on specific polymer grades can be obtained from manufacturers and distributors but are somewhat dependent on process history as well. Different grades of a particular polymer have been tailored for a specific application and processing method. For example, a bottle grade of high-density polyethylene (HDPE) has high molecular weight to provide the toughness needed for a bottle application and high melt strength necessary for the melt blowing process. An injection molding grade might have a lower molecular weight yielding lower melt viscosity and good flow properties.

1.3   Wood: structure and properties

Wood contains natural polymers such as lignin, cellulose, and various hemicelluloses but has very different properties from the synthetic polymers with which it is most often combined. The efficient structure and anatomy make it a stiff, strong, tough, and lightweight material that can efficiently perform functions such as moisture transport that are critical for survival of the tree. Its excellent material performance and low cost have made it a useful structural material for millennia.

From a polymer composite standpoint, wood is less expensive, stiffer, and stronger than many commodity synthetic polymers, making it a candidate for filling or reinforcing them. However, some of the same material behavior such as moisture sorption that serves it well in nature can be problematic in a composite material. Therefore, to effectively use wood as a filler or reinforcement in polymers, an understanding of its material behavior is important. We will first discuss the structure and anatomy of wood and then describe its behavior and that of the fillers and reinforcements made from them.

1.3.1   Structure and anatomy of wood

Wood anatomy

As with most natural materials, the anatomy of wood is complex. Wood is porous, fibrous, and anisotropic. Wood is often broken down into two broad classes: softwoods and hardwoods, which are actually classified by botanical and anatomical features rather than wood hardness. Figures 1.6 and 1.7 are schematics of a softwood and hardwood, respectively, showing the typical anatomies of each wood type. Softwoods (or gymnosperms) include such species as pines, firs, cedars, and spruces; hardwoods (or angiosperms) include species such as the oaks, maples, and ashes.

Raw materials for wood–polymer composites

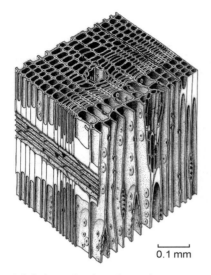

1.6 Schematic of a softwood.

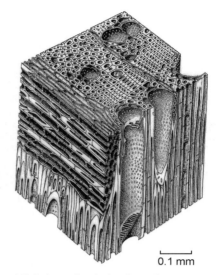

1.7 Schematic of a hardwood.

Wood is primarily composed of hollow, elongated, spindle-shaped cells (called tracheids or fibers) that are arranged parallel to each other along the trunk of the tree (Miller, 1999). The lumen (hollow center of the fibers) can be completely or partially filled with deposits, such as resins or gums, or growths from neighboring cells called tyloses (Miller, 1999). These fibers are firmly cemented together and form the structural component of wood tissue. The length of wood fibers is highly variable but average about 1 mm (1/25 in.) for hardwoods and 3–8 mm (1/8 to 1/3 in.) for softwoods (Miller, 1999). Fiber diameters are typically 15–45 μm.

Chemical constituents

The wood substance itself is a complex, three-dimensional, polymer composite made up primarily of cellulose, hemicellulose, and lignin (Rowell, 1983). These three hydroxyl-containing polymers are distributed throughout the cell wall. The chemical compositions of selected woods are shown in Table 1.3.

Cellulose varies the least in chemical structure of the three major components. It is a highly crystalline, linear polymer of anhydroglucose units with a degree of polymerization (n) around 10 000 (Fig. 1.8). It is the main component providing the wood's strength and structural stability. Cellulose is typically 60–90% crystalline by weight and its crystal structure is a mixture of monoclinic and triclinic unit cells (Imai and Sugiyama, 1998; Wada *et al.*, 1994). Hemicelluloses are branched polymers composed of various 5- and 6-carbon sugars whose molecular weights are well below those of cellulose but which still contribute as a structural component of wood (Pettersen, 1984).

12   Wood–polymer composites

Table 1.3 Approximate chemical composition of selected woods from Petterson (1984)

Species	Cellulose[a]	Hemicellulose[b]	Lignin[c]	Extractives[d]	Ash
Ponderosa pine	41	27	26	5	0.5
Loblolly pine	45	23	27	4	0.2
Incense cedar	37	19	34	3	0.3
Red maple	47	30	21	2	0.4
White oak	47	20	27	3	0.4
Southern red oak	42	27	25	4	0.4

[a]Alpha cellulose content as determined by ASTM D1103-60 (1977).
[b]Approximate hemicellulose content determined by subtracting the alpha cellulose content from the holocellulose content values from Pettersen (1984).
[c]Klason lignin content as determined by ASTM D1106-96 (2001).
[d]Solubility in 1:2 volume ratio of ethanol and benzene according to ASTM D1107-96 (2001).

Lignin is an amorphous, crosslinked polymer network consisting of an irregular array of variously bonded hydroxy- and methoxy-substituted phenylpropane units (Pettersen, 1984). The chemical structure varies depending on its source. Figure 1.9 represents a partial softwood lignin structure illustrating a variety of possible structural components. Lignin is more non-polar than cellulose and acts as a chemical adhesive within and between the cellulose fibers.

Additional organic components, called extractives, make up about 3–10% of the dry wood grown in temperate climates, but significantly higher quantities are found in wood grown in tropical climates (Pettersen, 1984). Extractives include substances such as fats, waxes, resins, proteins, gums, turpenes, and simple sugars. Many of these extractives function in tree metabolism and act as energy reserves or defend against microbial attack. Though often small in quantity, extractives can have large influences on properties such as color, odor, and decay resistance (Pettersen, 1984). Small quantities (typically 1%) of inorganic matter, termed ash, are also present in wood grown in temperate regions.

Cellulose forms crystalline microfibrils held together by hydrogen bonds and then cemented to lignin into the wood fiber cell wall. The microfibrils are aligned in the fiber direction in most of the cell wall, winding in a helix along the fiber axis. The angle between the microfibril and fiber axes is called the microfibril helix angle. The microfibril helix angle is typically 5–20° for most of

1.8 Chemical structure of cellulose (Pettersen, 1984).

Church on the Hill
Fundraising Dinner
Saturday, May 14, 2011

Guelph Knights Charities Trust

Ticket Sellers

Ticket #	First Name	Last Name	Address	Postal Code	Pho
1 - 10	Frank	Maine	71 Sherwood Drive	N1E 6E6	823-
11 - 20	Leon	Yaskowich	4 Hawthorne Pl.	N1E 1V8	824-
21 - 30	Pat	Moynihan	80 Burns Dr., Unit 16	N1H 6V9	265-
31 - 40	Gary	Calverley	80 Forest St.	N1G 1H9	824-
41 - 50	Bernie	Miller	87 Walser St., Elora	N0B 1S0	846-
51 - 60	Matt	Sarabura	120 Water St.	N1G 1A9	823-
61 - 70	Jerry	Matthews	24 Sherwood Dr.	N1E 1R6	822-
71 - 80	Brian	Pittam	270 Water St.	N1G1B7	83
81 - 90	Peter	Borrett	28 Elmira	N1H2C9	824
91 - 100	Mike	Kennedy	5A Hamilton	N1H 6J2	823-
101 - 110					

1.9 A partial softwood lignin structure (Pettersen, 1984).

the cell wall (Parham and Gray, 1984) and varies depending upon many factors, including species and stresses on the wood during growth.

1.3.2 Sources and production of fillers and reinforcements from wood

There are a number of methods of producing fillers and reinforcements from wood for use in polymer composites. However, most commercial methods result in either some type of fiber or particulate.

Though often used somewhat loosely, wood fibers more precisely refer to the spindle-shaped cells of wood and are the main structural element of wood on the macroscopic scale. Fibers can be separated from wood by various mechanical or chemical pulping methods, for example, which influence the final properties of the fibers themselves. These fibers offer good reinforcing potential because their high strength and reasonably high aspect ratio (i.e. length-to-diameter ratio) can allow efficient transfer of stresses to the fiber if fiber–matrix adhesion is promoted. However, the use of fibers in polymer composites still greatly lags that of wood flour owing to greater cost and increased processing difficulties when using conventional plastics processing methods.

1.10 Scanning electron micrograph of pine wood flour.

The term 'wood flour' is somewhat ambiguous, referring to wood reduced to finely divided particles approximating those of cereal flours in size, appearance, and texture (Reineke, 1966). Wood flour comprises fiber bundles, rather than individual wood fibers, with aspect ratios typically only about 1–5 (see Fig. 1.10). Though the low aspect ratio limits the reinforcing ability (Bigg *et al.*, 1988), mechanical performance of the composite is sufficient for many applications. Wood flour is also less expensive and is easier to feed and meter into conventional plastics processing equipment than fibers derived from wood. Wood flour particle size is often described by mesh of the wire cloth sieves used to classify them. Table 1.4 lists the US standard mesh sizes and their equivalent particle diameters. However, different standards may be used internationally

Table 1.4 Conversion between US standard mesh and particle diameter

Mesh US	Particle diameter (μm)	Mesh US	Particle diameter (μm)	Mesh US	Particle diameter (μm)
20	850	60	250	170	90
25	710	70	212	200	75
30	600	80	180	230	63
35	500	100	150	270	53
40	425	120	125	325	45
45	355	140	106	400	38
50	300				

(*International Sieve Chart*, 1997). Grades of commercially manufactured wood flours used as fillers in thermoplastics are supplied in different particle size ranges but these ranges typically fall within 180–840 μm (80–20 US standard mesh).

Wood flour is derived from various scrap wood from wood processors. Though there is no standard method of producing wood flour, the main steps in wood flour production are: (i) size reduction using various mills and (ii) size classification by screening or air classification (Reineke, 1966). Many different species of wood flours are available and are often based on the regional availability of clean, raw materials from wood-processing industries. The most commonly used wood flours for plastic composites in the United States are made of pines, maples, and oaks. Many reasons are given for species selection including slight color differences, availability, and familiarity. Additionally, some species that contain significant amounts of extractives that can be leached from the wood by water are avoided. This water can migrate to the surface and evaporate, leaving behind stains from the extractives. Extractives can also volatilize or darken during processing if the temperatures are too high.

1.3.3 Properties

Owing to its commercial importance, the properties of common wood species are readily available. However, the properties of fibers and particles derived from wood can be significantly different from the wood from which it is derived. Methods for producing wood-derived fillers and fibers as well as the high temperatures and pressures often found during composite processing influence attributes such as surface chemistry, density, and moisture content of the wood component in the final composite. For example, wood fibers produced by thermomechanical means lead to lignin-rich surfaces while those produced by chemical means lead to carbohydrate-rich surfaces (Stokke and Gardner, 2003). These changes in surface chemistry can affect adhesion with polymers, for example. Important properties of wood and fillers and fibers derived from them are discussed below.

Density

Though the density of the wood cell wall is about 1.44–1.50 g/cm^3 (Kellogg, 1981), the porous anatomy of solid wood results in overall densities of about 0.32–0.72 g/cm^3 (20–45 lb/ft^3) when dry (Simpson and TenWolde, 1999). Not surprisingly, production of fillers and reinforcements from wood result in materials with bulk densities that are significantly lower than wood. For example, the bulk density of wood flour depends on factors such as moisture content, particle size, and species, but typically is about 0.19–0.22 g/m^3 (12–14 lb/ft^3).

The bulk density of reinforcing fibers derived from wood is considerably lower, especially with long fibers, and is quite variable. This low bulk density and entanglement of the fibers make metering and feeding into conventional polymer processing equipment such as extruders difficult. Though methods have been developed that overcome these hurdles, they typically add cost.

As a filler in polymers, wood flour is unusual in that it is compressible. The high pressures found during many plastics processing methods can collapse the hollow fibers that constitute the wood flour or fill them with low molecular weight additives and polymers. The degree of collapsing or filling will depend on such variables as particle size, processing method, and additive viscosity, but wood densities in composites approaching the wood cell wall density (i.e. 1.44–1.50 g/cm^3) can be found in high-pressure processes such as injection molding. Consequently, adding wood to commodity plastics such as polypropylene, polyethylene, and polystyrene increases their density despite having densities higher than that of wood prior to compounding. However, even the density of compressed wood is considerably lower than those of common inorganic fillers and reinforcements, which are typically about 2.5–2.8 g/m^3 (Xanthos, 2005). This density advantage is important in applications where weight is important, such as automotive components.

Moisture sorption

The major chemical constituents of the wood cell wall contain hydroxyl and other molecular groups that attract moisture. Absorbed moisture interferes with and reduces hydrogen bonding between cell wall polymers and alters its mechanical performance (Winandy and Rowell, 1984). However, the interior of the crystalline cellulose is not accessible to moisture, which is important in maintaining rigidity in the tree even at high moisture contents (Tarkow, 1981).

The equilibrium moisture content of wood is affected by temperature and humidity and can vary as much as 3–4% depending on if it is approached from a higher or lower humidity (i.e. wood exhibits a moisture sorption hysteresis). Table 1.5 shows approximate equilibrium moisture contents for wood at different temperatures and humidities at a midpoint between the hysteresis curves.

The moisture sorption of fillers and reinforcements derived from wood are affected by the methods used to produce them. However, wood flour is produced mechanically and its moisture sorption properties are similar to that of solid wood. Wood flour usually contains at least 4% moisture when delivered, which must be removed before or during processing with thermoplastics. Even if dried, wood flour can still absorb moisture quickly. Depending on ambient conditions, wood flour can absorb several percent of moisture within hours (Fig. 1.11).

Moisture of up to about 30% can be adsorbed by the cell wall with a corresponding reversible increase in apparent wood volume. Volume changes of the wood component due to moisture sorption, especially repeated moisture cycling,

Table 1.5 Equilibrium moisture content for wood at different temperatures and humidities from Simpson and TenWolde (1999)

Temperature		Moisture content (%) at various relative humidities								
°C	°F	10%	20%	30%	40%	50%	60%	70%	80%	90%
−1.1	30	2.6	4.6	6.3	7.9	9.5	11.3	13.5	16.5	21.0
4.4	40	2.6	4.6	6.3	7.9	9.5	11.3	13.5	16.5	21.0
10.0	50	2.6	4.6	6.3	7.9	9.5	11.2	13.4	16.4	20.9
15.6	60	2.5	4.6	6.2	7.8	9.4	11.1	13.3	16.2	20.7
21.1	70	2.5	4.5	6.2	7.7	9.2	11.0	13.1	16.0	20.5
26.7	80	2.4	4.4	6.1	7.6	9.1	10.8	12.9	15.7	20.2
32.2	90	2.3	4.3	5.9	7.4	8.9	10.5	12.6	15.4	19.8
37.8	100	2.3	4.2	5.8	7.2	8.7	10.3	12.3	15.1	19.5

can lead to interfacial damage and matrix cracking (Peyer and Wolcott, 2000). As a result, many manufacturers of WPCs used in exterior applications limit wood flour content to 50–65% by weight and rely on the partial encapsulation of the wood by the polymer matrix to prevent major moisture sorption and subsequent negative effects.

Durability properties

The surface of wood undergoes photochemical degradation when exposed to UV radiation. This degradation takes place primarily in the lignin component and

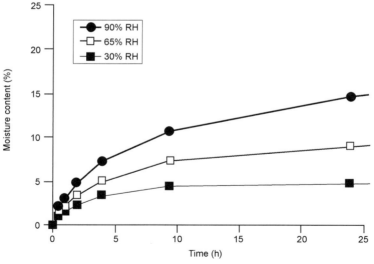

1.11 Moisture sorption of pine wood flour at several relative humidities (RH) and 26 °C (80 °F).

results in a characteristic color change (Rowell, 1984). Mold can form on moist surfaces of WPCs. Although mold does not reduce the structural performance, it can be an important aesthetic issue. In severe cases, where the moisture content of the wood flour in the composite exceeds about 25–30%, decay fungi can begin to attack the wood component leading to weight loss and significant reduction in mechanical performance. Wood is degraded biologically because organisms recognize the celluloses and hemicelluloses in the cell wall and can hydrolyze them into digestible units using specific enzyme systems (Rowell, 1984). Further information on the durability of wood and WPCs can be found in Chapter 7.

Thermal properties

The onset of degradation differs for the major components of wood with cellulose being the most thermally stable. Owing to its low thermal stability, wood flour is usually used as a filler only in plastics that are processed at temperatures lower than about 200 °C. Above these temperatures the cell wall polymers begin to decompose. High-purity cellulose pulps, where nearly all of the less thermally stable lignin and hemicelluloses have been removed, have recently been investigated for use in plastic matrices, such as nylon, that are processed at higher temperatures than most commodity thermoplastics (Sears *et al.*, 2001). Figure 1.12 shows thermogravimetric analysis curves showing weight loss of pine wood flour and high purity cellulose fiber as temperature is increased. Wood flour begins to degrade at significantly lower temperature than the cellulose fiber.

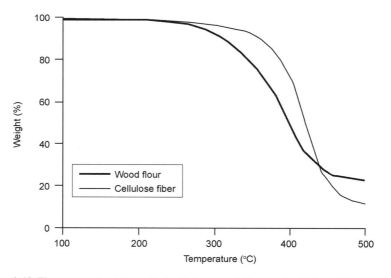

1.12 Thermogravimetric analysis of pine wood flour and cellulose fibers made from wood.

The thermal expansion of wood is less than that of the commodity plastics commonly used as matrices. Thermal expansion coefficients for wood are directional but are roughly (Kellogg, 1981):

$$\alpha = A \rho \times 10^{-6} \qquad 1.3$$

where α is the coefficient of thermal expansion (in K^{-1}), ρ is the specific gravity (oven-dry basis), and A is roughly 50–80 perpendicular to the fiber direction and about 5–10 times less in the fiber direction. This roughly yields an average of about $10–15 \times 10^{-6}/K$ if we assume a density of 1.5. This is well below that of common matrix materials for WPCs for which α is about $150–250 \times 10^{-6}/K$ (Osswald and Menges, 1996).

Mechanical properties

Reported mechanical properties for reinforcing fibers, particularly natural fibers, vary greatly. Variability is due to differences in preparation methods as well as to differences in species, growing conditions, age, and location. Also, differences in methods of determining mechanical performance are common and make direct comparisons of literature values difficult (Mark, 1967). However, some general statements regarding mechanical performance can be made.

Table 1.6 lists typical tensile properties for solid wood as well as various fibers used in polymer composites. The mechanical properties of wood are quite anisotropic with the largest values are found parallel to the grain, which corresponds to the direction of alignment of the wood fibers. The strength properties of wood pulp fibers are larger than those of wood itself. This is not surprising since the fiber is the main macroscopic structural element of wood. Strengths of air-dried pulp fibers tend to increase with removal of lignin (Mark, 1967). The mechanical properties of some natural fibers such as flax are considerably higher than those of wood, which is one of the reasons that flax

Table 1.6 Typical mechanical properties of wood and various reinforcements

Material	Tensile modulus (GPa)	Tensile strength (MPa)	Reference
Southern pine wood*	12–27 (∥) 0.7–1.4 (⊥)	100–200 (∥) 3–5 (⊥)	Kretschmann and Green (1996)
Southern pine sulfate pulp fiber	5–10	350–700	Mark (1967)
Flax fiber	26–107	750–1100	Mark (1967)
E-glass fiber	76	1500	Xanthos (2005)
Carbon fiber	230–340	3200–2500	Xanthos (2005)

* Properties are for uncompressed wood at 4% moisture content and densities between 0.4 and 0.7 g/cm³. (∥) and (⊥) refer to the properties parallel and perpendicular to the grain.

fibers are of considerable interest in polymer composites in applications in the automotive industry, for example (Sudell and Evans, 2004). Common synthetic reinforcing fibers such as glass and carbon fibers have greater mechanical performance than natural fibers. However, advantages such as low cost and low density of the natural fibers offer advantages in some applications.

The strength of wood flour would be expected to be less than that of wood fiber at the same conditions since it is easier to separate fibers than to break them. Also, the low aspect ratio of the flour and limited adhesion between the wood flour and plastic often lead to incomplete stress transfer that prevents optimal reinforcement of matrix. However, sufficiently useful property improvements (e.g. increased modulus, heat deflection temperature, and dimensional stability with changes in temperature), ease of processing, and low cost make wood flour desirable as a filler.

1.4 Sources of further information and advice

There are many introductory texts on polymer science including several used as references for this chapter (Billmeyer, 1984; Osswald and Menges, 1996). Additionally, practical information on various polymers can be obtained from polymer manufacturers and distributors as well as industry handbooks. The recent advent of electronic material databases offers further opportunity for comparisons of polymer properties.

Similarly, many introductory texts are available on wood science and technology (Wangaard, 1981; Fengel and Wegener, 1983). Another popular, readily available source of basic information on wood is the *Wood Handbook: Wood as an Engineering Material* (1999). As its use as a filler increases, handbooks are beginning to include information on wood-derived fillers and reinforcements alongside more traditional ones (Clemons and Caulfield, 2005). Suppliers are also a source of practical information on commercial grades.

1.5 References and further reading

ASTM D1103-60 (1977), 'Method of test for alpha-cellulose in wood', in *Annual Book of ASTM Standards*, Vol. 04.10, ASTM International, West Conshohoken, PA.

ASTM D1106-96 (2001), 'Standard test method for acid-insoluble lignin in wood', in *Annual Book of ASTM Standards*, Vol. 04.10, ASTM International, West Conshohoken, PA.

ASTM D1107-96 (2001), 'Standard test method for ethanol-toluene solubility of wood', in *Annual Book of ASTM Standards*, Vol. 04.10, ASTM International, West Conshohoken, PA.

Bigg, D M, Hiscock, D F, Preston, J R and Bradbury, E J (1988), 'High performance thermoplastic matrix composites', *Journal of Thermoplastic Composite Materials*, **1**, 146–161.

Billmeyer, F W (1984), 'The science of large molecules', in *The Textbook of Polymer Science*, 3rd edn, John Wiley and Sons, Inc., New York.
Carley, J F (ed.) (1993), *Whittington's Dictionary of Plastics*, Technomic Publishing Company, Inc., Lancaster, PA.
Clemons, C M and Caulfield, D F (2005), 'Wood flour', in *Functional Fillers for Thermoplastics*, M. Xanthos, ed., Wiley-VCH Verlag GmbH and Co. KGaA, Weinheim, 249–270.
Fengel, D and Wegener, G (1983), *Wood: Chemistry, Ultrastructure, and Reactions*, Walter de Gruyter and Co., Berlin.
Gordon, J E (1988), 'Composite materials', in *The New Science of Strong Materials (or Why You Don't Fall Through the Floor)*, 2nd edn, Princeton University Press, Princeton, NJ, 179.
Imai, T and Sugiyama, J (1998), 'Nanodomains of I_α and I_β cellulose in algal microfibrils', *Macromolecules*, **31**, 6275–6279.
International Sieve Chart (1997), Reade Advanced Materials, www.reade.com/Sieve/international_sieve.html
Kellogg, R M (1981), 'Physical properties of wood', in *Wood: Its Structure and Properties*, Wangaard, F F ed., Pennsylvania State University, University Park, PA, 191–223.
Kissel, W J, Han, J H and Meyer, J A (2003), 'Polypropylene: structure, properties, manufacturing processes, and applications', in *The Handbook of Polypropylene and Polypropylene Composites*, Karian, H G ed., 2nd edn, Marcel Dekker, Inc., New York.
Kretschmann, D and Green, D (1996), 'Moisture content–specific gravity relationships for clear southern pine', in *Proceedings of the International Wood Engineering Conference*, Louisiana State University, New Orleans, LA, 536–542.
Mark, R E (1967), 'Tests of mechanical properties', *The Cell Wall Mechanics of Tracheids*, Yale University Press, New Haven, CT, 28–34.
Miller, R B (1999), 'Structure of wood', in *The Wood Handbook: Wood as an Engineering Material*, General Technical Report FPL-GTR-113, USDA Forest Service, Forest Products Laboratory, Madison, WI.
Morton, J, Quarmley, J and Rossi, L (2003), 'Current and emerging applications for natural and wood fiber composites', in *7th International Conference on Woodfiber-Plastic Composites*, The Forest Products Society, Madison, WI, 3–6.
Nakagawa, S and Shafizadeh, F (1984), 'Thermal properties' in *Handbook of Physical and Mechanical Testing of Paper and Paperboard*, Mark, R E ed., Marcel Decker, New York, 271.
Osswald, T A and Menges, G (1996), *Materials Science of Polymers for Engineers*, Carl Hanser Verlag, New York.
Parham, R A and Gray, R L (1984), 'Formation and structure of wood', in *The Chemistry of Solid Wood*, Rowell, R M ed., American Chemical Society, Washington, DC, 1–56.
Pettersen, R C (1984), 'The chemical composition of wood', in *The Chemistry of Solid Wood*, Rowell, R M ed., American Chemical Society, Washington, DC, 76–81.
Peyer, S and Wolcott, M (2000), 'Engineered wood composites for naval waterfront facilities', *2000 Yearly Report of Office of Naval Research under Contract N00014-97-C0395*.
Reineke, L H (1966), 'Wood flour', *U.S. Forest Service Research Note FPL-0113*, USDA Forest Service, Forest Products Laboratory, Madison, WI, USA.
Rowell, R M (1983), 'Chemical modification of wood', *Forest Products Abstracts*, **6** (12), 363–382.

Rowell, R M (1984), 'Penetration and reactivity of cell wall components', in *The Chemistry of Solid Wood*, Rowell, RM ed., The American Chemical Society, Washington, DC, 176.

Sears, K D, Jacobson, R and Caulfield, D F (2001), 'Reinforcement of engineering thermoplastics with high purity cellulose fibers', in *Sixth International Conference on Woodfiber-Plastic Composites*, The Forest Products Society, Madison, WI, 27–34.

Simpson, W and TenWolde, A (1999), 'Physical properties and moisture relations of wood', in *The Wood Handbook: Wood as an Engineering Material*, General Technical Report FPL-GTR-113, USDA Forest Service, Forest Products Laboratory, Madison, WI.

Stokke, D D and Gardner, D J (2003), 'Fundamental aspects of wood as a component in thermoplastic composites', *Journal of Vinyl and Additive Technology*, **9** (2), 96–104.

Sudell, B C and Evans, W J (2004), 'The increasing use and application of natural fiber composite materials within the automotive industry', in *Seventh International Woodfiber-Plastic Composites (and other natural fibers)*, The Forest Products Society, Madison, WI, 7–14.

Tarkow, H (1981), 'Wood and moisture', in Wood: Its Structure and Properties, Wangaard, F F ed., Pennsylvania State University, University Park, PA.

Wada, M, Sugiyama, J and Okano, T (1994), 'The monoclinic phase is dominant in wood cellulose', *Mokuzai Gakkaishi*, **40** (1), 50–56.

Wangaard, F F (ed.) (1981), *Wood: Its Structure and Properties*, Pennsylvania State University, University Park, PA.

Winandy, J E and Rowell, R M (1984), 'The chemistry of wood strength', in *The Chemistry of Solid Wood*, Rowell, R R ed., The American Chemical Society, Washington, DC, 218.

Wood Handbook: Wood as an Engineering Material, General Technical Report FPL-GTR-113, 1999, USDA Forest Service, Forest Products Laboratory, Madison, Wisconsin, USA. Also available online at: www.fpl.fs.fed.us/documnts/fplgtr/fplgtr113/fplgtr113.htm

Xanthos, M (2005), 'Modification of polymer mechanical and rheological properties with functional fillers', in *Functional Fillers for Plastics*, Xanthos, M ed., Wiley-VCH Verlag GmbH and Co. KGaA, Weinheim.

2
Additives for wood–polymer composites

D V SATOV, Canada Colors and Chemicals Limited, Canada

2.1 Introduction

The process of bringing any product to market is initiated when the sponsors of the venture agree there is an identified market need and, upon commercialization, profit can be generated. Additive selection is just as important to the success of the venture as the choice of processing equipment and raw materials for the final product.

Additives play two roles in contributing to the success of the venture. The role seen by the marketplace is that of effecting end product performance attributes to meet consumer expectations, regulatory requirements, and the marketing objectives of the manufacturer. The role internal to the business enterprise is that of effecting the manufacturing process to minimize the unit cost of finished goods for greater profitability.

Having had the opportunity to spend years both as a user of chemicals as additives and modifiers in manufacturing, as well as promoting them to industry, I have developed the perspective that it is important to minimize the number and total volume of additives used in a product. Rarely is additive selection a major consideration from the start of a venture. More often additives and modifiers are included based on typical formulas and industrial norms, and are fine-tuned only after inadequacies have been identified. The single greatest reason industry uses more additives than necessary is that there is a lack of adequate development in coordinating or matching the finished product requirements with the particulars of the process technology and the primary material components; in the case of composites these primary materials being resin and natural fiber. Industry tends to use additives to make up for process deficiencies, a boon to the chemical industry. This is because process equipment is the primary initial investment and is an obvious cost to contain from the start, regardless of the fact that a capital cost can be amortized and written down. Subsequently any process inadequacy is addressed with additives, resulting in an operational cost that will continue until the process is upgraded.

It is the author's opinion that choosing which resin and fiber to use must be an integral part of the initial strategy. Considerations are: suitability for application, availability and cost of raw materials from chosen source, reliability of supply especially assuming demand will grow, and/or potential for alternative sources, and how choice fits into the marketing plan (such as environmental marketing). The equipment must be suitable for processing of the designated materials to make a part cost effectively. Considerations include the base process being considered, design features such as screw configuration, output potential, both as weight per unit time and for a given die, length per unit time, overall energy consumption, having adequate cooling capacity which takes account of projected growth, and other downstream equipment parameters such as specifications on the hauler, saw-speed, and packing.

In approaching optimized additive use, it is essential first to establish the part to be made, the required product attributes, the primary material components, and the process technology, and then design the additive package to ensure performance requirements are met and unit costs are optimized. Tables 2.1–2.3 summarize the functions, materials and dosages for polyethylene-, polypropylene-, and polyvinyl chloride-based composites.

Table 2.1 Polyethylene (PE)-based composite

Function	Material	Dosage range (%)
Matrix component	Polyethylene resin	Difference from total of other components to 100%
Matrix component	Natural fiber	30–60
Coupling agent	Maleated polyolefin	2–5
Lubricant(s)	Stearates/esters/EBS/other	3–8
Antioxidants	Phenolics/phosphites	0–1
Acid scavengers	Stearates/hydrotalcites	0–1
UV protection	HALS/benzophenones/benzotriazoles	0–1
Mineral filler	Talc	0–10
Biocide	Zinc borate	0–2
Density reduction	Microspheres/chemical or physical blowing agent	0–5
UV protection/aesthetics	Pigments	As required
Flame retardants/smoke suppressants	Various	As required

Additives for wood–polymer composites 25

Table 2.2 Polypropylene (PP)-based composite

Function	Material	Dosage range (%)
Matrix component	Polypropylene resin	Difference from total of other components to 100%
	Natural fiber	30–60
Lubricant(s)	Stearates/esters/EBS/other	3–8
Antioxidants	Phenolics/phosphites	0–1
Acid scavengers	Stearates/hydrotalcites	0–1
UV protection	HALS/benzophenones/benzotriazoles	0–1
Mineral filler	Talc	0–10
Biocide	Zinc borate	0–2
Density reduction	Microspheres/chemical or physical blowing agent	0–5
UV protection/aesthetics	Pigments	As required
Flame retardants/smoke suppressants	Various	As required

Table 2.3 Polyvinyl chloride (PVC)-based composite

Function	Material	Dosage range (%)
Matrix component	Vinyl	Difference from total of other components to 100%
	Natural fiber	30–40
Heat stabilizer	Various	1–3
Lubricant(s)	Stearates/paraffin/esters/OPE/EBS/other	3–8
Acid scavengers	Stearates/hydrotalcites	0–1
UV protection	HALS/benzophenones/benzotriazoles	0–1
Mineral filler	Carbon filler	0–10
Biocide	Zinc borate	0–2
Co-stabilizer	Zeolite	0–3
Density reduction	Microspheres/chemical or physical blowing agent	0–5
UV protection/aesthetics	Pigments	As required
Flame retardants/smoke suppressants	Various	As required

This table does not take into account most PVC processors use PHR versus %.

2.2 Lubricants and rheology control additives for thermoplastic composites

'Lubricant' is a generic term for many different chemicals that affect the rheology of molten thermoplastics, whether or not they contain fibers. The rheological effect of additives intended for other functions can be seen in unfilled thermoplastics and is even more pronounced in highly filled thermoplastic composites. Lubricants are used to improve the rheology of the total formula so it will process as required.

Rheology refers to how the melt behaves in processing. There can be a number of parameters affecting where in the process the material melts (or fuses if polyvinyl chloride): apparent viscosity, apparent pressure at different process points, anti-stick attributes on the metal part of the equipment, how melt flows into different zones of the extrusion die or mold, even energy draw, etc. Other terms for lubricant used interchangeably are wax, process aid, anti-stick, slip, release agent, flow modifier, etc. Most of these terms describe function while wax refers to appearance of many lubricant materials regardless of the actual chemistry.

The terms 'internal' and 'external' lubricant are often used and, depending on the perspective, can have different interpretations. The author uses the following definitions:

- internal lubricants affect viscosity and flow attributes because these additives are compatible with the resin of the melt, essentially lubricating resin molecules;
- external lubricants affect anti-stick and slip attributes as these molecules will be incompatible with the melt, thereby separating and migrating to the surface of the melt, hence lubricating between the melt and the metal of process equipment.

Wax products as a group have diverse chemistry. The most basic division is between oleo- and petroleum-derived products. Oleo-derived materials come from naturally occurring triglycerides (fats and oils), where the fatty acids and alcohols derived from triglycerides are subsequently saponified, esterified, or otherwise chemically reacted. Petroleum-derived materials are paraffin, microcrystalline wax, and polymer waxes. Polymer waxes are chemically the same as many of the thermoplastics polymers commonly used, but with much lower molecular weights so their behavior is wax-like, not resin-like. The best way to distinguish them is the melting point range.

All waxes are solid. However, though rarely desirable, many products of similar chemistry are available as liquids. If the compound blending system is more oriented to liquid additives, it is important to look for the same general chemistry, but select additives with low enough molecular weights and/or a more branched than linear structure, and/or containing hydrocarbon chains that are predominantly unsaturated.

Lubricant selection will vary with choice of the thermoplastic polymer, and possibly the fiber. However, there are also further considerations. The predominant composite in the market today is wood fiber in a polyolefin resin (polyethylene or polypropylene). The common wax lubricants are metal stearates (soaps), amides, and esters. Stearates are cost-effective lubricants that provide external lubricity, and the rheology is often balanced with amide wax for flow attributes. The most commonly used metal stearate in polyolefin-based composites is zinc stearate. The zinc content will be relatively consistent among manufacturers. There can be variation due to the manufacturer's saponification process and the choice of fatty acid to saponify. It is imperative that the formulator recognizes the likelihood of reaction between metal stearates and maleated coupling agents. This prohibits the use of stearate lubricants when maleated coupling agents are required. Alternative lubricant chemicals will work, as discussed in Section 2.3.

Amide waxes are *bis*-amides or mono amides, the former being most prevalent as lubricants in composites. The most common one is EBS, ethylene bis stearamide, though there are others. For reference, the mono amides are not used in composites, being mostly used to affect the slip and block properties of film products.

There are many families of ester, in that all have the ester function, though differing in the overall structure of the molecule. An ester is formed by the dehydrogenation of a hydroxyl functional group with a carboxylic acid functional group. The families of ester that can be dehydrogenated are simple esters, di-esters, glycerol esters, and esters of polyols other than glycerine. All can vary with choice of fatty acids. Where there are multiple hydroxyl sites on a polyol, an additive can vary by degree of esterification as well as selection of fatty acids. As an example, complex esters are made with a blend of mono- and di-functional carboxylic acids esterified to longer polyols. Since ester lubricants are the most varied type of lubricant chemistry, the author recommends depending on a reliable manufacturer with application knowledge to guide selection.

Generally esters and polymer waxes cost more than other lubricants. However, these materials make different contributions to melt rheology, and if the process or product requirements can be achieved only with these lubricants, the price will be easily justified.

PVC, polyvinyl chloride, or vinyl, is becoming more prevalent in the composite industry as it has superior physical properties to polyolefins, and in principle is a lower-cost thermoplastic. PVC as a resin has different additive requirements from other resins, including lubrication. It is in PVC that paraffin and functionalized derivatives of polymer waxes come into play. In the early days of PVC, unplasticized material was extruded on single-screw extruders, and the lubricant system was calcium stearate and EBS. When twin-screw extruder technology was introduced, alternative lower-cost lubricant systems came into being, remaining prevalent today, being calcium stearate and paraffin, often enhanced with ester and/or oxidized polyethylene wax. In PVC composites, all these lubricant chemistries remain applicable. There are now materials other than waxes used to provide rheology control. In PVC, certain oxidized polyethylene waxes and acrylic polymer process aids are used to shorten fusion time while addition of paraffin will delay fusion.

One last comment is appropriate regarding improved flow of PVC. As long as the composite is to be used with structural performance in mind, do not make the mistake of improving flow with a plasticizer. These materials are chemically similar to some ester lubricants and are lower in cost but there is a major difference in performance. Both internal lubricants and plasticizers will improve the flow of the PVC melt, but plasticizers will also make the end product flexible where the lubricants will not. Flexible PVC simply does not have the rigidity for structural requirements, and can also depress the heat distortion temperature below that which is acceptable.

Heretofore we have only reviewed various organic chemicals, yet there is a new inorganic technology on the market that affects rheology of thermoplastics, with a number of performance attributes that could justify the term process aid. Benefits include the potential for increased output, lower energy demand, and a variety of improvements to secondary process and product attributes with very low dosages of the additive, approximately 0.5% maximum. Off-white to colorless, these materials are natural aluminosilicate glasses (e.g. volcanic ash) specifically that are for the most part amorphous, i.e. with minimal crystalline content. The author subscribes to a minimum 87% amorphous constituent for adequate performance. This is the key characteristic that allows this mineral to be included in compounds in the processing of thermoplastics. Higher levels of crystalline silicates will give the additive sufficient hardness to be harmful to the metal surfaces of the processing equipment. Since these materials are so old in origin, unless there is an overriding parameter, they are extremely inert and will have no reactivity with polymers or additives.

The obvious rheological benefits mimic the internal lubrication previously discussed. However, there are additional characteristics to simply lubricating between polymer molecules. Although the mechanism determining how these materials function is still being elucidated, nonetheless their performance is well established. The benefits can be seen during evaluation of commercial processes. The improvements to productivity and product quality can be quantified to determine the cost effectiveness of these products, which are more expensive than most additives. These materials should be considered when processes are already otherwise optimized for productivity, energy consumption, overall unit cost, and product attributes.

A further attribute of these particular materials is that in the end product the particles will be locked into the polymer matrix. This can be a benefit when a process uses an organic additive that exhibits some problematic type of migration or blooming behavior. This occurs when the organic material, at a variety of possible kinetic levels, moves through the polymer matrix to the surface of the melt or part. If the material has extremely fast kinetic behavior, this can happen when the material is at melt stage. Even items with a low tendency for such behavior can migrate when overdosed into the formula. Many terms have been used to describe these phenomena: die drool, plate-out, exudation, calibrator fouling, etc.

As with any inorganic material from a ground deposit, there is natural variation in composition from deposit to deposit. The products being marketed in this category are likely to be different on that basis. Furthermore, the marketers of this technology all have proprietary processes for finishing the material into the final additive, and these also are not necessarily the same. It is recommended to consider the following in differentiating between product options: confirmation of complete inertness so as to be assured of no chemical interaction with other additives and the resin, ratio of amorphous glass to crystalline minerals where the latter needs to be minimized, proof that the production process provides for lot-to-lot consistency of particle size distribution and particle morphology suitable to the user needs. It will also be helpful that the manufacturer can provide data substantiating their performance claims. The reader should refer to Section 2.5 on mineral fillers for further general information on inorganic materials.

2.3 Coupling agents

Since there are many different types of composite, this implies there can be different chemistries of coupling agents. As the thrust of this text is woodfiber thermoplastic composites, further discussion will focus on those commonly used. Coupling agents, also known as compatibilizers, have the primary function in composites of improving the blend homogeneity of dissimilar or incompatible materials. Lack of homogeneity can prevent the development of satisfactory structural properties in the end product; hence the use of these materials improves physical properties. It has also been well substantiated that there is the further advantage of reducing water absorption by the fibers in application of the end product, the mechanism being enhanced encapsulation of the fiber by the polymer. Further benefit from encapsulation will be elaborated in Section 2.7 on biocides. Reduced water absorption minimizes the fiber swelling that can distort the physical dimensions of the end product.

Compatibilizing is a surface-active phenomenon. Essentially it comes down to the electrostatic nature of the two surface materials at the interface. Molecules constituting a surface can vary in polarity, from non-polar to various strengths of dipole moments. When they are widely different, there is a lack of compatibility, and a molecule is needed that can bridge the two surfaces, creating compatibility. Wood surfaces are electrostatic, while polyolefin resins are non-polar. Coupling agents are molecules with structures and functionality that ensure different parts of the molecule have different dipole moments. These dissimilar sections of the one molecule will be compatible with different components of the composite matrix.

The predominance of polyolefin resins in today's wood–polymer composite (WPC) market takes this discussion to the most common general chemistry of coupling agent, that of maleated polyolefin. These materials are manufactured by grafting maleic anhydride onto a backbone of polyethylene or polypropylene. The maleic functionality provides compatibility with the wood fiber surfaces, while the polymer backbone is compatible with the matrix resin.

Although the chemistries are the same, there is product differentiation in the market based on degree of maleation and molecular weight of the polymer. Mechanisms are promoted for both types. One theory is the polymer must be of 'resin' molecular weight, to provide maximum compatibility with the matrix resin, whereby benefit is derived from improved anchoring of the coupling agent to the matrix resin. The author suggests an alternative theory.

Since coupling is a surface phenomenon, the focus must be on the interface of the two materials. An important consideration is delivery of functionality to the surfaces. Regardless of resin and molecular weight, the maleated version of a given product will melt at higher temperatures than the resin itself. Hence it is probable that a maleated 'resin' will melt at a higher temperature than the matrix resin – therefore later in the conversion process – so allowing fibers to be first coated by matrix resin, i.e. prior to the coupling agent. For any polymer chemistry, the lower molecular weight of the wax-like version will melt at a lower temperature than the higher molecular weight variant. Hence the likelihood is that a maleated 'wax' will melt before the matrix resin, allowing the coupling agent to first coat the fiber. The net result from using maleated wax-based coupling agents is potentially superior delivery of functionality to the fiber–polymer interface, with the likelihood of improved cost–benefits.

Further to this 'interface' model, it is the author's experience that choice of coupling agent to minimize the compatibility of the coupling agent backbone with matrix resin further improves performance of the coupling agent, owing to enhancement of concentration at the interface. Maleated waxes can cost more than maleated resins on a unit weight basis; however, if performance testing shows that the wax versions provide better results, then less is needed for the required performance, providing savings on the additive.

An important note from Section 2.2 is that there is a strong contra-indication to the use of stearate-based lubricants in the presence of maleated coupling agents. If coupling agents are required for structural and water absorption parameters, do not use stearates. In PVC composites, coupling agents have little relevance. PVC is a polymer with greater physical properties, and the electronegativity from the chlorine atom on every monomer unit probably allows for dipole–dipole interaction at the polymer–wood fiber interface, negating the need for a coupling agent. There are similar expectations at the interface for any polymer resin being used as the matrix resin where there is a dipole moment in the monomer.

2.4 Stabilizers

Stabilizers function by preventing or minimizing the deleterious chemical reactions that result in the degradation of either the composite matrix or a component of the matrix. As there are numerous chemical reactions to prevent, and more than one way to stabilize any one system, the scope of this section is potentially enormous, hence the discussion will be necessarily streamlined. The focus is on the stabilization of the main matrix components, polymers and fibers, subsections being antioxidants, UV stabilizers, heat stabilizers for PVC, and a few miscellaneous types considering the demands of processing and field performance. Fortunately there is a large and well-defined body of knowledge regarding the stabilization of thermoplastic polymers, which is germane to the technology of thermoplastic composites. Nonetheless, performance needs to be confirmed by the formulator/processor since the cocktail of chemicals that can be introduced by the non-cellulose components of any fiber can have an effect.

2.4.1 Antioxidants

Antioxidants have two types of function: that of performing when the material is being processed, and that of performing when the end product is in use. Prior to further discussion, a process history must be established. Processing of a thermoplastic will cause some degradation. Correct matching of material and process with stabilizer chemistry and dose will keep this degradation minimized and virtually undetectable. Nonetheless, though undetectable, there will be some degradation, and degradation will start again when the end product is put to use. The main reason for minimizing undetectable process degradation is that whatever happens in the field, degradation will pick up from where it left off in processing. To better illustrate, think of the polymer chain as a zipper, and degradation as the unzipping action. Degradation in production will partially unzip the chain. In the field, degradation will start with both previously degraded and unaffected polymer chains. Degradation on the former group will start where it left off, i.e. the polymer chain that is partially unzipped will have the unzipping continue. It is for this reason that process antioxidant will have a beneficial effect to end product performance, hence both types are important.

Most polymers upon being heated by processing start to degrade slowly, except for PVC, which will degrade very quickly. Hence antioxidants are used for most polymers and heat stabilizer is used for PVC. Traditional antioxidant chemistries are phenolics, phosphates, and thioesters. In addition to their chemistry, another way to review antioxidants is through their functionality, meaning that while certain antioxidants will prevent degradation during processing, others will be more suitable to prevent degradation in field performance from deleterious effects of heat aging. The traditional chemistries are all valid, being well established for decades, and are now quite economical since patents have expired. The new chemistries being introduced to the market are subject to patents by the manufacturer, so can be priced higher; hence they require evaluation versus traditional materials to confirm which is the most cost effective.

One subset of additive that should be reviewed at this point is that of acid scavengers. Some processors consider these because of the chemical cocktail that can be present originating from fiber resins. There are a variety of possible chemistries: stearates are a low cost per unit weight material, and hydrotalcite-like products are high cost per unit weight materials. The benefit of the latter material, although expensive, is that if the fiber of choice causes need for acid scavenger, it does not have an adverse effect on coupling agent, and very little is needed for extraordinary performance; hence it can be much more cost effective.

2.4.2 UV stabilizers

The energy of UV radiation upon impacting an outdoor end product can initiate chemical reactions. Many possible chemical reactions will utilize the presence of oxygen and water. To counter the potential of UV-initiated degradation, three strategies can be employed: blocking/screening UV energy, absorption of UV energy, and finally stabilization, which is trapping radical species generated subsequent to degradation.

Blocking/screening is done with high-coverage/opacity materials such as titanium dioxide, carbon black, and many pigments. UV absorption is achieved with chemistries of benzophenones and benzotriazoles. UV stabilization is traditionally achieved with HALS – hindered amine light stabilizers. HALS are available as both monomeric and polymeric forms; however, most of the latter are not on the Domestic Substances List in Canada. The same comments under antioxidants are germane for UV chemistry regarding new patented versus traditional off-patent chemistries, and cost effectiveness.

Blocking/screening is an important concept. In WPCs the very high level of fiber can be an effective UV blocker, in that the UV radiation can only penetrate to just below the surface, so the bulk of the matrix is protected. Hence many processors achieve satisfactory product attributes without anything further than appropriate pigments used for aesthetics.

These chemistries should be used only for the prevention of polymer degradation, having no effect on preventing fade of wood fiber. Pigment systems also require separate considerations for UV protection, which are best addressed by pigment manufacturing or pigment masterbatch industry specialists.

2.4.3 PVC heat stabilizers

PVC heat stabilizers are almost always a metal ion with organic ligands. Rigid PVC is most commonly stabilized by tin in North America and, in the rest of the world, calcium–zinc is replacing lead. The archaic choice of cadmium is now rarely used. Tins are almost always liquid, and there is a variety in the market. Further to cost effectiveness and performance requirements, choice is often based on organoleptic considerations in that there can be odor issues in the manufacturing venue. Calcium–zinc products are available in both liquid and solid form. The cost effectiveness of these materials has evolved nicely in the last decade, with near parity on this parameter with lead and tin products.

Further to the primary heat stabilizer, there are various ways to enhance PVC stabilization with materials that act as co-stabilizers. These are additives that in small doses contribute to the cost effectiveness of the stabilizer system, yet on their own these same materials are not adequate primary stabilizers. Epoxidized oils and resins are most common, the performance arising from the epoxy functionality. There is also a well-established technology based on synthetic zeolite, which traps chlorine ions released by degradation, preventing these ions from contributing to further propagation of degradation chain reactions. The materials are synthetic zeolite, which eliminates natural variations. However, the formulator should note that many synthetic zeolite products are detergent grade, and not suitable for polymer processing due to excessive grit.

All the stabilizer chemistry heretofore is used to protect polymer from degrading. There is a new patent-pending technology, which purportedly claims that use will lead to less process degradation of fiber, and will result in improvements to composite physical properties, with further benefits to overall processing as well. At the time of writing this chapter, the author has not seen any performance data; however, more information will be available in the future. This material is similar to the zeolites previously discussed as co-stabilizers in PVC, and hence the same cautionary statements apply.

2.5 Fillers

In keeping with the subject of this text, the fillers to be discussed are those that are natural fibers, predominantly wood fiber. As these are matrix components, and therefore covered in Chapter 1, this section will focus on other materials that are used as fillers, further to the natural fiber content. The reason these other materials are considered as additional constituents in the formula is that after determining the maximum level of the chosen natural fiber that can possibly be incorporated, as limited by the process and the end product attributes, we still seek to replace resin with less costly materials.

The primary perspective of fillers is that they are mineral based, low cost, and added at as high a level as can be accepted by the system, so conversion costs are not inflated, and required end product attributes are not undermined, all with the intention of reducing material costs. The following commentary on fillers is from a number of perspectives as well as financial consequences, such as morphology and functionality.

Besides the possibility of any ground deposit becoming a mineral filler, there are those materials that can be used for the same purpose, but are essentially waste product, for example, ground tire rubber or ground thermoset-based tire bumpers. Even these fit into the 'lower the material cost' perspective. Let us first address a major consideration of cost for any low-cost material. By definition, these materials have a low value, and hence a major factor in the price to the processor is the freight from source to the processing plant: whether paid separately or included in the price, this cost is present. While sometimes a given mineral is most suitable to a given polymer or application, the intention should be to identify the 'closest' source of any acceptable product. It generally costs the same to extract and grind any mineral, but shipping costs make a difference, and these are usually differentiated by distance. This is a good rule-of-thumb, but it does not account for other marketing or economic parameters, such as a sole deposit of a particular mineral having been developed that has a niche application, or some other distinctive consideration.

One important differentiating factor between like minerals is the content of crystalline silicates. These are present at different levels in various ore bodies, are species with great hardness, which will contribute to wear and tear of plastic processing equipment, and, depending on the end product, can affect the end properties. Since crystalline silicates are from ore bodies, and if the closest of two ore bodies has more crystalline silicates, then the decision to be made is between the lower cost of this material versus the possible costs of using a higher crystalline content on machine maintenance.

Another differentiating mineral property to consider is heavy metal content. As previously alluded to in earlier sections of this chapter, regarding functional materials that are inorganic, there are similar considerations to inorganic fillers. For the end product to be weatherable, under-the-hood, or otherwise exposed to various heat histories and UV radiation, and/or oxygen and/or moisture, caution is required as higher heavy metal content under these conditions will contribute to increasing resin degradation, hence lowering longevity of end-product performance. Exposure to heat and/or UV light can have an adverse effect on long-term resin stability in the end product since it contains inorganic additives with greater heavy metal content.

Filler morphology alone can be the subject of complete texts, so for simplicity think in terms of shape and effect. Filler particles can be amorphous, spherical, platelet, or fibrous. Fibrous and platelet types will have reinforcing attributes to bolster certain physical properties; the amorphous type will be the opposite and will weaken the product. Spherical types can go either way depending on particulars of the particle size distribution and the caliber of the dispersion in the matrix. Fibrous fillers can be the most difficult to process; however, if using them for effect, the proper selection of processing additives can mostly compensate. The reinforcing properties can be interpreted through a single specification for each type: if platelet shaped, consider aspect ratio (area to thickness), and if fibrous, consider dernier (ratio of length to diameter).

Other key attributes are particle size distribution, color, hardness, specific gravity bulk density, and oil absorption character. Particle size distribution can be the limiting parameter on formula content of the additive. Consider two products of the same mineral, same cost into the plant, same general properties, same average particle size, but differing in the actual particle size distribution from which the same average particle size is derived. At either end of the distribution are the smallest particles known as 'fines' and largest particle sizes known as 'top-cut'. Fines have more surface area for the same weight of material than larger particles. At the same dosages of our two conjectural products, the one with more fines will cause higher melt viscosities, which generally will lower output or increase energy draw, hence dose is at the highest possible level for melt viscosity still to be acceptable for good processing, and less can be used. Alternatively, the top-cut consideration is based on the fact that weaknesses of impact resistance can always be associated with irregularities in the matrix homogeneity (more of this in Chapter 11). When comparing the two conjectural fillers, again at the same dosage, the one with the greater amount of top-cut will result in an end product with lower impact resistance; hence less of this filler can be added to the end product to meet this specification. The conclusion is to not dismiss more expensive fillers if the particle size distribution is narrower, as it is still lower in cost than resin and increased dosage could actually result in lower cost of end product because more can be added with good conversion rates and/or without jeopardizing end product physical properties.

The most common mineral filler in polyolefin-based composites is talc. The platelet is purported to aid processability as the particles align with the vector of polymer processing, and also adds reinforcing character. Carbonates are the rule in PVC, though they are amorphous. The most common carbonate is calcium carbonate; however, the dolomitic version, calcium magnesium carbonate is similar in inertness, and slightly different in crystal structure. The differentiating factor should be total net contribution of cost to end product.

Other minerals will have special attributes to consider, depending on application. Most notable are barytes, chemically barium sulfate. All barytes fall into one of two categories, either natural or processed, sometimes called synthetic. The difference is that the natural baryte is simply the ore as it is found, ground to specification; the processed is where the barium sulfate is dissolved, the crystalline silicates and heavy metal-containing species are therefore separated, and the barium sulfate is precipitated from solution in very high purity, often with outstanding control of particle size distribution. The benefits of lower crystalline silicates, lower heavy metals, and narrow particle size distribution already have been review. The distinguishing attribute of this mineral is the high specific gravity of approximately 4.4, which varies by purity. Use will enhance density of finished articles, improving attributes of sound and vibration dampening, and there can be other performance considerations.

2.6 Density reduction additives

Though Chapter 11 discusses foaming, this section is included to present further considerations. In brief, density can be reduced by a number of options. Physical blowing agents are the use of uncompressed gas released into the matrix during mid-processing. Ultimately this is the low-cost scenario, although it is the most capital-intensive option and requires reliable knowledge by the process personnel. Chemical blowing agents are solid materials that are dispersed with other additives into the compound and, upon exposure to heat of processing, decompose to release a gas as one of the decomposition products. This option is far less capital intensive, but often pre-existing process equipment such as dies need to be redesigned, and, again, require dependable processing expertise. The third option for density reduction is really not foaming, but achieves the same objective using a recently developed additive technology. These materials are more expensive, but there is no requirement for new capital expense or extra processing competence. These materials are generically known as microspheres.

Microspheres are one of two types: either they are resilient to the effects of shear and pressure, or they are not. The latter are mostly ceramic or glass, and are easily broken by the dynamic environment of polymer processing. They are more used in coatings, and hence are not pertinent to this discussion. The resilient material in the market today is a polymer shell encapsulating gas. This is either expanded or compressed, and we are most interested in the latter. These materials vary in the polymer of the shell, choice of compressed gas and therefore the temperature of expansion, and the diameter of the shell after expansion. The benefit is that the cell structure forms, while the gas remains contained. Therefore the matrix does not have to be managed to contain the gas and forms a satisfactory cell structure, eliminating equipment and specialized process capability. The selection process is based on first using the temperature profile of the process, and also the final shell size, to approximate the correct product and dosage. After initial testing, further refinement of product choice can be made.

2.7 Biocides

Biocides for these composites prevent microbial species from feeding on the organic matter of the natural fiber. There are two primary types of microbial intrusion to avoid: those that occur on the surface of the composite and which are often attributed to mildew, and those that occur sub-surface and are mostly fungal. Mildew on surfaces is unsightly, while fungal rot will undermine the structural integrity of the composite by ingesting the fiber.

Microbial growth requires moisture and time, so antimicrobial additives are not required for products designed for short or temporary service periods, or that have adequate enhancement of resistance to water absorption from coupling agents, or otherwise have no chance of any exposure to moisture. These materials should be considered for products that will be exposed to moisture, or should not be exposed to moisture, but can be, for example, sub-floor components.

There are both organic and inorganic products on the market. The organic products are often proprietary molecules, and generally have a higher cost per unit weight with low dose requirements. Manufacturers can present data on performance versus typical dosages so cost effectiveness can be ascertained. The inorganic type is a much simpler molecule, essentially zinc borate, made from reacting zinc oxide with boron. The rule-of-thumb dosage is 2% maximum. It is important to note that not all zinc borates are the same; quality of performance can vary for a number of reasons. A different purity of the zinc oxide reactant, and the caliber of the manufacturing process to fully complete reaction, can cause some versions to have lower zinc borate content diluted with unreacted raw materials. Regardless, lower-quality zinc borate, while lower in cost, has a less satisfactory performance in the composite, often because the inadequacies lead to susceptibility of leaching to the composite surface, causing another type of aesthetic inadequacy.

An interesting but less well-documented phenomenon is that while composites are not a target of wood-consuming insects, insects will eat through a composite to access a wood substrate or core. It has been established that zinc borate-containing composites are resistant, preventing insects from accessing the wood. One last point is that not all commercial trade names of zinc borates in the market are registered with government agencies (varying with each country) as functional biocides. Only the use of registered products allows for marketing and label claims by manufacturers.

2.8 Product aesthetics additives

Additives that affect end product aesthetics fall into two categories: those that affect characteristics of the part surface, by influencing interaction between the molten compound and the die by modifying process rheology, and those that alter appearance simply by being present, usually pigments. Appearance can also be affected by choice of the filler, whether fiber or mineral, i.e. different materials look different. In summary, depending on what is sought, look to pigments, rheology modification, or possibly alternative fillers.

The last parameter on aesthetics from an additive perspective is the understanding that the UV additive system that protects the integrity of pigment longevity needs to be distinguished from the UV additive system that is meant to protect long-term resin integrity. The former is appearance while the latter is structural.

2.9 Flame retardants and smoke suppressants

Here is another topic worthy of its own textbook and if this functionality is important to you, I strongly recommend you seek out one. The following is more a synopsis of key considerations. Flame retardant (FR) chemistry falls into only a few families and functions; however, before the additives can be chosen, it is important to understand why a certain additive should be used. The expense of incorporating such additives is only undertaken when essential, and this is when the article in question, if not flame retarded, can potentially contribute to propagation of a fire in which human life or assets can be lost. If so, then there are codes and standards for performance that need to be met. First, the standard the product must meet must be confirmed. Often, reference to the name of the test used to determine the performance is confused with the standard itself, so ensure you acquire the correct information. Details will vary by application, e.g. automotive versus roofing versus wall panels, and the standards will be different. Knowledge of the test is also important, since any given particular formula will generate results that differ when the test procedure is a variable.

Once the standard is identified, the actual material constituents of the part need be considered. A roof made of slate will meet UL790 with no additives as it is stone and has no fuel content, but one made from a polymer will require additives, furthermore as the polymer varies, so will the FR system, i.e. one from PVC will have different requirements from one of polypropylene. Remember the objective is to make the part, regardless of material, meet the standard. The two most basic data points are which polymer and how much inorganic (filler) is contained. The latter will not contribute to the fuel content, and the former is a variable of its own, in that different resins burn differently and present different challenges to flame retardancy. Some standards require limits on the smoke that can be generated in a burn, and therefore sometimes a smoke suppressant is required. There can also be other parameters of limited universality for which special solutions are needed.

Once the aforementioned details have been established, a well-versed FR professional can assist in nominating a suitable system. Primary FR chemistries are halogenated, phosphated, metal hydroxide, plus a small number of miscellaneous chemistries with niche applications. Halogenated FRs are generally the most cost effective, especially the brominated versions. Chlorinated products when providing a further functionality to flame retardancy can be cost effective. For example, in PVC, chlorine is already part of the resin and in the system at no extra cost. The cost effectiveness of halogens is maximized when combined with a synergist. The function of these systems is in the gas phase of the flame to extinguish the fire.

The most common synergist is antimony trioxide despite its often being very expensive when at the top of the pricing cycle. One interesting note is that antimony trioxide can enhance smoke generation. If this is a problem, other products can be used in combination with it: zinc borate or a molybdenum derivative. These are primarily smoke suppressants, but they are also halogen synergists, though not as efficient as the antimony trioxide. In combination, equal flame retardancy with lower smoke can be achieved. Metal hydroxides at low levels in certain polymer systems can contribute to incremental improvement of smoke suppression, although whether this is also in the case of thermoplastic wood composites remains to be ascertained.

Phosphate chemistry FRs act as char formers after the burn has started, physically separating the fuel from the necessary oxygen in the atmosphere, and extinguishing the flame. Most phosphate FRs are liquid esters and most suitable for PVC, rubber, and latex matrices. Since they also plasticize, parts cannot be structural. Note that many tests are for complete structures, not the individual components, such as the UL790 roof test where a section of roofing is burnt. So where a composite can be used for a roofing component such as a tile or frame, a non-composite roofing membrane can be flame retarded with a phosphate ester, such as a PVC product, and as a part of a whole roof construction, together contribute to the total FR performance and adherence to the standard.

Ammonium polyphosphate is a solid phosphate FR that can be used in composites. As this material will foam when decomposing, it is referred to as an intumescent FR. There are also other chemistries of intumescent FRs, some of which are based on melamine chemistry. Intumescent products all form particularly strong chars so that there is a certain structural characteristic to the char that prevents collapse after the fire. The only important consideration in processing with intumescent FRs is to ensure the peak process temperature is below the decomposition temperature.

Metal hydroxides of all FRs are the lowest cost per unit weight, but for most applications can be the least cost effective owing to the large quantities required. Applications where these are best used will be where part surfaces are mostly horizontal. Metal hydroxides when exposed to fire will decompose at a characteristic temperature into water and metal oxide. The primary mechanism for function is that instead of the heat energy being used to burn the fuel in the part, it is absorbed by the endothermic decomposition of metal hydroxide. When the part has a horizontal surface, the decomposition product of metal oxide provides further functionality by reflecting heat energy so it is not available to the fuel, and this cannot occur on a vertical burn. The formation of water accounts for only approximately 15% of the additive performance. Unlike the other chemistries of FR, there are basically only two types of metal hydroxide, though these have many versions varying with particle size distribution and whether coated or not, and types/loading of the coating. Magnesium hydroxide is more efficient on a unit weight basis than aluminum trihydrate, but costs more. The real cost effectiveness of the former is based on the lower loading requirement of what is generally a high loading chemistry. The inorganic materials, like mineral fillers, at high levels increase melt viscosity and can inhibit conversion efficiency.

Magnesium hydroxide, with approximately 15% lower requirement for equal performance to aluminum trihydrate, will incrementally allow for more efficient conversion and therefore lower unit costs. Staying with our roofing example, a roofing tile will meet UL790 with 49% aluminum trihydrate but 42% magnesium hydroxide. If the two products have identical particle size distribution, use of the latter will provide for best possible processability.

Other miscellaneous FR chemistry exists but is likely not pertinent to a discussion on composites. Nonetheless, the following are three examples:

1. Certain sulfur derivatives can be the most cost-effective chemistry to FR polycarbonate.
2. Some standards require no flaming drips. If polyolefin, addition of high oil absorption talc stops drips of the molten polymer.
3. Some standards require that after the flame has been extinguished, the ember must not glow. If this needs to be addressed, a low percentage of zinc borate will prevent glow.

2.10 Future trends

Development of new chemistry is infrequent; however, industry often generates new additive product technologies from using existing chemistries in novel ways. It is the author's experience that new technologies can present new paradigms that allow for either new capabilities or improved conversion scenarios. Either way, meeting market expectations and securing new profits can result. Assessments of claims are required, and the onus is on the manufacturer to generate supporting data that can be scrutinized.

2.11 Conclusion

Additives can be used wisely and cost effectively to optimize performance and profit. Due diligence is required with an understanding of all the parameters. The catch is that for any one objective there is often more than one solution available; one must identify the option that provides the best overall answer, after considering all the possible ripple effects of any option.

3
Interactions between wood and synthetic polymers

K OKSMAN NISKA, Luleå University of Technology, Sweden
and A R SANADI, University of Copenhagen, Denmark

3.1 Introduction

The structure and properties of the wood fibre–polymer matrix interphase play a major role in determining the mechanical and physical properties of the composite material. In general, the interphase or adhesion between the fibre and the matrix has little effect on the composite stiffness. However, the adhesion and interphase play a very important role in determining properties such as strength and toughness, and long-term properties such as creep and moisture stability. Adhesion can be improved by using coupling agents or compatibilisers. The fibre surfaces can vary greatly which results in varying interaction with the polymer matrix – this is particularly true with wood fibres (WF) due to their natural variability. Regions of both low and high surface energy may exist on the same fibre: some sites may be inert, while others provide sites for specific interactions with polymer molecules. Furthermore, the surface of the fibre may be smooth or rough, and fibre modifications may further enhance surface area.

Figure 3.1 shows scanning electron micrographs of fracture surfaces of a wood–polymer composite (WPC) and the difference between poor and good interfacial adhesion in polyethylene-wood fibre (PE/WF) composite is clearly represented here. Figure 3.1(a) shows the microstructure of the WPC when no compatibiliser was used. As can be seen, the wood fibre surface is smooth and clean and the PE matrix is rougher. Furthermore, a gap or space between the fibre and matrix can be seen, indicating poor adhesion. Figure 3.1(b) shows the microstructure of the PE/WF composite when a compatibiliser was added. As shown in the figure, it is more difficult to discern the wood fibres since they are partially covered by the compatibiliser. Furthermore, there is no gap between the wood fibre and PE matrix, indicating good adhesion. Considerable work has been going on studying the interface, adhesion and stress transfer in wood and natural fibre composites and several groups around the world are working this area. Several special issues of the journal, *Composite Interfaces* (2000, 2005), have been devoted entirely to the interface, interphase and adhesion in cellulose-based composites, and more are planned in the near future.

42   Wood–polymer composites

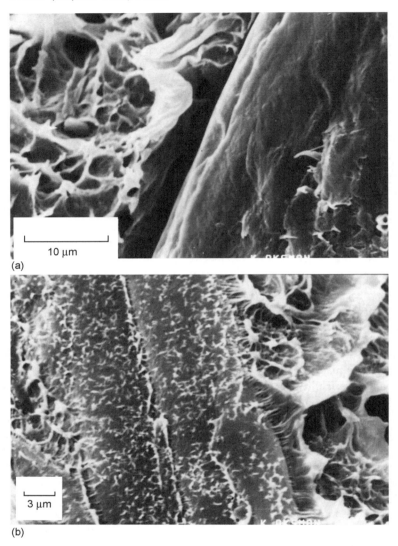

3.1 Example of (a) poor and (b) good adhesion between the wood and PE matrix (Oksman and Bengtsson, 2007).

3.2   The interface and interphase in composites

The interface is a two-dimensional surface between two phases while the interphase is defined as the three-dimensional bulk regions between the fibre and the bulk of the polymer matrix. Poor interaction between the polymer and fibre decreases the adhesion between them. Figure 3.2 shows a schematic description of the interface and interphase in a cross-section of a WPC. From the figure it

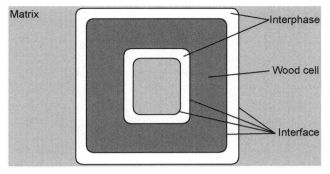

3.2 A schematic description of an interface and interphase in a cross-section of WPC.

can be seen that two interfaces exist, one between the interphase and the wood fibre and another between the interphase and the polymer matrix.

There are several scenarios of failure which can occur in a composite. In the simplest of cases, adhesive failure can occur in the fibre–interphase *interface* (I), the interphase–matrix *interface* (II) or cohesive failure of the interphase (III), shown in Fig. 3.3. However, in general, failure in WPCs is usually more complex. This is due to the uneven nature of the surfaces discussed previously.

3.3 Wetting, adhesion and dispersion

Wetting and dispersion are of great importance in achieving good adhesive interaction between two phases. Wetting can be defined as the extent to which a liquid makes contact with a surface. In wood–thermoplastic composites it is characterised by the degree of direct interfacial contact between the wood and polymer surfaces. Adhesion is sticking together two surfaces so that stress can be transmitted between them and can be quantified by the amount of work is required to pull the two surfaces apart. The degree of dispersion describes the

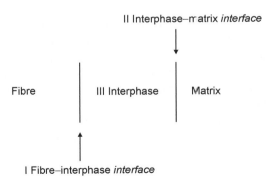

3.3 Schematic of the three distinct failure modes; adhesive (I and II) and cohesive (III).

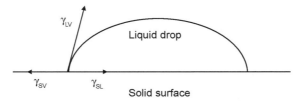

3.4 Contact angle of a liquid drop on a solid surface.

degree of mixing in a colloidal system; good dispersion would correspond to colloidally stable individual particles and poor dispersion to particle agglomeration.

Wetting, dispersion and adhesion are three concepts that are closely related but not synonymous. Wetting is a necessary but not sufficient condition to achieve good dispersion and adhesion in composites. Even with good wetting, dispersion may not be adequate and neither does it guarantee good fibre–matrix adhesion.

Wetting is generally described by Young's equation, as the contact angle of a liquid on a surface equilibrium.

$$\gamma_{SV} = \gamma_{SL} + \gamma_{LV} \cos\theta \qquad 3.1$$

where the γ_{SV}, γ_{SL} and γ_{LV} are the surface energies or surface tensions of solid/vapour, solid/liquid and liquid/vapour interfaces respectively, and θ is the contact angle. The contact angle of a liquid drop on a flat surface is shown in Fig. 3.4.

Ideally for complete wetting, θ should be zero. For systems where θ is high, more work is necessary to penetrate and wet out the wood fibre surface. Equation (3.1) indicates that θ can be reduced by and wetting improved in three ways: by raising γ_{SV} or lowering γ_{SL} and/or γ_{LV}. Reduction of surface tension in the polymer melt can be achieved by using surfactants. However, changing the surface energetics of the system such as increasing the fibre surface energy and/or lower polymer surface energy can result in improved wetting. This can be achieved by the use of suitable surface treatments or by addition of interfacial active additives. Wetting is a precondition for adhesion, since the latter requires contact between the phases. Unwetted areas are very weak and do not contribute to adhesive strength. Furthermore such regions constitute defects around which stress concentrations can occur and thus provide sites for initiation of interfacial failure. The work of adhesion, W_A, is the work per unit area that is required to separate two phases reversibly, and will be lower if the wetting is poor since the interfacial area is reduced. There is a direct relationship between W_A and wetting. Thus:

$$W_A = \gamma_{SV} + \gamma_{LV} - \gamma_{SL} \qquad 3.2$$

and combined with equation (3.1) yields

$$W_A = \gamma_{LV}(1 + \cos\theta) \qquad 3.3$$

so that W_A is maximised when θ is zero. However, the work of adhesion is not fully described by equation (3.2) because the fracture process is generally irreversible. In practice, W_A constitutes only one fraction of the work required. A major contribution to the energy required to pull the interface apart is from energy dissipation via viscoelastic processes occurring during fracture in either or both phases in the interfacial region. Thus the observed fracture energy is strongly dependent on their viscoelastic properties, on the rate of fracture and on the nature of interaction between the phases. For example the presence of covalent bonds or molecular entanglements between the phases will increase the work required to separate them. The adhesion theories can be divided to mechanisms which are showed schematically in Fig. 3.5.

3.3.1 Chemical bonding

Covalent bonding occurs between chemical groups on the wood fibre surface and the compatible chemical group in the matrix polymer. Chemical bonds are formed between the fibre surface and some molecules in the matrix. The strength of the interface depends on the type of bond and the number of bonds. In WPCs, groups such as silanes can result in chemical bridging between the matrix and fibre surface. A siloxane bridge is formed on the surface of the fibre, while the other end of the silane interacts with the polymer (Bengtsson and Oksman, 2006). Maleated polyolefins are also used to create a chemical bond on the fibre surface, while the other end of the anhydride grafted polymer molecules physically entangle with the matrix polymer (Oksman *et al.*, 1998).

3.3.2 Electrostatic bonding

Difference charges between two surfaces, such as one surface having a negative charge and the other surface a positive charge can result in forces of attraction. Again, the strength of the interface will depend on the intensity and number of the bonds. These forces include ionic, hydrogen bonding, acid–base interactions, dipole–dipole and dipole-induced dipole interactions, polar interactions and van der Waals forces. However, a more recent way of thought has been developed by Folkes (1964) where the work of adhesion is split into two parts, dispersive (*d*) and polar (*p*).

$$W_A = W_A^d + W_A^p \qquad 3.4$$

The polar component can be rewritten as

$$W_A = W_A^d + W_A^D + W_A^h + W_A^i + W_A^{ab} \qquad 3.5$$

In this case, the polar component has been expanded to separate the dipole–dipole interactions (*D*), dipole-induced dipole (*i*) interactions, hydrogen bonding

46 Wood–polymer composites

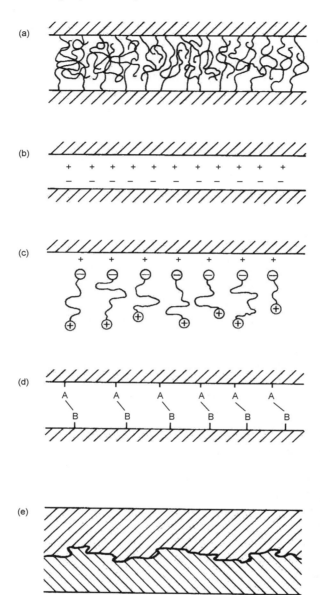

3.5 Different adhesion mechanisms: (a) bond formed by molecular enlargement following interdiffusion, (b) bond formed by electrostatic attraction, (c) cationic groups at the end of molecules attracted to an anionic surface resulting in polymer orientation at the surface, (d) chemical bond between groups A on the surface and groups B on the other surface, (e) mechanical bond formed when liquid polymer wets a rough solid surface (from Figure 3.3 page 41, *An introduction to composite materials*, Derek Hull, Cambridge University Press, 1982).

(*h*) and acid–base interactions (*ab*). A more recent way of thought of Folkes expression is to rewrite it to include the Lifshitz–van der Waals (LW) forces and acid–base interactions:

$$W_A = W_A^{LW} + W_A^{ab}$$ 3.6

The LW forces include all types of dipole interactions and dispersive forces. The acid–base interactions according to the Lewis acid/base theory involve the sharing of an electron pair, and includes hydrogen bonding.

3.3.3 Mechanical bonding

Mechanical bonding can occur due to the physical interlocking of two surfaces. Frictional shear can be significant in such cases, and this may be enhanced due to induced processing stresses such as resin shrinkage during cooling of thermoplastics or curing of thermoset-based resins. Roughness will increase the work of adhesion because the surface area is increased. It can also provide adhesion by a mechanical interlocking effect. This kind of mechanical adhesion can be formed if one or both phases are liquid and stress transfer is possible even if there is no interaction at molecular level other than weak secondary bonds, and W_A is low. WPCs are formed when the matrix polymer has low viscosity, and therefore this mechanism may make a significant contribution to the stress transfer efficiency especially if no compatibilisers or coupling agents are used. Although in some cases the mechanical stress transfer can be high, there is no or little improvement in properties such as rate of water absorption into the wood fibre or the long-term composite properties such as creep.

3.3.4 Interdiffusion

It is possible to form interactions between two polymers entirely due to molecular entanglement between two types of molecules. The level of interaction and adhesive depends on the amount of molecular entanglement. Molecular entanglement depends on the level of diffusion and a minimum degree of polymerisation is necessary for the adhesive strength to be high. The use of maleated polyethylene (MAPE) results in chemical bonding between the anhydride moiety and the wood fibre surface, while the other end of MAPE results in interdiffusion with the PE matrix (Oksman *et al.*, 1998). A minimum polymer chain length is necessary to obtain optimised stress transfer through molecular entanglement and therefore the molecular weight of the MAPE is an important consideration for adhesive strength.

Clearly the stronger the molecular interaction between the wood fibres and the matrix, the greater is the resulting adhesive strength and better the stress transfer efficiency. The strongest adhesion will occur when covalent bonds are formed at the interface between the fibre and coupling agent or the matrix itself.

Interfacial modification can be designed to produce this kind of interaction. Hydrogen bonding and other acid–base interactions, dipole–dipole interactions and dispersion forces across the interface also provide a means to improve adhesion. The advantage of covalent bonding is clearly seen from experiments by Sanadi *et al.* (1994) on recycled wood fibre–polypropylene (WF/PP) composites. Maleic anhydride grafted polypropylene (MAPP) which forms covalent bonds with the fibre and acrylic acid grafted polypropylene (AAPP), which forms only non-covalent (acid–base and LW forces) interaction were compared as compatibilisers. Properties when using the MAPP were significantly better than when the AAPP was used as compatibiliser.

3.4 Techniques to evaluate interfacial interactions and adhesion

There are a number of methods for studying the properties and interfacial phenomena in composites. Some surface analysis methods directly characterise the chemical nature of fibre surface while others use indirect methods to obtain information of surface characteristics. Model single fibre tests such as pull-out and single fragmentation test give ideas on the interfacial shear strength and are useful in comparing different interfaces, particularly when the same fibre and matrix systems are used.

3.4.1 Surface analysis

Chemical characterisation not only gives information on the chemical structure of the fibre surface but can also give information on the type of chemical interactions occurring between the fibre surface and a coupling agent or compatibiliser. Electron spectroscopy for chemical analysis (ESCA), also known as X-ray photoelectron spectroscopy (XPS), and Fourier transform infra-red spectroscopy (FTIR) are among the techniques used to gain information on the surface properties of wood and cellulose-based fibres, before and after surface treatment. Techniques such as inverse gas chromatography can give information on the acid–base interaction potential of untreated and treated wood fibres.

ESCA is a highly surface-specific technique and can provide information on elemental composition on the fibre surface up to a depth about 10 nm. Spectra are obtained by irradiating a beam of X-rays onto the surface and plotting the binding energy of the electrons escaping the surface versus the number of electrons detected in an ultra-high vacuum. Each element produces a characteristic binding energy values and these peaks are related to the electron configuration of the electrons within the elemental atom. The binding energy and intensity of the peaks provide information on the surface modification such as the identity of the elements and their quantity within about 10 nm of the sample surface. This method was used to get information about surface characteristics of untreated cellulose fibres and fibres grafted with MAPP (Felix and Gatenholm, 1991). The treated fibres showed an increase in the peak intensity at about 285 eV, indicating increased number of C–C bonds, while the O/C and the O/(O–C=O) ratios were decreased significantly. The change in C–C bond numbers and the presence of a different O/C ratio indicated that the MAPP was concentrated on the fibre surface, thus making the fibre surface more hydrophobic since the amount of O atoms on the surface was reduced.

Infra-red (IR) spectroscopy works on the principle that chemical bonds have specific frequencies at which they vibrate. Diatomic molecules have a bond that may stretch, while complex molecules have many bonds, and this leads to infrared absorptions at different characteristic frequencies. These absorptions at specific frequencies are related to a particular bond. To obtain information on the type of bonds on the surface, a beam of IR light is passed through the sample, and the energy absorbed at each frequency is noted. IR wavelengths frequently used for analysis range from 2500 nm to 16 000 nm. The transmittance and absorption spectra obtained at a particular frequency allow us to identify certain bonds present. Sample preparation for solids usually involves making a very finely ground sample mixed with a special salt (e.g. potassium bromide), and this is made into a compacted translucent pellet. FTIR spectroscopy is a measurement method for collecting infrared spectra over a wide range of frequencies simultaneously, and using a mathematical Fourier transform of the signal results in a spectrum identical to that conventional IR spectroscopy, albeit at a much faster rate. A newer IR technique is to use attenuated total reflect Fourier transform infra-red spectroscopy (ATR-FTIR) which allows samples to be used in their original state. In case of solids, the sample is pressed directly against an ATR crystal.

Oksman *et al.* (1998) used FTIR to show evidence of grafted anhydride on wood fibres. Maleic anhydride grafted styrene–ethylene–butylene–styrene (SEBS-MA) was used to improve the adhesion and toughness of WF/PE composites. Figure 3.6 shows the IR spectra using a diffuse-reflectance technique. The spectra of three samples are given: (A) SEBS-MA, (B) SEBS-MA with 40 wt% WF and (C) WF. Absorption bands at 1865 and 1786 cm^{-1} are characteristic of cyclic anhydride, where as the bands 1713 and 1602 cm^{-1} are typical frequencies for maleic acid carbonyls and aromatic ring stretching vibrations. From these spectra the authors concluded that maleic anhydride moiety grafted on the fibre surface but was also was present in its maleic acid form. Thus from the evaluation of the type of chemical bonds present it is possible to gather information on the type of molecular interactions present in the system. Mohammed-Ziegler *et al.* (2006) used many spectroscopic techniques including ATR-FTIR to study the surface of silane-modified wood composites and also found this technique useful in studying the interactions.

Electron probe microanalysis (EPMA) can be used to get information on the presence of select elements in ppm. EPMA works by irradiating a small volume

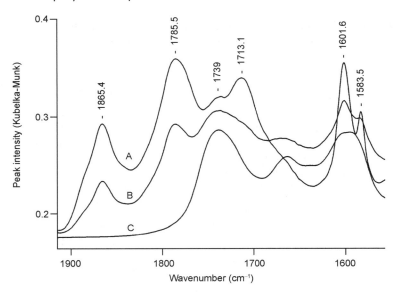

3.6 FTIR spectra of (A) SEBS-MS, (B) SEBS-MA with 40 % WF and (C) WF (Oksman *et al.*, 1998).

of a sample with a focused electron beam (usually 5–30 eV) and collecting X-rays induced and emitted by various elements. Since the wavelengths of the X-rays are characteristic of an elemental species, the sample can be easily identified. For example, this technique, coupled with a scanning electron microscope (SEM) has been used to obtain information the location of silicon present in a WPC (Bengtsson and Oksman, 2006), when silane was used in HDPE-based WPC. WFC samples were embedded in an epoxy matrix and polished before coating with a thin layer of carbon to prevent a charge build up. Figure 3.7 shows an SEM-EPMA scan of WFC with and without silane coupling agents – the dark areas indicate the presence of silica. Very little silica was present in samples without the silane, in Fig. 3.7(a) while a significant amount was seen is Figure 3.7(b). Secondly, an interesting observation was that the silica was located close to the wood fibres, suggesting that the silane thermodynamically segregates to the wood fibre surface during processing of the composites, and thereby indicating the effectiveness of surface modification when using silane.

Inverse gas chromatography (IGC) has been used to study the surface energetics and acid–base characteristics of polymers. The term 'inverse' is used since the technique characterises the stationary phase of the column rather than the injected probe. In IGC, the stationary phase (in this case wood fibres) is of interest. In addition to *n*-alkanes (where the interactions are only dispersive forces), liquid probes of known acid–base characteristics can be used. The interaction between the probe and the wood fibre column determines the resid-

Interactions between wood and synthetic polymers 51

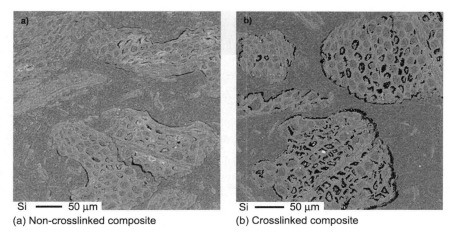

(a) Non-crosslinked composite   (b) Crosslinked composite

3.7 SEM-EPMA mapping of silicon in a polished cross-section of WF/PE composites (a) without silane coupling agent and (b) with silane coupling agent (from Bengtsson and Oksman, 2006).

ence time of the probe liquid in the column and the shape of the chromatograph. Bleached sulphite wood pulp fibres, with and without surface modification, were used in a study by Felix and Gatenholm (1993a) to observe changes in the surface characteristics. Fibres were modified by grafting alkenyl succinic anhydride (ASA) on the fibre surface. Probes such as n-hexane, n-heptane, n-octane and n-nonane provided only dispersive interaction with the wood fibre surface. Chloroform and ethyl acetate were used as the acidic probes, while acetone and diethyl ether were used as the basic probes. From their IGC study the authors concluded that the ASA blocked a significant amount of acid–base interactive sites on the fibre surface. Rials and Simonsen (2000) also studied the surface characteristics and adhesion potential of wood and thermoplastics using IGC. Another method of evaluating the surface energetics of modified fibres is the Wilhelmy plate technique and this has been used effectively by Park *et al.* (2006) on many interfacial modifications in jute and hemp plastic composites.

3.4.2 Microscopy

Many different microscopy methods can be used to study the interactions between the fibre and matrix. SEM is a technique commonly used in material science. The theory behind this technique is that when an electron beam targets a sample volume, there is emission of electrons and other electromagnetic radiation that are detected to produce an image of the sample surface. Secondary electrons and backscattered electrons from the sample are detected and the observed image of the sample displayed on a screen. Conventional SEMs need a very high vacuum to scan the sample, and this can result in changes on the sample due to, for example, dehydration. Environmental SEM (ESEM) has been developed that can scan samples in a low-pressure gaseous environment and high humidity and the specimen need not be coated with a conductive material as in conventional SEM. However, the resolution is lower.

SEM is extensively used to observe the fracture of composites, and gives insights into the quality of fibre–matrix adhesion and the type of interface/interphase failure in composites. Figure 3.1(a) is a good example of interphase behaviour and shows failure at the fibre surface interface of a WF/PE composite where the surface of the fibre is clean with no indication of polymer adhering to it. Furthermore, the physical gap between the fibre and interphase also indicates poor fibre–matrix bonding. Figure 3.1(b) shows a WF/PE composite where a maleated polymer has been added. The presence of some polymer adhering to the fibre surface and the lack of physical gap between the fibre and interphase provides evidence of a good fibre–matrix adhesion resulting in good stress transfer. Comparing the lengths of fibres pulled out of the fracture plane between systems with different interfaces can give insights into the effect of fibre surface or interphase modification. Smaller pull-out lengths indicate better adhesion when the same fibre–matrix combinations are used. Figure 3.8(a) shows a typical fracture surface of a bleached cellulose fibre in PP showing fibre pull-out and the fairly 'clean surface' suggest relatively poor bonding. Figure 3.8(b) shows a significant decrease in pull-out lengths when 3% by weight MAPP was added to the composite.

Optical microscopy has been used in conjunction with a hot stage to get information on crystallisation behaviour. Although isothermal crystallisation studies are not truly representative of the composite in real situations, they do give insights into how induced transcrystallisation can affect interphase and thereby influence composite properties. Increase in the thickness of the transcrystalline layer, improved the interfacial shear strength, as measured by the single fibre fragmentation test, by 40–100% (Felix et al., 1994).

Transmission electron microscopy (TEM) has been used in WFC composites to observe samples at a higher resolution than can be obtained using an SEM. In TEM a beam of electrons is transmitted through a thin sample. The darker areas represent those areas where fewer electrons are transmitted due to that area being thicker or denser than a lighter area. Additional sample preparation is needed when using the TEM. This technique has been used to get high-resolution images of WF/PE composites with a SEBS-MA compatibiliser (Oksman et al., 1998). Figure 3.9 shows a TEM micrograph of the sample. SEBS-MA is expected to be concentrated in the interphase region between the wood fibre and PE matrix, shown as a dark region in the picture. The micrograph also suggests that the compatibiliser is not uniformly distributed around the wood fibre, with some regions showing a thicker interphase.

Confocal microscopy is a technique that is valuable in getting high-resolution images and 3D reconstruction. Since images are produced point by point and reconstructed using a computer, the image produced tends to be blur-free. In one

Interactions between wood and synthetic polymers 53

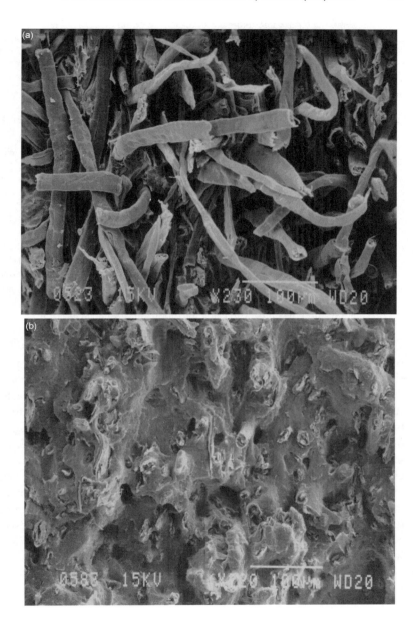

3.8 Scanning electron micrograph of the fracture surfaces of 20% by weight cellulose-PP composites: (a) uncoupled and (b) coupled using 3% by weight MAPP (courtesy Craig Clemons, USDA Forest Products Laboratory, Madison, WI, USA).

54   Wood–polymer composites

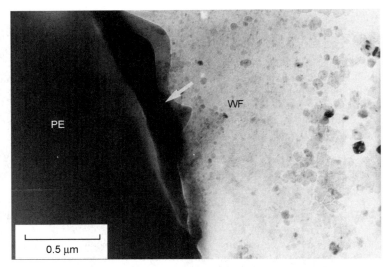

3.9 Transmission electron micrograph of a LDPE/WF composite with 5% SEBS-MA as compatibiliser. The dark region, between the fibre and PE, indicated by the arrow shows the area where compatibiliser is concentrated, suggesting the efficacy of using the modifier (Oksman *et al.*, 1998).

such technique, confocal laser scanning microscopy, a laser beam is focused within a specimen and the emitted fluorescent light is detected and the light signal is transformed into an electrical signal which is recorded by a computer. As this technique depends on fluorescence, a sample may have to be treated with fluorescent dyes. However, the amount of dye can be very low so disturbance to the sample is minimal. The confocal laser technique yields better images than other confocal techniques such as confocal spinning disk microscope, but the image framing rate is very slow. The laser confocal technique has been used effectively by Grigsby and Thumm (2007) in observing WFC and fibre swelling of composites. Information such as potential matrix fracture at the interface/interphase due to water swelling of the fibre in composites may also be observed using this technique. Kazayawoko *et al.* (1999) used a confocal microscopy to observe changes in the dispersion and distribution of wood fibres in PP, with and without compatibilisers.

3.4.3   Single fibre pull-out and microbond tests

The pull-out test has received considerable attention, because of the possibility of obtaining valuable information about the fibre–matrix interface/interphase and the ability to distinguish between failure modes. Information on the debonding energies, average and maximum shear strength, and frictional shear components can be obtained by careful control on the test (Piggott, 1991; Penn and Lee, 1987; Sanadi *et al.*, 1993).

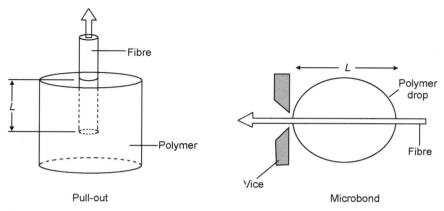

3.10 Schematic of micromechanical interfacial shear strength test samples: pull-out (left side) and microbond tests.

Figure 3.10 show schematics of pull-out and microbond tests. A single fibre is embedded in a polymer cylindrical block (typical pull-out test) or a droplet or bead of polymer (micro-drop test). Force is applied to the end of the fibre and pulled out of the matrix. The average load measured prior to debonding by the fibre from the matrix is recorded. The interfacial debonding stress, τ, is calculated from the following equation:

$$\tau \frac{F}{\pi dl} \qquad 3.7$$

where F is the maximum load at debonding, d is the fibre diameter and l is the embedded length. In the case of wood and natural fibres, the fibres are not uniform and circular, but have complex shapes and a more appropriate way is to use the perimeter in the equation instead of the diameter. This equation can then be modified using the perimeter, P, to give:

$$\tau = \frac{F}{Pl} \qquad 3.8$$

No information is available using wood flour or fibres using the pull-out test, possibly due to the difficulty in obtaining long and straight fibres to perform the test. Wood dowels have been used to gather information on surface modification and the effect on the adhesion in case of low molecular weight polyethylene, and clear differences between uncoated, maleated polypropylene coated and polyethylene acrylic acid coated systems were observed using the pull-out test (Sanadi et al., 1993).

However, the microbond test was used by Liu et al. (1994) to get the interfacial shear strength (IFSS) of wood fibres and polystyrene. It must be noted that their experimental technique describes the wood fibres as greater than 5 mm in length, and possibly in the range greater than 7 mm. Their results are interesting! IFSS increased from about 3 to 10 MPa with only fibre acetylation. This is a significant increase with only secondary electrostatic interactions (mainly acid–base work of adhesion) enhancement and possibility of increased mechanical bonding, but no possibility of covalent bonding.

3.4.4 Single-fibre fragmentation test

The single fibre fragmentation (SFF) test was adapted to polymers after Kelly and Tyson (1965) observed fragmentation of tungsten fibres in copper. Figure 3.11 shows a schematic of a single fibre embedded axially along the length, in a dog bone specimen, and a tensile is applied to it. The load builds up in the fibre through stress transfer from the matrix. As the stress increases, the fibre breaks, first at its weakest point. As the load increases, the fibre progressively fractures, until all fragment lengths are below the fibre critical length (see Fig. 3.11). In other words, the fracture stress in the fibres is unable to be reached since not enough stress can be transferred to the fibre. Kelly and Tyson (1965) used the critical length to obtain the IFSS and developed the equation:

$$\tau = \frac{\sigma_f d}{2l_c} \qquad\qquad 3.9$$

where τ is the IFSS, σ_f is the fibre strength, d is the fibre diameter and l_c is the fibre critical length. Drzal et al. (1982) used the SFF test in carbon fibre polymer

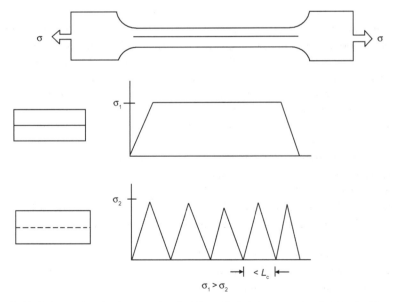

3.11 Schematic of micromechanical interfacial shear strength: single fibre fragmentation test.

composites. They concluded that the fibre fragments are not all the same length, but are a distribution of fragment lengths and the two-parameter Weibull distribution fit their data. They then used the following equation to determine the IFSS:

$$\tau = \frac{\sigma_f}{2\beta} \Gamma\left(1 - \frac{1}{\alpha}\right) \qquad 3.10$$

where α and β are the shape and scale parameters, respectively, and Γ is the gamma function. Own *et al.* (1986) suggested that the bimodal log-normal distribution was a better fit for their data.

One disadvantage of the test is that the failure strain of the matrix has to be *significantly higher* than the failure of the fibre. This limits the wood fibre composites to matrices such as PE and PP, but this test may not be successful for brittle polymers such as polystyrene where the failure strain of the polymer is closer to that of the fibre.

Few results are available with wood fibres owing to the difficulty in obtaining longer lengths and relatively straight fibres needed for proper handling to set up the test. Thuvander *et al.* (2001) used single wood pulp fibres in a vinyl ester (a thermoset polymer) and studied the IFSS using the SFF test. They concluded that the processing of the wood pulp fibres can change the results of the IFSS. With increased time in a PFI mill, the variability in the fibre strain to failure decreased, and they associated this with increased fibrillar debonding of fibres that have been processed for a longer time.

Quite a few authors have used other cellulosic and plant-based fibres to obtain IFSS using this technique (Felix and Gatenholm, 1994; Herrara-Franco and Valadez-Gonzalez, 2005). Felix and Gatenholm (1994) successfully used the test to evaluate the effect of the transcrystalline interphase on model isothermally crystallised cotton–PP systems, while Herrara-Franco and Valadez-Gonzalez (2005) used the test effectively to distinguish between different surface treatments and silane modification on henequen fibres in a high-density polyethylene (HDPE) matrix. In the tests by these researchers, the fibre was placed between two sheets of polymer, which was then pressed in a hot press. After pressing specimens were cut out of the pressed films, and then analysed using the SFF test. Felix and Gatenholm (1994) obtained IFSS between cotton fibres and polypropylene, which varied from 6.1 to 12 MPa, depending on the type and thickness of the interfacial layer.

3.4.5 Interfacial studies using laser Raman spectroscopy

A newer technique that is getting attention is laser Raman spectroscopy. The advantage of this technique is that IFSS estimates can be obtained at low strains, and furthermore very long fibre lengths are not a requisite. The principle is that Raman frequencies of molecular vibrations in fibres are stress dependent (Galiotis, 1991). Calibration curves can be developed that plot Raman frequency versus strain from this phenomenon. A great deal of information can be gathered on the interface and this has been successfully conducted on carbon and Kevlar in an epoxy matrix (Galiotis, 1991; Schadler *et al.*, 1992). Using the strain-induced shifts in Raman spectra of cellulose fibres such as hemp (Eichhorn *et al.*, 2000), information such as the maximum interfacial shear stress of a hemp single fibre embedded in a droplet of epoxy was obtained (Eichhorn and Young, 2004). Tze *et al.* (2007) used the test to get data on the maximum interfacial shear stress data of regenerated cellulose (lyocell) fibres in PS and compared several different surface treatments.

3.4.6 Other techniques

Mechanical tests give indirect, but very useful, information on how the interface affects composite properties, such as tensile and flexural strength and impact toughness, both notched and un-notched. Tensile and flexural tests and different types of impact tests are examples. Improvements in strength and failure strain using the tensile and flexural test, for example, give insights into adhesion and stress transfer efficiency (Sanadi *et al.*, 1994; Oksman and Clemons, 1998), with all other factors being equal.

Dynamic mechanical thermal analysis (DMTA) has been used effectively to evaluate changes of interphase modification on dynamic properties. DMTA measures the amplitude and phase of the displacement of a specimen in response to an oscillating force. In purely elastic materials the stress and strain are in phase so that the response of one caused by the other is instantaneous. In viscous material the strain lags the stress by a 90° phase shift. In viscoelastic material, there is a lag between the responses of the strain to the stress. The storage (E') and loss (E'') moduli of the composite can be evaluated using the DMTA over a wide temperature range and a wide range of oscillating frequencies. Tan δ, the damping factor can also be calculated. A variety of testing fixtures are possible including tensile and bending modes of operation.

Figure 3.12 shows a DMTA scan of PP/EPDM-MA (polypropylene/ethylene propylene diene terpolymer) with and without addition of WF and MAPP (Oksman and Clemons, 1998). The elastomeric tan δ peak located at $-38\,°C$ is shifted about 5 °C downwards and broadens with the addition of the WF and MAPP. The peak shift to lower temperatures indicates increased mobility of EPDM molecular motion through increased free volume near the filler surface.

Improvement in interactions between the fibre surface and polymer such as through chemical bonding results in changes in polymer molecular mobility at the interphases. This can result in, for example, the broadening of the glass transition and changes in the peak temperature of the transition, and can be observed in the tan δ or E'' spectra. Felix and Gatenholm (1993b) studied the effect of different molecular weight MAPP compatibilisers that were grafted on to cellulose fibres prior to composite preparation and this can be seen in Fig.

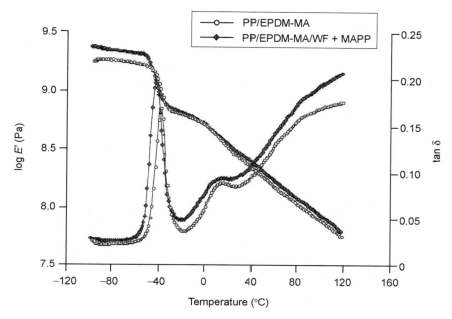

3.12 DMTA spectra of PP/EPDM-MA and PP/EPDM-MA plus WF and MAPP (from Oksman and Clemons, 1998).

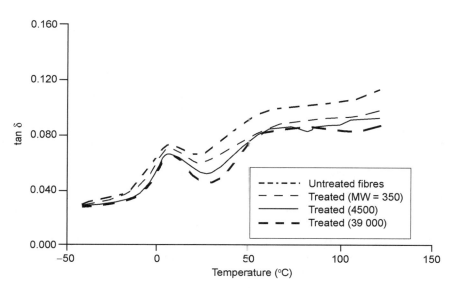

3.13 DMTA curves of 40% by weight cellulose–PP composites. The numbers 350, 4500 and 39 000 indicate the molecular weight of the MAPP grafted on the fibres (from Felix and Gatenholm, 1993b).

3.13. The higher the MW of the compatibiliser, the higher the peak temperatures of the relaxation process and the broader the peak. The higher peak temperatures suggest reduction in molecular mobility, and this is likely to be at the interphase region. Similar broadening and increase in peak temperatures with improved adhesion and stress transfer efficiency have been observed with other lignocellulosic fibres such as kenaf-PP systems (Sanadi and Caulfield, 2000).

Differential scanning calorimetry (DSC) is a technique in which the difference in the amount of heat required to increase the temperature of a sample and reference are measured as a function of temperature. Both the sample and reference are maintained at very nearly the same temperature throughout the experiment. The basic principle underlying this technique is that when the sample undergoes a physical transformation such as phase transitions, a different amount of heat will need to flow to it than the reference to maintain both at the same temperature. DSC may also be used to observe more subtle phase changes, e.g. molecular transitions such as the glass transitions. Changes are crystallinity and changes in the glass transition temperature with the addition of coupling agents can give information on the effect of interphase modification. DSC can also be used to obtain information on the effectiveness of crosslinking agents such as silane coupling agents in WPC.

3.5 Improving interface interactions in wood–polymer composites

Improvement of the interfacial adhesion between the wood fibre and thermoplastic matrix is important and many researchers have studied this phenomenon. Different methods have been tested and the basic role of these different modifications have been to reduce fibre–fibre interaction by facilitating better wetting and dispersion, and to improve the adhesion and stress transfer between the matrix and the fibres (Dalväg *et al.*, 1985; Sain and Kokta, 1993a,b; Felix *et al.*, 1995; English *et al.*, 1996; Oksman, 1996; Lu *et al.*, 2005; Angles *et al.*, 1999; Balasuriya *et al.*, 2002; Wang *et al.*, 2003; Danyadi *et al.*, 2006; Bengtsson and Oksman, 2006; Correa *et al.*, 2007). Several means to try to improve the adhesion have been used, some with more success than others. Gauthier *et al.* (1998) have covered some of the many surface modification agents used in early work with thermoplastics.

This section will address several of the modification techniques used by various authors in the field of study. Two distinct paths have been tried in order to achieve the above goals. The first method is to modify the fibre surface without changing any component of the matrix. This involves an initial stage of fibre surface modification prior to the fibres being compounded with the matrix polymer. The second method is to modify the matrix by adding dispersing, coupling agents and/or compatibilisers and crosslinking agents in the compounding stage in an extruder or other plastic processing equipment. This second method has a better chance of commercialisation since other than additional material cost, no additional processing step is needed for this technique.

3.5.1 Surface modifications

Plasma modification of the fibre surface has been studied to enhance adhesion and stress transfer. Felix *et al.* (1994) used cold plasma treatment to change the surface properties of the fibres to enhance acid–base interactions in composites. They used two gases, ammonia and nitrogen, to provide basic groups on the fibre surface. They also used methacrylic acid (MAA) to provide acidic sites on the surface and also the creation of plasma polymer on the fibre surface. Neither of the two gases proved effective with PP since no additional adhesion mechanisms were added to the system since PP being non-polar allows only dispersive interactions. However, the MAA showed improvement, although no new acid–base interactions were developed. They explained the improvement in properties with interdiffusion of the PP molecules into the plasma polymer deposited on the surface. In case of a PS matrix which is basic in nature, the MAA (acidic surface) was effective since both acid–base interactions increased and inter-diffusion was also possible, as in the case of PP. In case of a chlorinated polyethylene polymer matrix which has an acidic surface, both nitrogen and ammonia improved properties were effective in improvement of properties due to enhanced acid–base interactions since the gases enhanced the basic nature of the fibre surface.

Another method that has been used to modify the fibre surface is to first deposit and then graft suitable molecules on to fibre surface, prior to the plastic compounding stage. This is usually done by soaking the fibres in a solution of the molecules, and then drying/heating the fibres in an oven to achieve the molecular graft. In general, the additional processing step will *increase* the cost of the composite and therefore is *seldom used in industry*. Some of the molecules that have been used are silanes, isocyanates, MA grafted polymers, etc. Depositing or grafting (e.g. dispersing agents, coupling agents or compatibilisers, with or without crosslinking agents) not only improves wetting and dispersion, but can also result in improved mechanisms of adhesion such as the possibility of molecular entanglements between the deposited molecules and the matrix.

3.5.2 Improved interactions using additives

Many types of molecules have been used to modify the fibre–matrix interaction, either by initial deposition and then grafting or by addition of suitable molecules in the compounding step. Non-reactive treatments change the surface energy of the fibres by thermodynamically segregating to the fibre surface without reacting with the polymer matrix. This leads to changes in fibre–fibre interaction, leading to improved dispersion, distribution, easier processing, and generally also improved impact properties. In WPC systems these non-reactive additives can be different waxes and stearates. Dalväg et al. (1985) used dispersing agents such as stearic acid, paraffin wax, and polyethylene waxes with PE and PP. Stearic acid resulted in the highest improvement in failure strains (no data on strength for the three dispersing agents were reported by these authors). Kokta et al. (1990) used stearic acid to improve the dispersion of the fibres in PVC and this led to about a 20% increase in strength and ultimate elongation at 20% by weight fibre loading. Reactive surface treatments result in a chemical reaction between the added couplings/compatibilising agent and the fibre surface. This will result in improved fibre–matrix adhesion and thus lead to improved mechanical properties. A wood fibre contains several surface hydroxyl groups that can be utilised for chemical bonding and hydrogen bonding interactions with the polymer matrix to create strong adhesion at the wood fibre interface. Sain and Kokta (1994) tested bismaleimide as an additive to enhance better interaction and thermal stability. They concluded that maleimide reaction with cellulose resulted in improved both the mechanical and thermal properties of WPC.

There have been several studies on coupling agents such as silanes and isocyanates as coupling agents (Kokta et al., 1989, 1990; Raj et al., 1989; Girones et al., 2007; Karmarkar et al., 2007) on different wood fibre–polymer systems. These agents chemically react with the fibre surface and the other end, then interact with the matrix to improve the stress transfer from the matrix to the fibres. Both silanes and isocyanates react with the OH group on the fibre surface to form primary chemical bonds. Different types of interactions are possible between the coupling agent on the fibre surface and the matrix and these improved adhesion and thus composite properties. Such interactions could include one or more of the following: (a) covalent interaction, (b) acid–base interactions, and (c) molecular entanglement (interdiffusion) between the compatibiliser and matrix polymer.

Raj et al. (1989) used both isocyanates and silane coupling agents in cellulose fibre–linear low-density polyethylene (LLDPE) composites, by depositing these molecules onto the fibres from a dilute solution. The fibres need to be dried before compounding with plastic in an extruder. They reported better properties with use of isocyanates as compared with silanes, and improvement in strength and ultimate failure strain using either of the coupling agents over the uncoupled systems. Xanthos (1983) and Bataille et al. (1989) used silane coupling agents with dicumyl peroxide as the catalyst on cellulose fibre–PP composites, and reported some improvement in strengths. In all the above cases with isocyantes or silanes they suggested that the coupling agent bonded covalently to the OH groups on the fibre, but no covalent bonds are created with the polymer matrix.

A newer method is to incorporate coupling agents during the compounding stage, and the advantage is that no pre-treatment of the fibres is necessary.

Bengtsson and Oksman (2006) used a combination of silane and dicumyl peroxide (an initiator), wood fibres and HDPE in a one-step process in the extruder and this resulted in much improved properties: they suggested that silane forms covalent and hydrogen bonds to the fibre surface, and analytical evidence also indicated the presence of silane–PE cross-links. In earlier work, Bengtsson et al. (2005) also suggest that a siloxane bridge is formed from the surface of the fibre, while the other ends of the silane crosslinks with the HDPE. As Gauthier et al. (1998) explain, these are genuine bifunctional coupling agents where one side or group reacts with the hydroxyl group on the fibre surface, while the other group reacts with polymer molecules.

Geng et al. (2005) used an isocyanate in combination with stearic anhydride, along with the wood fibre and HDPE. No preheating or curing step was used prior to the compounding stage, and according to the authors, the reaction between wood and isocyanate takes place in the extruder and results in significant property enhancement.

The most commonly used and, it appears, most effective compatibilisers for wood fibres and polyolefins are maleic anhydride modified thermoplastics such as MAPP or MAPE (Dalväg et al., 1985; Felix and Gatenholm, 1991; English et al., 1996; Oksman et al., 1998; Angles et al., 1999; Balasuriya et al., 2002; Wang et al., 2003; Lu et al., 2005; Qui et al., 2005; Correa et al., 2007). Here grafting is possible either using a pre-treatment phase or can also occur in the extruder stage during compounding. Unlike the case of crosslinking silane mentioned earlier, these maleated polyolefins do not form a covalent bridge between the modifier and polymer. In these cases, the maleic anhydride grafts to the OH groups on the fibre surface, and the other end molecularly entangles/diffuses with the polymer matrix. Acid–base interactions are possible between any reacted and unreacted MA groups (maleic acids) and the fibre surface. Another advantage is that flexible molecules with the maleic anhydride moiety can be used to optimise both stress transfer and impact toughness (Bengtsson and Oksman, 2006).

Table 3.1 shows how the properties of a WPC composite, with 50% wood flour, is affected by addition of 2 wt% MAPP. The flexural strength of the composites increased from 46 to 59 MPa (~28% increase) upon addition of MAPP and the tensile strength increased from 22 to 30 MPa (36% increase). The addition of

Table 3.1 The influence of MAPP additive in WPC composite (fibre content 50%)

Type of composite	Flexural strength (MPa)	Bending modulus (GPa)	Tensile strength (MPa)	Tensile modulus (GPa)	Impact energy (kJ/m^2)
WF/PP	46	4.0	22	5	8
WF/PP/MAPP	59	4.2	30	5	10

MAPP did not affect the modulus in the same level but the flexural modulus was slightly improved. The impact strength improved from 8 to $10\,kJ/m^2$.

Danyadi et al. (2006), using several techniques including acoustic emission, concluded that using MAPP resulted in wood flour fracture during deformation. They also concluded that fibre–matrix interface debonding was the main failure process when no MAPP was used. This suggests that addition of MAPP improved the stress transfer sufficiently so the length of the wood particles were more than the fibre critical length (l_c). On the other hand, Hristov et al. (2006) indicated that using microdeformation tests with samples deformed *in situ* while under a high-voltage electron microscope, no fracture of wood particles was observed with the addition of the MAPP coupling agent. Manchado et al. (2003) suggested that MA grafted compatibilisers may reduce the interfacial stress concentrations and prevent fibre–fibre interactions which can lead to premature composite failure. Bledzki and Faruk (2003) found that the use of MAPP reduced the damping index significantly, and this again is an indication of better adhesion and improved stress transfer efficiency. According to the authors, damping index is a measure of ratio of dissipated energy (loss energy) to the stored (elastic energy), and gives an indication of impact characteristics during failure. The loss energy is a measure of energy for irreversible deformations such as matrix and interfacial cracking, debonding, delamination, etc.

The molecular weight of the MAPP and the amount of anhydride grafted are both factors to be considered for good stress transfer. Enough anhydride is needed to provide sufficient covalent bonding to the wood fibre surface, while a minimum molecular chain length of the MAPP is needed to provide good entanglement or interdiffusion between the MAPP and the PP matrix (Sanadi et al., 1994). In general, in the production of MAPP, higher anhydride content results in chain scission and a decrease in MAPP molecular weight. Thus MAPP polymers optimised for both amount of anhydride and molecular weight are necessary for good stress transfer to provide strong fibre surface–MAPP interface bonding and MAPP–PP entanglement.

Stress–strain curves of wood fibre–PP composites with different coupling agents and compatibilisers can be seen in Fig. 3.14. Although in experiments by Felix et al. (1994) the coupling agents/compatibilisers were deposited and grafted on the fibres before composite processing, the stress–strain curves give a good indication of how the molecular weight of the MAPP changes composite properties. The numbers at the end of the curves indicate the weight average molecular weight of both MAPP which have about the same percentage of MA graft (about 6% by weight). From the curves, the importance of the entanglement can be gauged – the MAPP with longer molecules permits enhanced interdiffusion and entanglement and thereby result in improved properties. Lu et al. (2005) studied the reaction between MAPE and wood fibres using FTIR and ESCA and also indicated that the anhydride group reacts with the hydroxyl groups on the surface to create an ester. The failure and deformation of WFC,

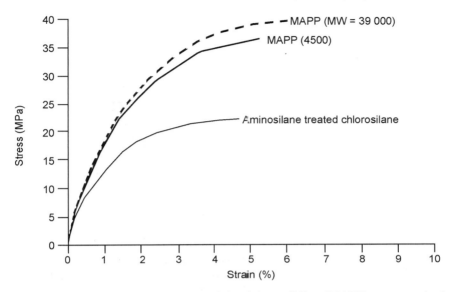

3.14 The effect of molecular weight of compatibiliser (MAPP) on composites' mechanical properties. Stress–strain curves of WF/PP composites using different compatibilisers/coupling agents (from Felix *et al.*, 1994).

with and without MAPP, have been studied by Danyadi *et al.* (2006); addition of MAPP changes the mechanism of failure from fibre debonding to fibre fracture, indicating that adhesion improved sufficiently to ensure that particles were longer than the fibre critical length.

Karmarkar *et al.* (2007) have used a m-isopropoenyl-dimethyisocyanate (m-ID) grafted on to PP as a compatibiliser in PP–wood fibre composites and have seen considerable improvement in properties. A 5% by fibre weight of m-ID grafted PP used as a compatibiliser in a 50% wood fibre–PP composite resulted in a 85% increase in flexural strength. An interesting observation using m-ID PP was that they also observed an increase in tensile modulus which is not usually seen when using MAPP.

Nachtigall *et al.* (2007) have shown some very interesting results using a novel coupling agent. They grafted a vinyltriethoxysilane on to PP to use as a compatibiliser in wood flour–PP composites. Strength improved using these compatibilisers in a 30% by weight wood flour–PP composites. They achieved better properties using these silane-modified PP as compared with MAPP. However, it should be pointed out that the MAPP they used had a very low degree of MA functionality.

Geng *et al.* (2005, 2006) have used combinations of different materials as couplings between WF and PE. A combination of polyaminoamide-epichlorohydrin/stearic anhydride was used as a compatibiliser and they found that this combination was superior to MAPE regarding the composite strength and water resistance (Geng et al., 2006). They also used a poly(diphenylmethane diisocyanate) and stearic anhydride as a novel compatibiliser for WF/PE composites (Geng et al., 2005). They concluded that this combination resulted in better properties than MAPE compatibilised composites. This latter technique is interesting since they also reported data on the reaction between the isocyanate and fibre occurring in the extruder, rather than a pre-treatment stage.

3.6   Interphase effects on other properties

The interphase in composites also plays an important role for toughness and impact properties as well as for long-term properties such as creep and water absorption. Figure 3.15 shows how the impact properties of WPC can be improved by using different additives, impact modifiers. The different modifiers were 1 wt% MAPP, 2 wt% ethylene-propylene-diene terpolymer (EPDM), 2 wt% SEBS-MA. The WF content in the composites is 50 wt% SEBS-MA was shown to act as a good impact modifier for WPC and the improvement in impact strength upon addition of SEBS-MA was larger than when MAPP and EPDM were used as impact modifiers. The elastomeric properties of the SEBS provide a flexible interlayer that is well bonded to the fibre surface that results in the improved impact properties. Theoretical analysis of a flexible interlayer suggest that assuming there is good adhesion, the flexibility (or compliance) and the thickness of the interlayer are both important factors in determining impact improvement and stress transfer efficiency and composite properties can be tailored using those two parameters (Oksman and Clemons, 1998). Sain and Kokta (1994) studied the effect of epoxy-coated WF on the toughening of WPC. The crosslinking of epoxy resin resulted in enhanced energy absorption properties of the polypropylene composites.

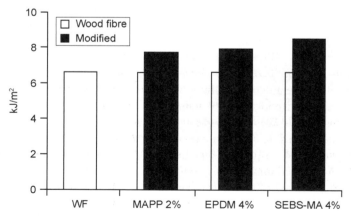

3.15 The effect of different modifiers on the un-notched Charpy impact strength of WPC.

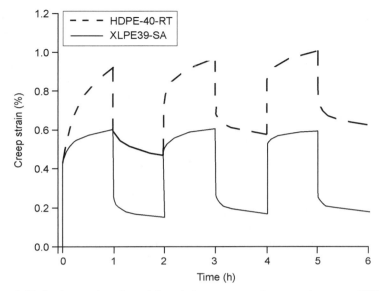

3.16 Strain as a function of time during creep cycling experiments at 60°C in WF/HDPE composites. (Uncrosslinked system, HDPE and crosslinked system, XLPE) The applied stress during loading was 2 MPa (from Bengtsson and Oksman, 2006).

Creep behaviour is dependent on several factors including crystallinity, fibre–matrix adhesion and the interphase properties, and polymer chain entanglements. Bengtsson and Oksman (2006) studied the effect of silane crosslinking on several properties and results indicate that the interphase modification can have a profound effect on short-term creep properties. Figure 3.16 shows the strain as a function of time during creep experiments. The silane crosslinked samples indicated by XLPE have a lower creep response than the uncrosslinked samples (HDPE). The lower creep response can be related to reduced viscous flow due to crosslinking as well as improved adhesion between the matrix and wood flour.

Dynamic creep studies using the dynamic mechanical thermal analyser by Feng *et al.* (2001) of lignocellulosic (kenaf)–PP composites show that in some cases the interphase has a profound effect on creep. They observed that creep improvement when using the MAPP can be significant when lower molecular weight copolymer matrices were used.

The fibre–matrix interface/interphase can also play a significant role in the rate of water absorption in WPC and the deterioration of properties due to water effects. Dalväg *et al.* (1985) have shown that the MAPP has a profound influence in composite properties after testing samples that were boiled in water for 24 h and then tested. Without MAPP the WF/PP composite tensile strength decreased from 19 to 12 MPa. When MAPP (6% by weight) was used, the tensile strength was unchanged at 26 MPa after the 24 h boiling test. A similar

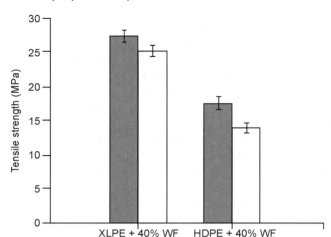

3.17 Tensile strength before (dark) and after boiling test of crosslinked (XLPE) and non-crosslinked (HDPE) composites (from Bengtsson *et al.*, 2005).

test was made for crosslinked composites (Bengtsson *et al.*, 2005). The average moisture uptake was 5.9% for crosslinked and 10.3% for non-crosslinked. Figure 3.17 shows the mechanical properties before and after the boiling test. The tensile testing after boiling is decreased for both materials but the loss is more significant for non-crosslinked composites (-20.6%) compared with 7.8% for the crosslinked.

3.7 Conclusions

The fibre–matrix interface and the interphase are important factors in determining a variety of properties of WPC composites, and therefore should be seriously considered when developing WPCs for any application. In WPC, there have been several techniques used to modify the surface/interface/interphase. From data available in literature, in general, the most effective method is to covalently bond the coupling agent to the fibre surface and then have a strong interaction (such as polymer entanglement) or covalent bonding to the bulk matrix. Covalent bonding and polymer entanglements provide stronger bonds as compared with secondary adhesive mechanisms such as acid–base interactions or Lifshitz–van der Waals forces. The type of interaction between the coupling agent and matrix depends on the type of adhesive bonding possible.

The improvements in interfacial adhesion in WPC have gained in interest during the last 20 years. Many groups around the world are studying means to improve interaction, adhesion and stress transfer in WPC. It is obvious that there is still a lot of potential to improve, optimise and tailor properties using new techniques and/or new coupling agents or compatibilisers.

Two major techniques have been used to modify the fibre–matrix interaction so as to improve adhesion and stress transfer efficiency. The first is to change the fibre surface by depositing a suitable molecule on the fibre surface and then heating or curing the fibres. The fibres are then compounded with the plastic in an extruder. The second method is to use a compatibiliser such as maleated polyolefins or crosslinking agents such as silanes in the compounding step so that any chemical reaction to enhance adhesion occurs in the extruder. In the case of maleated polyolefins, the anhydride functional group reacts with the fibre's surface hydroxyl groups, while the other end of the compatibiliser molecule interdiffuses with the matrix polymer, creating a strong interphase. In the case of silane when used with suitable initiator such as a peroxide, a covalent chemical bridge between the fibre surface and the polyethylene matrix can occur, creating a true 'coupling reaction'. The second method, of a single step reaction in the extruder, has much greater commercial potential since no additional pre-treatment and drying/curing step is necessary prior to the compounding stage.

3.8 References and further reading

Angles, M. N., Salvido, J. and Dufresne, A. (1999), *J. Appl. Polym. Sci.*, **74**, 1962–1977.
Balasuriya, P. W., Ye, L., Mai, Y. W. and Wu, J. (2002), *J. Appl Polym. Sci.*, **83**, 2505–2521.
Bataille, P., Ricardo, L. and Saphieha S. (1989), *Polym. Comp*, **10**, 2, 103–108.
Bengtsson, M. and Oksman, K. (2006), *Comp. Part A*, **37**, 752–765.
Bengtsson, M., Gatenholm, P. and Oksman, K. (2005), *Comp. Sci. Technol.*, **65**, 1468–1479.
Bledzki, A. J. and Faruk, O. (2003), *Appl. Comp. Mater.*, **10**, 365–379.
Composite Interfaces (2000), Guest Editors: Sanadi, A. R. and Rials, T., Special Issues on 'Interfaces in Cellulose-Polymers', **7**, 1 and 2.
Composite Interfaces (2005), Guest Editor: Dufresne, A., Special Issue on 'Natural Fiber Composites', **12**, 1.
Correa, C. A., Razzino, C. A. and Hage Jr., E. (2007), *J. Thermoplastic Comp. Mater.*, **20**, 323–339.
Dalväg, H., Klason, C. and Strömvall, H. E. (1985), '*Intern J Polym Mater*, **9**, 38.
Danyadi, L., Renner, K., Szabo, Z., Nagy, G., Mocso, J. and Pukanszky, B. (2006), *Polym. Adv. Technol.*, **17**, 967–974.
Drzal, L. T., Rich, M. J. and Lloyd, P. F. (1982), *J. Adhes.*, **16**, 1.
Eichhorn, S. J. and Young, R. J. (2004), *Comp. Sci. Technol.*, **64**, 767–772.
Eichhorn, S. J., Hughes, M., Snell, R. and Mott, L. (2000), *J. Mater. Sci. Letters*, **19**, 721–723.
English, B., Clemons, C., Stark, N. M. and Schneider, J. P. (1996), Waste-wood derived fillers for plastics, *General Technical Report, FPL-GTR-91*, Madison: US Department of Agriculture, Forest Service, Forest Products Laboratory.
Felix, J. M. and Gatenholm, P. (1991), *J. Appl. Polym. Sci.*, **42**, 609–620.
Felix, J. M. and Gatenholm, P. (1993a), *Nordic Pulp and Paper Res. J.*, **8**, 2000–2003.
Felix, J. M. and Gatenholm, P. (1993b), *J. Appl. Polym. Sci.*, **50**, 699–708.

Felix, J. M. and Gatenholm, P. (1994), *J. Mat. Sci.*, **29**, 3043–3049.
Felix, J. M., Gatenholm, P. and Schrieber, H. P. (1994), *J. Appl. Polym. Sci.*, **51**, 285–295.
Feng, D., Caulfield, D. F. and Sanadi, A. R. (2001), *Pol. Comp.*, **22**, 506–517.
Folkes, F. M. (1964) in *Contact Angle, Wettability and Adhesion, Advances in Chemistry*, **43**, Amer. Chem. Soc., Washington DC, p. 99.
Galiotis, C. (1991), *Comp. Sci. Technol.*, **42**, 125–150.
Gauthier, R., Joly, C., Coupas, A. C., Gauthier, H. and Escoubes, M. (1998), *Polym. Comp.*, **19**, 287–300.
Geng, Y., Li, K. and Simonsen, J. (2005), *J. Adhes. Sci. Technol.*, **19**, 987–1001.
Geng, Y., Li, K. and Simonsen, J. (2006), *J. Appl. Polym. Sci.*, 99, 712–718.
Girones, J., Pimenta, M. T. B., Vilaseca, F., de Carvalho, A. J. F., Mutje, P. and Curvelo, A. A. S. (2007), *Carbohydrate Polym.*, **68**, 537–543.
Grigsby, W. and Thumm, A. (2007), 'Evaluation of WPC fibre swelling behavior observed by fluorescent microscopy', *9th Inter. Conf. on Wood and Biofibre Plastics Conference*, Forest Products Society, Madison, WI, USA.
Herrara-Franco, P. J. and Valadez-Gonzalez, A. (2005), 'Fibre-matrix adhesion in natural fibre composites', in *Natural Fibers, Biopolymers, and Composites*, eds. A. Mohanty, M., Mishra and L. T. Drzal, CRC Press, 177–230.
Hristov, V., Krumavo, M. and Michler, G. (2006), *Macromol. Mater. Eng.*, **291**, 677–683.
Karmarkar, A., Chauhan, S. S., Modak, J. and Chanda, M. (2007), *Comp. Part A*, **38**, 227–233.
Kazayawoko, M., Balatinez, J. J. and Matuana, L. M. (1999), *J. Mater. Sci.*, **34**, 6189–6199.
Kelly, A. and Tyson, W. R. (1965), *J. Mech. Phys. Solids*, **13**, 329.
Kokta, B. V., Raj, R. G. and Daneault, C. (1989), *Polym.-Plast. Technol. Eng.*, **28**, 247–259.
Kokta, B. V., Maldas, D., Deneault, C. and Beland, P. (1990), *Poly. Comp.*, **11**, 84–89.
Liu, F. P., Wolcott, M. P., Gardner, D. J. and Rials, T. G. (1994), *Comp. Interfaces*, **2**, 419–432.
Lu, Z. J., Negulescu, I. and Wu, Q. (2005), *Comp. Interfaces*, **12**, 124–140.
Maldas, D. and Kokta, B. V. (1990), *Polym.-Plast. Technol. Eng.*, **29**, 1119–1165.
Manchado, L. M. A., Arroyo, M., Biagiotti, J. and Kenny, J. M. (2003), *J. Appl. Poly. Sci.*, **90**, 2170–2178.
Mohammed-Ziegler, L., Horvolgyi, Z., Toth, A., Forsling, W. and Holmgren, A. (2006), *Polym. Adv. Technol.*, **17**, 932–939.
Nachtigall, S. M. B., Cerveira, G. S. and Rosa, S. M. L. (2007), *Poly. Testing*, **26**, 619–628.
O'Reilly, J. A., Cavaille, J. Y., Paillet, M., Gandini, A., Herrera-Franco, P. and Cauich, J. (2000), *Polym. Comp.*, **21**, 65–71.
Oksman, K. (1996), *Wood Sci. Technol.*, **30**, 197–205.
Oksman, K., Lindberg, H. and Holmgren, A. (1998), *J. Appl. Poly. Sci.*, **69**, 201–209.
Oksman, K. and Clemons, C. (1998), *J. Appl. Polm. Sci.*, **67**, 1503–1513.
Oksman, K. and Bengtsson, M. (2007), 'Wood fibre thermoplastic composites; processing, properties and future developments', in *Engineering Biopolymers; Homopolymers, Blends and Composites*, Eds. S. Fakirov and D. Bhattacharyya, Hansa Publisher, 655–671.
Own, S. H., Subramanian, R. V. and Saunders, S. C. (1986), *J. Mater. Sci.*, **21**, 11, 3912–3920.
Park, J. M., Quang, S. T., Hwang, B-S. and DeVries, K. L. (2006), *Comp. Sci. Technol.*, **66**, 2686–2699.

Penn, L. S. and Lee, S. M. (1987), *J. Comp. Technol. Res.*, **11**, 23–30.
Piggott, M. R. (1991), *Comp. Sci. Technol.*, **42**, 57–76.
Qui, W. L., Zhang, F. R., Endo, T. and Hirotsu, T. (2005), *Polym. Comp.*, **26**, 448–453.
Rials, T. G. and Simonsen, J. (2000), *Comp. Interfaces*, **7**, 81–92.
Raj, R. G., Kokta, B. V., Maldas, D. and Daneault, C. (1989), *J. Appl. Poly. Sci.*, **7**, 1089–1103.
Sain, M. and Kokta, B.V. (1993a), *J. Adv. Polym. Technol.*, **12**, 2, 167–183.
Sain, M. and Kokta, B.V. (1993b) *J. Appl. Polym. Sci.*, **48**, 2181.
Sain, M. and Kokta, B.V. (1994), *J. Appl. Polym. Sci.*, **54**, 1545–1559.
Sanadi, A. R. and Caulfield, D. F. (2000), *Comp. Interfaces*, **7**, 1, 31–43.
Sanadi, A. R., Subramanian, R. V. and Manoranjan, V. S. (1991), *Polym Comp.*, **12**, 377–383.
Sanadi, A. R., Rowell, R. M. and Young, R. A. (1993), *J. Mater. Sci.*, **28**, 6347–6352.
Sanadi, A. R., Young, R. A., Clemons, C. and Rowell, R. A. (1994), *J. Reinf. Plast. Comp.*, **13**, 54–67.
Schadler, L. S., Laird, C., Melanitis, N., Galiotis, C. and Figueroa, J. C. (1992), *J. Mater. Sci.*, **27**, 1663–1671.
Thuvander, F., Gamstedt, S. E. and Ahlgren, P. (2001), *Nordic Pulp Paper Res. J.*, **16**, 1, 46–56.
Tze, W. T. Y., O'Neill, S. C., Tripp, C. P., Gardner, D. and Shaler, S. M. (2007), *Wood and Fiber Sci.*, **39**, 184–195.
Valadez-Gonzalez, A., Cervante-Uc, J., Alayc, R. and Herrera-Franco, P. J. (1999), *Comp.: Part B*, **30**, 309–320.
Wang, Y., Yeh, F-C., Lai, S-M., Chan, H-C. and Shen, H-F. (2003), *Polym. Eng. Sci.*, **43**, 4, 933–945.
Xanthos, M. (1983), *Plast. Rubber Proc. Appl.*, **3**, 223–228.
Young, R. A., Sanadi, A. R. and Prabawa, S. (1997), *Proceedings of the 4th Int. Conf. on Wood-fibre Plastics Composites*, Forest Products Society, Madison, WI, USA, p. 94.

4
Manufacturing technologies for wood–polymer composites

D SCHWENDEMANN, Coperion Werner & Pfleiderer GmbH & Co. KG, Germany

4.1 Introduction

Wood fibres and polymers are among the most important materials of our time. Both materials have advantages as well as disadvantages, especially with respect to durability, mechanical properties, swelling, thermal resistance and their potentially limited availability as a long-term resource. The compounding process offers the opportunity to combine these materials, although the compounding system must comply with special requirements in order to obtain excellent properties.

Wood fibre requirements for the manufacturing system

The production technology is being developed by the plastic industry, so the machinery must be adapted to wood's special requirements. Wood fibre is a natural product produced from different tree species. This is reflected in the variation in material properties as well as the moisture content. When using wood fibre in conjunction with polymers it is very important to define and specify the qualities of wood fibre or flour.

Figure 4.1 is a schematic diagram of a wood composite compounding line. This chapter is structured in accordance with this schematic. To start with, raw material handling is discussed, than different feeding systems are shown. Compounding technologies present several extrusion systems. The final section is devoted to the downstream equipment and its behaviour relating to wood–polymer composites (WPCs).

4.2 Raw material handling

Wood fibre may differ greatly in its shape, with anything from fine powders to large chips. The amount of moisture may also vary dramatically. The shape and

Manufacturing technologies for wood–polymer composites

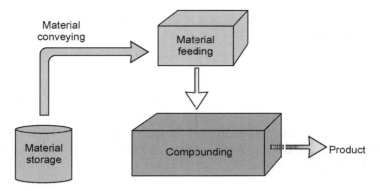

4.1 Schematic material flow of a compounding system.

moisture content both influence flow-ability, the most important factor in material handling. If materials consist of single particles and the particles have the capacity to move relative to each other then they are characterised as bulk materials. Wood fibre complies with these requirements and is therefore categorised as a bulk material.

Bulk material handling systems are very common in both the plastic and the agricultural industries. Bulk materials may sometimes change their properties similar to solid materials or to liquids (Fig. 4.2). The particle size may also vary from fine powder to much larger particles.

The properties of a bulk material are defined by:

- specific bulk weight;
- particle shape and size;
- elasticity of the particles;
- friction of the particles.

No general bulk properties are available with regards to wood fibre, because the properties vary widely. For example, bulk density varies typically between 70 and 350 kg/m. The fibres must be tested and characterised specifically.

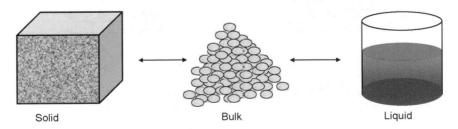

4.2 Characterisation of bulk material.

4.2.1 Material storage, transportation and conveying

Bulk material: wood fibre

Wood fibre is stored in bags, cartons, big-bags and silos. With larger-scale production, big-bags or silos are very common. Wood fibre is already used in many industries, so the storage, transportation and conveying of the fibre are well proven and state-of-the-art. Nevertheless, some basic details must be noted. Wood fibre is specified as an explosive material due to the formation of dust during manipulation, so the system and the machine set-up for handling the fibre must fulfil specific requirements.

A detailed discussion of the ATEX (**AT**mosphères **Ex**plosibles) issues would fill a complete chapter, but the hazard triangle (Fig. 4.3) provides a basic overview of explosive goods. The situation may become critical if all the three conditions specified in the triangle exist together: at least one must be eliminated to avoid an explosion.

Bulk materials: polymer

Typically, the polymer is conveyed from the storage area to the feed equipment via a pneumatic conveying system (Fig. 4.4). The particles are transported by the flow of air.

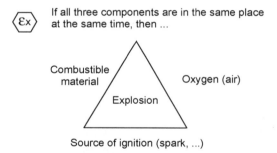

4.3 Hazard triangle (K-Tron, 2007[1]).

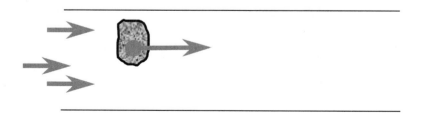

4.4 Pneumatic transportation of a particle or pellet by the air flow.

Manufacturing technologies for wood–polymer composites

Pneumatic conveying

When designing a pneumatic system, pressure loss in the system is important. It is influenced by the:

- throughput rate;
- conveying length and height;
- number of bends;
- bulk material properties.

Wood fibre is sometimes also conveyed via pneumatic systems, but in general mechanical conveying systems are used, such as:

- screw systems;
- belts;
- or other special systems, as a chain of receptacles.

Mechanical conveying

Mechanical systems have advantages for the conveying of wood fibre, e.g. very low dust formation means that normally the dust explosion risk is lower. However, they have a limited conveying distance.

4.2.2 Feeding systems

Feeding systems are basically classified according to method:

- Continuous feeding.
- Discontinuous feeding of batches.

Classification according to feeding principle is also possible, e.g. volumetric and gravimetric feeding. The volumetric principle calculates the feed rate theoretically using the bulk density. If there is a change in the property, the feeder does not react. This principle is used if only one feed stream is required or if the bulk properties are very homogeneous. In the gravimetric system, loss in weight is measured per time sequence (Fig. 4.5), which is why this feeder is also called a differential feeder. The differential feeder determines the feeding mass flow by dividing the loss in weight ΔG with the time Δt, in which the loss in weight was measured. Gravimetric feeding systems are normally used for wood–plastic compounding.

The following systems are used for the feeding of bulk materials:

- loss-in-weight feeding;
- weight belt feeding;
- flow metering.

76 Wood–polymer composites

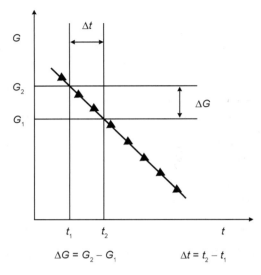

4.5 Loss in weight per time sequence (K-Tron, 2007[1]).

Loss-in-weight feeding

Figure 4.6 shows the control principle of a loss-in-weight feeder. The complete system (hopper, screw, gear box, motor, etc.) is placed on a gravimetric weight cell which measures the loss in weight during the process. The control system compares the actual mass flow with the set-point mass flow. If the material flow is too low, the speed is increased, and if it is too high, the speed of the metering unit is decreased.

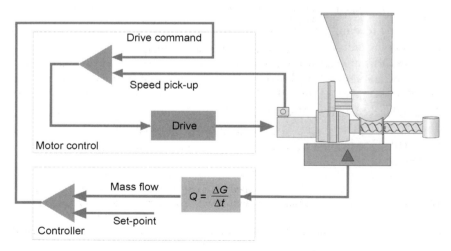

4.6 Loss-in-weight control principle (K-Tron, 2007[1]).

Manufacturing technologies for wood–polymer composites

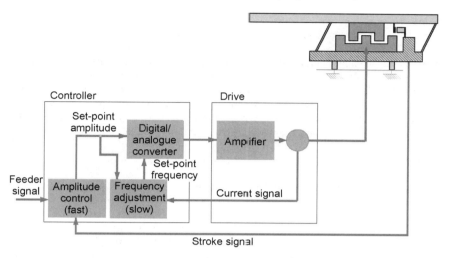

4.7 The vibration control system (K-Tron, 2007[1]).

The most commonly used differential feeder is the *screw feeder*. The material is conveyed by a single or a twin screw. Single screws are used for free-flowing bulk materials and twin screws for powders. Mixing elements or activated walls are available as options for activating the material in the hopper. The material is stored in the hopper. When a certain level is reached, the control switches in a volumetric mode and the hopper is refilled. With regard to wood fibre, screw feeders can also be used for smaller throughput rates. However, should the throughput increase, the system has some limitations; owing to the very low bulk density (between 0.1 and 0.2 kg/l) the volume stream is quite high, so the hopper must be large or it must be frequently refilled.

Vibration feeders are primarily used for shear-sensitive or abrasive bulk materials or for materials with a low melting point. In principle, the system is similar to screw feeders, except that the material metering is initiated by vibration (Fig. 4.7).

As the complete system is on a weight scale in a loss-in-weight feeder, *refilling* is an important factor. If the hopper reaches a defined minimum level, the feed equipment switches to volumetric mode during the refill. The hopper is loaded very fast by a refilling unit – usually a pneumatically loaded pre-hopper opened by a valve, although pre-feeding screws or vibration systems are also in use. During the refilling sequence, the material flow must be continuous and constant. Figure 4.8 illustrates the different steps of a refill sequence.

Belt feeder

In contrast to screw and vibration feeders, belt feeders (Figs 4.9 and 4.10) do not normally use differential calculation (a differential type is available for special

78   Wood–polymer composites

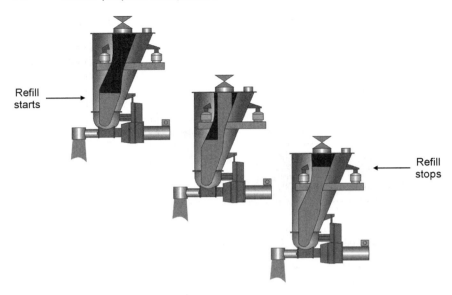

4.8 Refilling of a loss-in-weight feeder (K-Tron, 2007[1]).

applications). With the belt feeder, the weight of the material on the belt is measured per time sequence. If the mass flow is too low the belt speed is increased, if it is too high it is decreased. The hopper is independent of the control, which means that the refilling cycle of the hopper is not as critical to the accuracy of the feeding. If high throughput rates are required with a very low bulk density, as is the case with wood fibre, this system provides the plant designer with some advantages, including with regard to the explosive behaviour of the wood fibre.

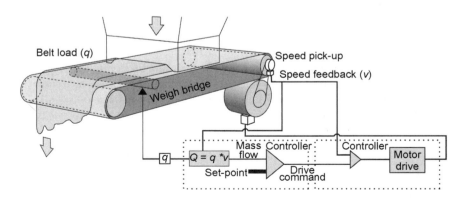

4.9 Belt feeder control principle of a single scale system (K-Tron, 2007[1]).

Manufacturing technologies for wood–polymer composites 79

4.10 Weight belt feeder with a pre-feeding system (photo by Coperion Werner & Pfleiderer[2]).

4.3 Compounding technologies

Various systems are currently used for the compounding of WPCs. Typical machines come from the plastics industry and the extruder is the most important system used in the production of WPC. The most commonly used systems for the compounding of WPC are (Fig. 4.11):

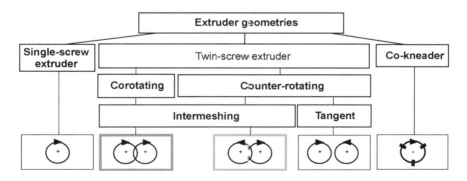

4.11 Overview on extruder systems.

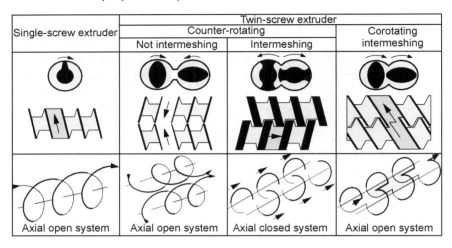

4.12 Conveying mechanisms of the extruder systems (Saemann[3]).

- counter-rotating twin-screw extruder (conical and parallel); and
- corotating twin-screw extruder.

But other extruder systems are also in use, such as:

- single-screw extruder;
- co-kneaders;
- planetary rolling extruder;
- multi corotation screw extruder;
- counter-rotating twin screw, not intermeshing.

Some of these systems are presented below, with attention being paid to their processing parameters as they relate to WPC production. Their conveying mechanisms are given in Fig. 4.12.

Table 4.1 shows a characterisation of the extruder systems specified. Each system has advantages and disadvantages. The single screw and the counter-rotating twin screw have excellent pressure build-up functionality which is why they are primarily used for profile extrusion. The corotating twin screw has the best mixing and degassing functionality and the highest flexibility – the primary reasons for its widespread use in the compounding industry. The single-screw and the counter-rotating twin-screw extruders often use a hot–cold mixer for the agglomeration of the polymer and the wood, which is why this system is presented in brief.

4.3.1 Hot–cold mixers

Hot–cold mixers are used mainly for the processing of polyvinyl chloride (PVC). In the WPC field, the machines are used as a pre-agglomeration system for

Manufacturing technologies for wood–polymer composites

Table 4.1 Characterisation of the different extruder systems[4]

	Single-screw extruder	Counter-rotating twin-screw extruder	Corotating twin-screw extruder
Conveying mechanism	Drag flow	Forced conveying solid flow	Drag flow (forced flow)
Screw speed range	60–250 rpm	25–80 rpm	100–1 200 rpm
Viscosity range	Small	Large	Large
Residence time range	Large	Small	Small
Mass and heat exchange	+	++	++
Mixing:			
Dispersing	+	++	++
Mixing	+	+	+++
Self-cleaning	+	+++	+++
Degassing	+	++	+++
Pressure build up	++	+++	+
Flexibility	+	+	++

+++ Very good, ++ Good, + Satisfactory.

polypropylene (PP) and wood, because the product-conveying capabilities of the single screw and counter-rotating twin screw systems are not sufficient for the intake of wood fibre with a low bulk density. All ingredients, such as wood, polymer and additives, are fed into the mixer as a batch. During the mixing procedure (5–10 min), the material is heated to 110–130 or 160–180 °C depending on the product.

The machine design parameters are:

- volume of the mixer;
- mixing tool speed;
- mixing tool configuration;
- power of the drive.

The processing steps are:

- feeding of all components into the mixer;
- heating of the mixture by friction and additional jacket heating, up to the melting point of the polymer;
- discharging the mixture, to take place when defined agglomerate size has been reached;
- discharging into cooler mixer. Cooling and post-cutting will be achieved in the cooler mixer down to 80 °C.

4.3.2 Single screw and counter-rotating twin screw

The single-screw extruder (Fig. 4.13) is the most commonly used extruder for all types of profile and sheet extrusion as well as for the melting of plastic on injection moulding machines. The product intake is direct from the hopper, so no feeders are required. The output rate is directly related to the screw speed: any increase in screw speed leads to an increased throughput. A variable speed drive is required to control the output rate. The melting is done by a compression section on the screw. The compression ratio is a typical feature of the screw design. Some basic screw designs are shown in Fig. 4.14.

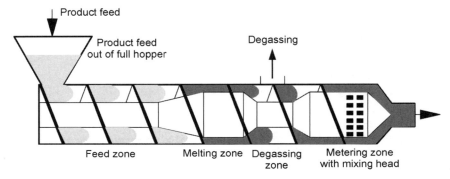

4.13 Schematic of a basic set-up of a single-screw extruder (Saemann[3]).

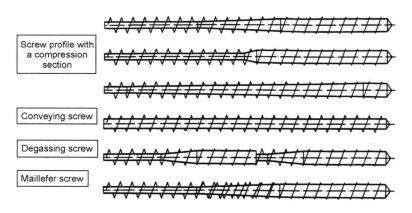

4.14 Basic screw designs of a single-screw extruder (Saemann[3]).

4.3.3 Counter-rotating twin screw

The counter-rotating twin screw is primarily used for PVC (Fig. 4.15). Process tasks are the melting and pressure build-up for the profile extrusion. The product intake capability is quite good so the machines are typically equipped with a

Manufacturing technologies for wood–polymer composites 83

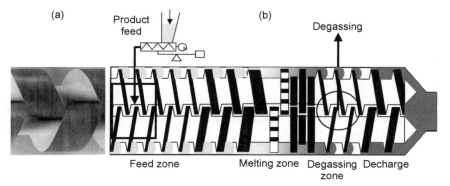

4.15 Schematic of a counter rotating, intermeshing twin-screw extruder (Saemann[3]).

feeder system. The output rate is then independent of the screw speed, which means that fixed speed or variable speed drives are possible. Both machine systems are looked at in detail in other chapters of this book.

4.3.4 Corotating twin screw

The corotating twin screw (Fig. 4.16) is used worldwide for all types of compounding in the plastic industry. The first machine of this concept shown by Coperion Werner & Pfleiderer in 1953 was named ZSK (German: *Zwei Schnecken Kneter* – twin-screw kneader). The machine is available today in everything from small lab sizes to huge production machines which can compound up to 100 tonnes of polyolefin per hour.

The product-conveying capability is excellent, so feeders are required. The output rate is then independent of the screw speed, which means that fixed speed or variable speed drives are possible. The variable drives now available are state-of-the-art.

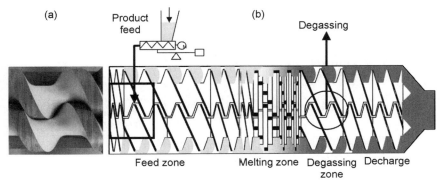

4.16 Schematic of a corotating, intermeshing twin screw extruder (Saemann[3]).

84   Wood–polymer composites

Machine design parameters

The geometry of the ZSK or any corotating twin screw is defined by three parameters:

- centreline distance;
- outer diameter to inner diameter ratio (D_o/D_i);
- specific torque (power/volume ratio defined as torque/centreline distance cubed M/a^3).

As shown in Fig. 4.17, the ZSK screw profile is closely intermeshing. This means that the crest of one screw traces the root and flank of the mating screw. The gap between the two elements is the minimum required for mechanical safety. This closeness results in one screw wiping the associated screw. The D_o/D_i ratio defines the free volume of the extruder: the higher the number, the greater the free volume but the smaller the available shaft diameter for torque transmission (Fig. 4.18). Increasing the D_o/D_i ratio also results in a lower average shear rate machine to its volume. Increasing the specific torque allows the ZSK to operate at higher percentage fill. This translates into higher rates at lower rpm, which, in turn, give lower material discharge temperatures.

Modular construction for an optimised processing section

The processing section of the ZSK comprises modular components (barrel section, shaft and screw bushings) that are assembled in the optimum configuration to meet required unit operations. The barrel sections and screw elements have a modular building block design (Fig. 4.19). The elements are classified as:

- conveying elements;
- return conveying elements;
- mixing elements;

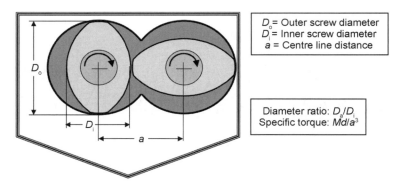

4.17 Basic corotating twin-screw definition parameters.

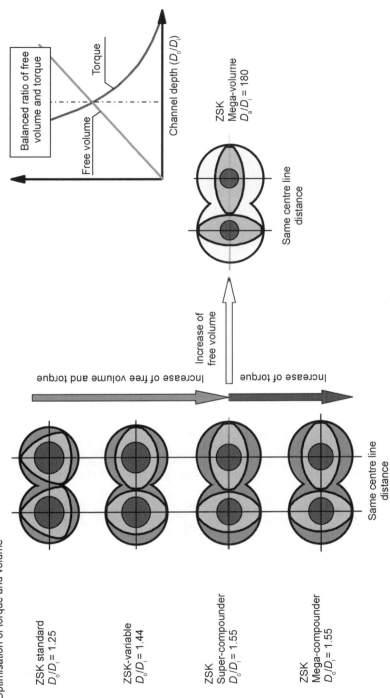

4.18 Balanced ratio of free volume and torque.

4.19 Building block principle of the ZSK.

- kneading blocks;
- special elements.

The conveying elements (Fig. 4.20) are defined by the pitch and the kneading blocks (Fig. 4.21) by the angle between each kneading disc.

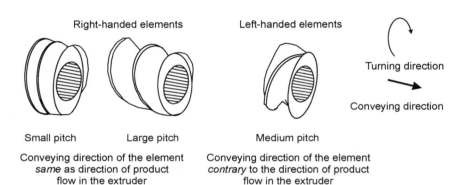

4.20 Functional principle for conveying elements.[3]

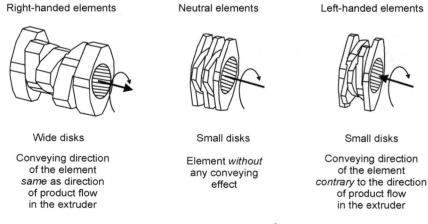

4.21 Functional principle for kneading blocks.[3]

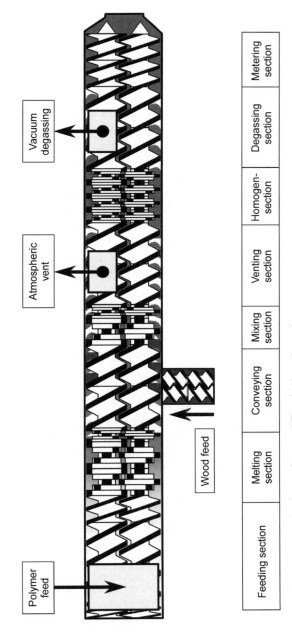

4.22 Processing section layout for wood fibre fed into the melt.

Figure 4.22 shows a typical processing section and the unit operations required for wood fibre incorporation into a polymer matrix. The system is very similar to the setup for polymer-compounding machines using filler (e.g. compounding system for production of PP and talc).

The polymer and additives are fed into the first barrel of the ZSK. The next section is the melting section in which the polymer pellets are melted using shearing energy introduced by kneading blocks. The corotating twin screw extruder transmits most of the energy via the screws from the main drive and not via the barrel heater. After the melting zone, the wood fibre is fed to the machine by a side feeding system. The fibre is incorporated into the melt by mixing and kneading elements. To degas the moisture content of the fibres, the mix then passes through an atmospheric and vacuum vent port. A pressure build-up zone is located at the end of the processing section.

The barrels of the ZSK have been designed with a rectangular cross-section to ensure uniform heat flow from the outer to inner barrel walls and therefore minimum temperature variation around the figure of 8 bore. Nevertheless, the temperature of the melt is primarily influenced by the screw configuration. The surface to throughput ratio is the smallest for the corotating twin-screw extruders in comparison with other machine systems when running at the most economical capacity.

As mentioned in Section 4.2.2, wood fibre has a very low bulk density. When it is fed into the machine, a great deal of air enters with the fibre. It is therefore vital that the fibre feed is vented. If the air cannot exit the system, it flows backwards in the conveying direction and blocks the fibre feed. This problem is

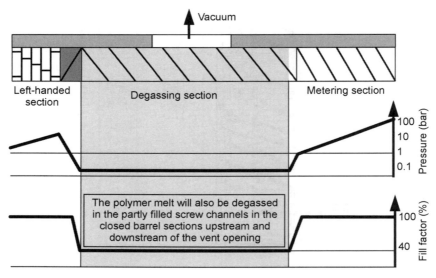

4.23 Principle of degassing on a corotating twin screw (Saemann[3]).

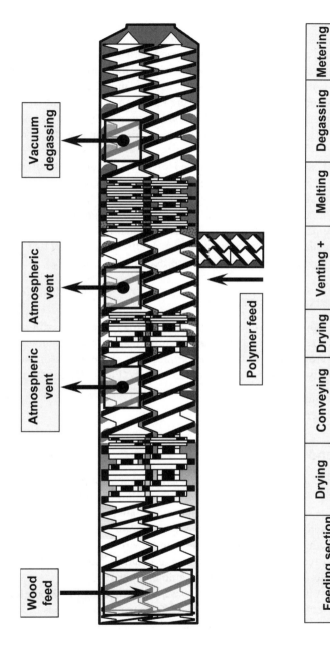

4.24 Processing section layout for wood fibre fed first.

avoided by the addition of a right vent on the side feeder hopper, a backward venting section on the twin screw and a balanced screw configuration.

Venting and degassing are key to the successful compounding of wood fibre polymers (Fig. 4.23). The moisture content of the wood acts as a physical foaming agent. At a venting port, the foamed melt is pushed out of the screws and closes the vent port. An operator must clean the port frequently, which is difficult at a vacuum vent port because the vacuum must be switched off to clean it, resulting in undefined production during the cleaning. To avoid this problem, Coperion Werner & Pfleiderer has developed a twin-screw side degassing unit, which pushes back the material and keeps open the atmospheric and vacuum venting ports.

Figure 4.24 shows a schematic set up of a wood fibre and polymer compounding line. The line has two atmospheric and one vacuum venting port. It is also possible to feed the wood fibres in the first barrel and dry the fibres in the corotating twin screw. The feed port atmospheric vent ports are placed downstream and the polymer is fed by a side feeder. Then the components are mixed and melted together. This process is used especially in North America for the in-line compounding and profile extrusion of decking materials.

4.4 Pelletising systems

In the plastic industry, it is very common to use pellets in converting machines such as injection moulding machines or profile extrusion machines. Several technologies are available for transforming the compounded material into pellets. Generally, two types of pelletising system (Fig. 4.25) can be identified:

- cold face cutter;
- hot face cutter.

4.4.1 Strand pelletising

The strand pelletising system (Fig. 4.26) is one of the simplest solutions for the production of pellets. The melt is pressed through a die with several holes and the strands (like spaghetti) are cooled in a water bath (Fig. 4.27). This solution is still used for many plastic compounding applications where production rates are below 1000 kg/h. It is a very simple and inexpensive installation and the cleaning effort required for a product change is quick and minimal. The system is therefore also used to produce small amounts per batch.

Regarding WPCs, the system is used in particular for laboratory-scale machines. However, at a filling level above 50% wood, the strands become very brittle and it is difficult to keep the pelletiser running. Often the operator is occupied in bringing the strands back to the pelletiser.

The strand pelletising system typically comprises:

Manufacturing technologies for wood–polymer composites 91

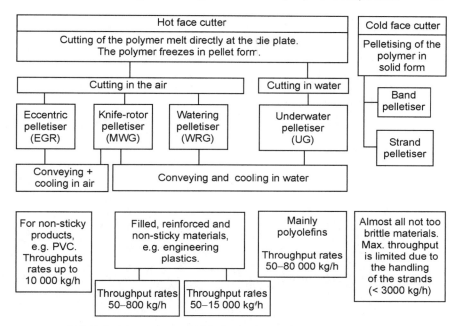

4.25 Pelletising systems (EGR, Exzentrische Granulierung, see Section 4.4.2; MWG, WRG, Wasser-Ring-Granulierung, see Section 4.4.2; UG, (Saemann[3]).

- a die head, directly mounted on the extruder;
- a water bath for cooling the strands;
- an air blowing dewatering unit (optional);
- a strand pelletiser.

4.4.2 Hot face pelletising

Head pelletising systems were developed to avoid the problems faced with strand pelletising systems. The most common systems are:

- underwater pelletisers;
- water ring pelletisers;
- dry cutting systems.

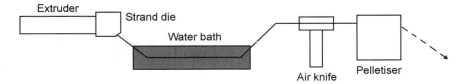

4.26 Schematic of a strand pelletising system.

92 Wood–polymer composites

4.27 Strand pelletising system with water bath.

The systems are presented below, with the impact of WPC on the systems shown in a matrix comparison at the end.

Underwater pelletiser

Underwater pelletising (Fig. 4.28) is currently the most commonly used hot face pelletising system in the plastics compounding field. An underwater pelletiser primarily comprises the following components:

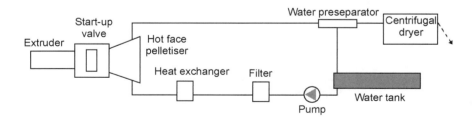

4.28 Schematic of an underwater pelletising unit.

Manufacturing technologies for wood–polymer composites 93

4.29 Underwater pelletiser system including a water treatment system.

- start-up valve;
- pelletiser with cutting chamber, die and cutter head;
- water treatment system including centrifugal dryer.

In the underwater pelletising system (Fig. 4.29), the extruder conveys the molten plastic through the start-up valve into the die plate. In the die plate, the melt stream is divided into a circle of strands and pressed into the cutting chamber through which the process water flows. The cutter head cuts the polymer strands into pellets directly in the water flow and the pellets are conveyed out of the cutting chamber immediately. The pellets flow towards the centrifugal dryer with the process water. Here the pellets and the water are separated from each other. The dry pellets flow out of the dryer continuously. The water is filtered, tempered and conveyed back into the production process. With regard to WPCs, the underwater pelletiser is suitable for fibre contents up to 60%.

Water ring pelletiser

In general, the water ring pelletiser (Fig. 4.30) comprises components similar to those of the underwater pelletiser:

- start-up valve;
- pelletiser with cutting chamber, die and cutter head;
- water treatment system including centrifugal dryer.

The major difference is that the die is not in contact with the water. The melt is also divided into a circle of strands and pressed into the cutting chamber. The strands are cut by knives rotating around the centre of the circle. The hot pellets

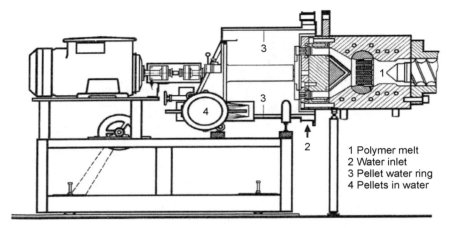

4.30 WRG 400 water ring pelletiser with integrated non-stop screen pack changer (Coperion Werner & Pfleiderer).

pass through the air to a water ring on the exterior of the chamber. The pellets are conveyed out of the chamber immediately and are dried in a centrifugal dryer, similar to the one used for the underwater pelletiser. With regard to WPCs, the WRG (German: *Wasser-Ring-Granulierung*) is suitable for wood fibre contents ranging from 40% to 70%.

Dry cutting systems

As is evident from the name, a dry cutting system is a pelletiser that does not use any water to cool or convey the pellets. These systems are now used primarily for the pelletising of PVC or in the food and animal feed industries. Two systems are detailed below.

The *eccentric pelletiser* (Fig. 4.31) is commonly used for the pelletising of all types of PVC, elastomers and special compounds. The product, which is discharged through the die plate in cylindrical strands, is cut into pellets by a knife rotor. The length of the pellets can be adjusted as required by changing the speed of the knife rotor. The knife rotor is arranged eccentric to the die plate. The unit comprises:

- rear wall mounted directly on extruder;
- hood connected to the rear wall with a lifting hinge and quick-release locks;
- die head mounted directly on the extruder flange or connecting flange;
- heater to heat the die plate;
- knife rotor drive, mounted on the rear wall with integrated knife shaft bearing assembly and infinitely adjustable frictional wheel geared motor;
- knife rotor with knife holder, knives and fastening for precise setting.

Manufacturing technologies for wood–polymer composites

4.31 EGR – eccentric pelletiser (Coperion Werner & Pfleiderer).

With regard to WPCs, the EGR (German: *Exzentrische Granulierung*) is suitable for fibre content in excess of 60% for PP–wood, greater than 50% for PE–wood and for PVC–wood.

The centric *pelletiser* on air generally comprises the same components as the underwater units in a schematic view. These units are used typically if the pellets cut are not sticky and the risk of agglomerates is very low. Regarding WPCs, the centric pelletiser on air is suitable for a wood fibre content in excess of 70%.

4.5 Profile extrusion

Most market activity is currently in WPC extrusion for decking applications. This is also featured in detail in another chapter, so the application is only shown in Fig. 4.32 in combination with a corotating twin-screw extruder, i.e. in-line compounding and profile extrusion.

4.6 Injection moulding

The injection moulding process is a wide area, so this chapter only provides an overview of the technologies available. Manufacturing of injected wood–plastic parts can be classified into two sections:

- standard injection moulding with pre-compounded pelletised material;
- in-line compounding and injection moulding in one step.

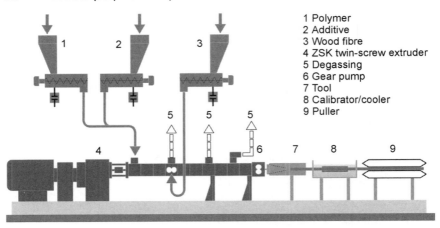

4.32 In-line compounding and profile extrusion.

4.6.1 Standard injection with pre-compounded wood–polymer composites

Most injection moulding machines today are equipped with single screws. The molten plastic pushes the screw back and after a certain volume of melt has built up, the screw rotation stops and the melt is pushed via a transfer movement of the screw into the mould installed in the clamping unit of the machine. Owing to the degassing nature of the single screw, the pre-compounded or pre-agglomerated WPC must have a very low moisture content in order to avoid surface defects on the moulded part.

In order to obtain a good injected WPC, some factors must be considered:

- Resin quality:
  - homogeneity is important,
  - the moisture content must be very low (below 0.15% measured with infra-red test equipment).
- Injection moulding unit and processing:
  - screw design not too aggressive,
  - degassing section,
  - temperature profile adapted to the WPC,
  - injection speed adapted to the WPC to reduce the shear rate.
- Hot runner design:
  - manifold design and melt channel layout must be modified to give lower shear stress on the material,
  - heater layout and location in the hot runner, no areas causing overheating.
- Mould design:
  - part geometry has an influence on the shear rate during injection,
  - venting of the mould.

Manufacturing technologies for wood–polymer composites

Most injection-moulded WPCs are currently produced using a compounding and pelletising process followed by a separate injection moulding process.

4.6.2 In-line compounding

New in-line compounding processes are now being introduced in which the compounder is piggy-backed on to a shooting pot of the injection unit (Fig. 4.33). This concept is the combination of a two-stage injection unit with a corotating twin screw. The two-stage injection units were used at the outset of the injection moulding industry and are still in use today for the production of shear-sensitive materials such as polyethylene terephthalate (PET), with very short cycle times (Fig. 4.34).

PP, additives and wood fibre are fed into the corotating twin screw. These materials are compounded and degassed in the extruder and fed by a transfer channel to a shooting pot. If the shooting pot is filled with a defined portion, the filling is stopped and the valve is changed to the injection position. The melt is pushed by the shooting pot hydraulics via the machine nozzle and hot runner into the mould. These systems are very economical for the injection of larger parts with a shot weight of 1 kg or more. The thermal stress to the material is reduced because the material is molten only once. The critical factor of moisture in the WPC pellet is also avoided.

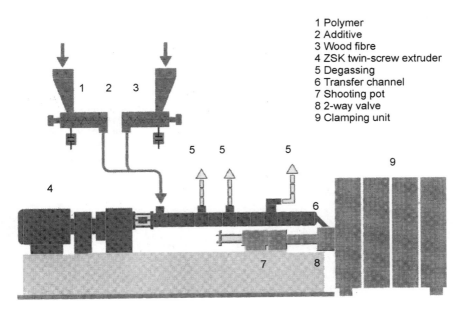

4.33 Schematic for in-line compounding and injection moulding.

98 Wood–polymer composites

4.34 In-line compounding and injection moulding (Husky[5]).

4.7 Sheet extrusion

From a manufacturing point of view, sheet extrusion of WPCs is classified as sheet extrusion on a double belt press or on a roll stack. The roll stack is used for sheet thicknesses of less than 6 mm and the double belt press for sheet thicknesses in excess of 6 mm.

4.7.1 Roll stack

The roll stack comprises three cooled rolls. Depending on the pressure loss required by the die, a melt pump is needed to build up the pressure. The melt is fed directly in a gap between the two lower rolls. The sheet is wound half around the middle and the upper roll, than the sheet is cooled down on rollers. Pullers support the movement on the rollers. Finally, the sheets are cut to length (Figs 4.35 and 4.36).

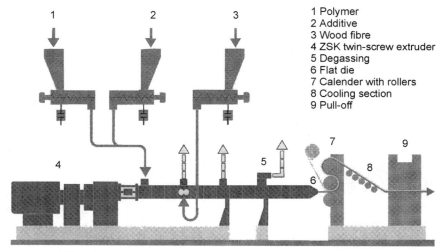

4.35 Schematic set-up of in-line compounding and sheet extrusion on a roll stack.

Manufacturing technologies for wood–polymer composites

4.36 In-line compounding and sheet extrusion on a roll stack (Isokon[6]).

4.7.2 Double belt press

The double belt press is used for thicker WPC sheets, because the sheet is not bent during the cooling. Double belt presses are also used for special applications with surface layers (Fig. 4.37). The melt is normally fed to the table at the beginning of the press by a die system. Two process options are available:

- heating and cooling;
- cooling.

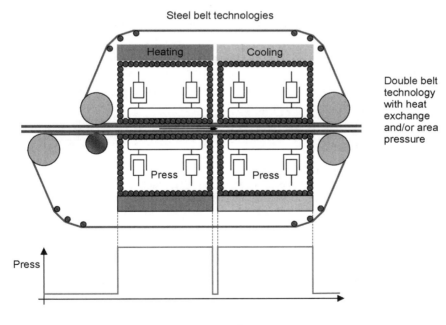

4.37 Schematic layout of a double belt press (SPS[7]).

The cooling double belt press only cools the material. With the heating section in front, some bonding actions may be obtained by keeping the melt at temperature and pressing at the same time. The process is also suitable for lamination.

4.8 Future trends

The compounding of polymers and wood fibre as well as natural fibre is, on the one hand, an old story because the naturally sourced materials were already used as a filler at the outset of the polymer industry. On the other hand, however, new technologies are now available for using these materials to a higher quality or property.

A very interesting field of technology for this material combination is all in-line processes, because the material is subjected to the thermal stress only once during the production process. In-line technologies are very interesting from a commercial viewpoint, especially with regards to higher throughput rates.

Compounding machines, especially the corotating twin screw, offer a high flexibility for combinations in formulations not used today. For example, combinations featuring glass fibre or other synthetic fibres or filler are also possible. Also a polymer blend could be used to give a more elastic characteristic.

Nevertheless, it is very important to analyse the melt properties of the WPC. In the field of injection moulding in particular, many melt flow analyses are carried out in advance to optimise the filling of the mould. If parameters are missing, simulation is impossible and sometimes the WPC material is not chosen because of that.

From the perspective of manufacturing technologies, many processes have been presented in recent years and the machine manufacturers have adapted their equipment for the production of WPC. The machinery is ready to produce parts made from the new material – WPC.

4.9 References

1. K-Tron, Switzerland, Info Module 2007.
2. Coperion Werner & Pfleiderer GmbH & Co. KG, 2006.
3. Saemann H.-J., Process Training Twin Screw Technology, Coperion Werner & Pfleiderer GmbH & Co. KG, 2005.
4. Berghaus, U., Direktextrusion mit Doppelschneckenextrudern (chapter 13, table 13.2), Der Doppelschneckenextruder, Düsseldorf 1998.
5. Husky, Luxemburg, Injection Molding Training Document, 2006.
6. Isokon, Slovenia, 2006.
7. Sandvik Process Systems, Fellbach, Germany, Process Presentation, 2007.

5
Mechanical properties of wood–polymer composites

M SAIN and M PERVAIZ, University of Toronto, Canada

5.1 Introduction

Mechanical and physical properties of wood–plastic composites (WPC), such as stiffness, strength, impact resistance, and density, play an important role in deciding the suitability of these products in various applications. Recent advances in catalyst technologies for polymerization of polyolefin resins and process engineering have made WPC a material of choice for different applications as shown in Fig. 5.1.

WPCs can be labeled as true composite materials, possessing properties of both major ingredients. The key mechanical properties such as strength and stiffness of these composite materials lie between those for polymer and wood. The structure morphology plays a vital role in defining most of the functional attributes of WPC. The excellent moisture resistance of polymers compared with wood directly relates to molecular structure of plastic material used, making WPC more durable and attractive.

A wide range of applications take advantage of the functional performance that WPC offers. For example, semi-structural building products, such as decking, siding, and roofing, take advantage of improved thermal and creep performance compared with unfilled plastics. Similarly, automotive applications rely on a lower density of WPCs compared with inorganic filled thermoplastics, whereas household consumer products depend on the aesthetics.[2]

5.2 Mechanical performance of wood–polymer composites

5.2.1 Anisotropic nature of wood fibers and role of polymer

Since the growth of a tree is subjected to various constantly changing parameters such as moisture, soil conditions, and growing space, the wood properties vary considerably. The physical structure, mechanical properties, density, and aspect ratio of wood change from species to species on a large scale. Although wood is the most common and abundantly available low-cost fiber source, its mechanical

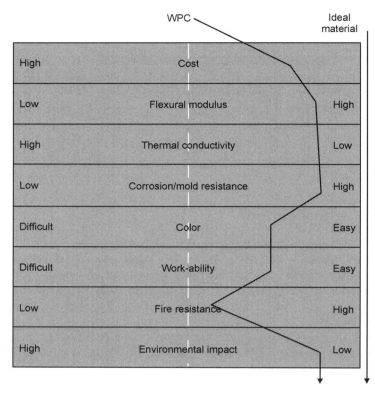

5.1 Combining the best features of wood and plastics, WPC functional properties are ideal for many applications.[1]

properties are often considered inadequate for use as an engineering material. Also important are some inherent shortcomings arising from wood's own basic components such as cellulose, hemicellulose, and lignin. All these building blocks of wood contain hydroxyl groups and other oxygen-containing groups that facilitate moisture absorption through hydrogen bonding.[3,4]

The hydrophilic characteristic of wood is the single drawback that is responsible for its dimensional changes due to swelling and shrinking and also makes it vulnerable to insect and fungal attack. Polymers, both thermoplastic and thermoset, have been successfully incorporated into wood not only to reduce its water absorbing tendency significantly but also to modify other functional properties. The wood particles in WPC are completely coated with plastic and moisture can penetrate only into the exposed sections of wood fibers and is not transmitted across the plastic boundaries, resulting in extremely moisture resistant composite product with little risk of swelling or fungal attack. In fact, filling of hollow centers of cells, also known as lumens, with polymer during compounding of WPC is the key in optimizing the mechanical properties of the product.[1,5]

5.2.2 Interfacial shear strength and fiber length

The interface in any fiber–matrix composite system is responsible for transmitting stresses from the polymer to the fibers. This stress transfer efficiency largely depends on the fiber–matrix interface and mechanical properties of the fibers and polymer. When stress is applied, the fiber–matrix bond ceases at the fiber ends and matrix deformation is experienced as shown in Fig. 5.2. The overall strength of a composite always requires a good interfacial bonding. Further, the mechanical properties of any fiber-based composite largely depend on the length of the reinforcement fiber used. The *critical fiber length*, l_c, at which maximum fiber load is achieved at the axial centre of the fiber, plays an important role in dictating the overall stress transfer from the matrix to the reinforcement medium. As shown in Fig. 5.2, there are three different possibilities when a stress equal to σ_f is applied to a fiber; if fiber length is just equal to critical fiber length, a maximum fiber load is achieved only at the axial centre of fiber and as the fiber length increases, the fiber reinforcement becomes more effective. Alternately, for fiber length less than l_c, no reinforcement is experienced.

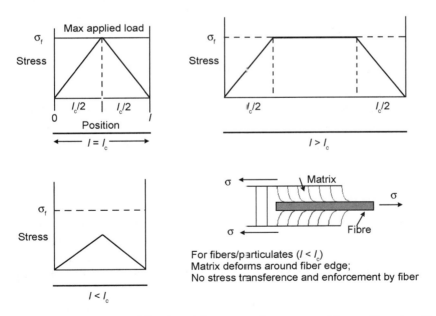

5.2 Influence of fiber length in a composite: stress-position profiles when fiber length is equal, greater and less than critical length under stress equal to fiber strength. Also shown is matrix deformation pattern when fiber–matrix bond fails at fiber ends.[6]

5.3 General mechanical properties of wood–polymer composites and test methods

Although the standards for WPC products in Europe and North America are voluntary, product conformance to certain standards is always advantageous from the marketing point of view as well as aiding proper identification and ensuring uniform product quality. Overall performance of WPC products comprising physical and mechanical properties is usually evaluated through laboratory test methods.

5.3.1 Physical property tests

The most common physical properties, such as moisture content, density, and dimensional changes, are referred to as American Society for Testing and Materials (ASTM) standard D1037. Water absorption behavior of WPC products and subsequent effects on composite dimensions are determined through the *water soak test* in which representative specimens are conditioned in a control environmental chamber prior to immersion in distilled water.

Thermal insulation is another important physical characteristic relating to heat flow through a composite product. A test method ASTM C177-92, the 'Standard Test Method for Steady-State Heat Flux Measurements and Thermal Transmission Properties by Means of Guarded-Hot-Plate Apparatus' is used to measure thermal conductivity of 25 mm thick samples.[7]

5.3.2 Mechanical properties and test methods

The mechanical properties of WPCs very much depend on the moisture content of the sample at the time of test. Samples tested dry are conditioned in a climate chamber to a constant weight and moisture at $20 \pm 3\,°C$ and relative humidity of $65 \pm 1\,°C$. The most common procedures to evaluate mechanical properties of cellulose fiber composite materials are described below.

Flexural strength and modulus

The ability of a WPC to resist deformation under load is defined as flexural strength and three-point static bending test, ASTM D790, is a usual procedure to measure this property as shown in Fig. 5.3. In this test geometry, the WPC sample experiences compressive stress at the load-bearing side and tensile stress at the opposite end.

The mathematical expression to calculate maximum surface stress, S, at failure is represented as:

$$S = 3PL/2bd^2$$

Mechanical properties of wood–polymer composites

5.3 Test assembly for static bending of WPC.

where

S = maximum stress at mid-span loading (MPa),
P = bending load at a given point (N),
L = length of span (mm),
b = width of sample (mm),
d = thickness (depth) of sample (mm).

The flexural modulus, a ratio between stress and strain within elastic region, is computed from the slope of a representative straight-line segment of load deflection curve obtained from flexural strength testing.

Tensile strength

Tensile strength, the measure of maximum amount of tensile stress to a composite to the point of failure, can be determined both parallel and perpendicular to the face of sample according to ASTM D-638. Parallel-to-face tensile strength, the resistance of WPC to be pulled apart parallel to its surface, is the more frequent test method employed by industry and research institutes as shown in Fig. 5.4. The specimens used in this method are of standard dumb-bell-shaped and tested under defined ambient conditions and machine speed.

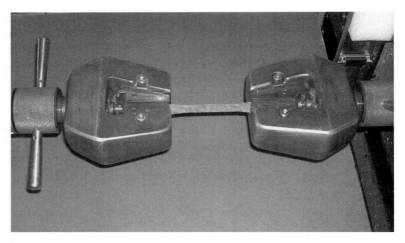

5.4 Test assembly for tensile testing of WPC.

The perpendicular-to-face orientation test procedure measures internal bond strength of composite material. In this method, a 50 mm square sample is bonded with an adhesive to metal alloy (aluminums or steel) loading blocks of same dimension and tensile strength is calculated as:[7]

$$IB = P/bL$$

where
  IB = internal bond strength (kPa),
  P = maximum load (N),
  B = sample width (mm),
  L = sample length (mm).

5.3.3 Secondary mechanical properties and test methods

Apart from the usual strength and modulus properties, there are a few other commonly used test procedures to evaluate the mechanical performance of WPC products in more detail for engineering applications.

Impact resistance

Toughness is a measure of the energy that a WPC product will absorb before breaking and is usually measured by Izod and Charpy impact tests according to ASTM standard D256. Impact samples are cut according to nominal size of $64 \times 12.7 \times 4.2 \, mm^3$. The specimens are clamped into a pendulum impact test fixture with the notched side facing the striking edge of the pendulum. The Izod notched test assembly comprising a typical impact tester is shown in Fig. 5.5.

Mechanical properties of wood–polymer composites 107

5.5 Test assembly for measuring Izod notched impact energy (Department of Chemical Engineering, University of Toronto, Toronto, Canada).

The quantity usually measured in this test, the energy absorbed in breaking the specimen in a single blow, is expressed as impact energy (J/m) and is calculated by dividing the energy (J) by thickness of sample. However, the ISO impact strength is measured in kJ/m^2 by dividing impact energy by area under the notch.

Hardness and dent resistance

Hardness and dent resistance measurement techniques are used to evaluate the resistance of WPC products to indentation and any deformation that may occur when these composites are struck by other moving or stationary objects. Although there is no specific test method solely designed for WPC, one of the best techniques to measure the ability of wood and its composites to endure indentation is the Janka test (ASTM D143). In this method, the force required to push a 11.28 mm diameter (0.44 inch) steel ball into WPC decking to half its diameter is measured. The results are usually reported either in kilogram-force (kgf), pound-force (lbf) or newtons (N).

Another method of hardness measurement of WPC products is using a Rockwell hardness tester according to ASTM E18. In this method, a hardened ball of usually 6.35 mm diameter and 588.4 N of force (Rockwell scale L) is used (Fig. 5.6). The Rockwell number represents the difference in depth penetration between two loads.

5.6 Rockwell hardness testing apparatus.

A method for impact resistance measurement of engineered WPC structures is known as the *falling ball impact resistance*. In this method a square sample of $304 \times 304\,\text{mm}^2$ or $152 \times 152\,\text{mm}^2$ held between supports is subjected to sudden localized load dropped by a 50 mm diameter ball from increasing heights. The procedure continues until the sample fails and the recorded height serves as an index of resistance to impact.[7,8]

Fastener holding strength

Fastener resistance, either screw or nail, is a key property of WPC composite panels used in structural applications such as sheathing and flooring The standard procedures for this test follow ASTM D1037; however, there are some individual standards as well pertaining to distinct type of products. The screw holding strength can be measured from the face or edge side of the sample which is 76 mm wide, 102 mm long, and approximately 19 mm thick. Number 10 Type

AB 25 mm sheet-metal screws are threaded down the sample up to 17 mm depth. As far as nail withdraw resistance is concerned, there are three types of methods: direct nail withdrawal, nail-head pull-through, and lateral nail resistance.

5.4 Critical parameters affecting mechanical properties of wood–polymer composites

As the WPC comprises three main ingredients – plastic, wood filler, and additives in different concentrations – it is imperative to anticipate a number of parameters affecting mechanical performance of these products. The plastic, typically a thermoplastic polymer such as polypropylene, polyethylene, or polyvinyl chloride, constitutes 30–70% of the mass of WPCs and the wood filler, typically 40 mesh saw mill residue, constitutes an additional 30–70% of the total mass of WPCs.[9] Further, these composite products contain about 5% additives of different types to facilitate manufacturing process and enhance physical appearance and durability of the products.

The quality of raw materials, both wood and plastic, is the single most important factor dictating ultimate performance of WPC products. Source and type of raw materials, whether virgin or recycled, can also have a profound effect on the strength and durability of WFCs. Similarly, modifications of the filler/polymer, method of blending main ingredients, manufacturing methods, and process conditions have their own significance in deciding the suitability of any WPC product for its intended use.

5.4.1 Effect of filler and polymer

Wood's morphology, density, and mechanical properties change from species to species and wood fibers can be separated from wood chips through a variety of chemical and mechanical methods. The pulping process has significant effect on the mechanical properties of wood fibers even within the same species and an average tensile strength of 0.5–1.5 GPa and a modulus of 10–80 GPa of wood fiber can be found.[3] Among the commonly used polymers, mostly thermoplastics, are high-density polyethylene (HDPE), low-density polyethylene (LDPE), polypropylene (PP), polystyrene (PS), and polyvinyl chloride (PVC). The filler content itself is a deciding factor for mechanical properties of fiber-reinforced composites.

A trend in increase of tensile modulus with an increase in fiber content is shown in Fig. 5.7. A significant difference between HDPE and PP is also observed where the latter has shown better modulus of elasticity.[3,10–12] It may be noted that Chemi-thermo mechanical pulp (CTMP) in this study was aspen whereas hardwood was 60% beech and 40% birch wood. The flexural properties of WPC also depend greatly on the type of filler and plastic, as well as their respective weight ratio. In a research study by VTT Processes, Finland, different

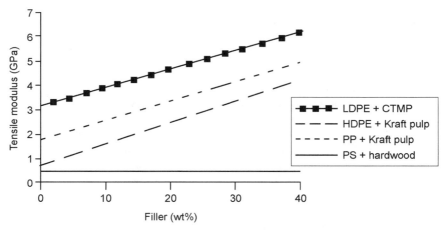

5.7 Effect of filler content and type of raw materials on tensile modulus of wood plastic composite.[3,10–12]

types of wood fibers and polymers were blended to manufacture WPC through their Conex® wood extruder.[13] Some of the results of this study involving mixed compressed wood pellets (spruce, pine), PP, PE, and recycled PE are presented in Fig. 5.8. For each blend, 2% lubricant and 2% coupling agent (g-MAH) were also added. It is interesting to note the difference between spruce-based and mixed pellet composites. The bending strength and modulus of spruce-based samples is higher, which might be attributed to its longer fiber length. PE, understandably, has shown poor values and recycled PP is also responsible for marginal decrease in bending strength of WPC.

As already mentioned, WPC properties are strongly influenced by wood content. Increasing the wood fraction increases the notched impact energy,

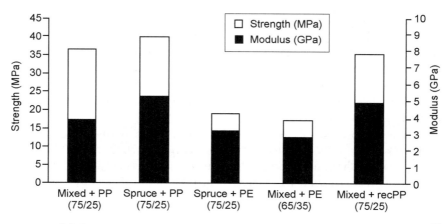

5.8 Flexural properties of WPC with different combinations of filler/polymer.[14]

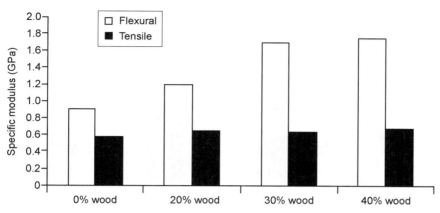

5.9 Specific flexural complex modulus (1 Hz) and specific tensile modulus of the WPCs manufactured from yellow birch and post-consumer PP/PE.[16]

flexural strength, and elastic modulus while decreasing melt index, tensile strength, as well as tensile elongation at break.[14] In another study,[15] WPCs were manufactured through injection molding consisting of a blend of post-consumer high-density PE and PP in a ratio of 85 : 15 and yellow birch fibers ranging in sizes between 425 and 1168 m. Three different wood contents (20, 30 and 40%), were blended with polymer. The results are shown in Fig. 5.9. It is observed from this study that specific complex moduli increased significantly with wood content; however, specific optimal value of Young's moduli was achieved at around 20–30% wood content.

The type of filler, softwood or hardwood, is another criterion to gauge the WPC performance especially in terms of impact strength and moduli of rupture and elasticity. Wood fiber–PP combination is probably the most extensively used formulation for general WPC products. Bledzki et al.[4] investigated the effect of type of wood and content on impact and strength properties. In this study hardwood and softwood fiber were arranged from J. Rettenmaier and Sohne GmbH & Co., Germany, with an average particle size of 150–500 mm. PP was used as the matrix and test samples were prepared by injection molding at molding temperature of 150–180 °C. The summary of results of mechanical properties concerning the effect of hardwood and softwood as filler is shown in Figs 5.10–5.12. It is interesting to note a significant drop in elongation at break and impact strength as fiber content is increased for both hardwood and softwood composites. However, an increasing trend is observed for tensile stiffness as fiber content is increased. As far as the difference in hardwood and softwood is concerned, better stiffness values for softwood WPC is presumably due to more lignin content in this kind of wood, 28% compared with 20% for hardwoods, which acts like thermoplastic, giving stiffer products.[5,16] In general, hardwood-based thermoplastic composites show better values for strength and elongation characteristics due to a higher cellulose content.

112 Wood–polymer composites

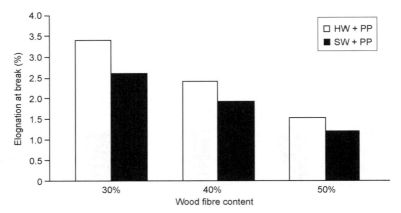

5.10 Tensile elongation at break of hardwood and softwood–PP composites.[4]

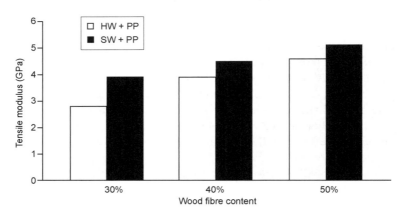

5.11 Tensile stiffness of hardwood and softwood–PP composites.[4]

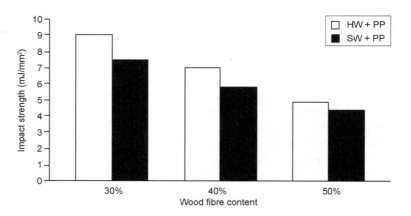

5.12 Charpy impact strength of hardwood and softwood–PP composites.[4]

5.4.2 Optimization of mechanical properties of wood–polymer composites: interfacial modification

Although there are several techniques available to overcome the compatibility issue between wood and polymers, the widely used way to improve wet-ability between filler and matrix is to use coupling agent having bi-functional groups. In this section the effect of maleic anhydride modified polypropylene (MAPP), the most commonly used coupling agent, on mechanical performance of WPC will be discussed. Generally, mechanical performance of WPCs improves significantly when these coupling agents are used under optimized conditions. The effects of MAPP copolymer were investigated at concentration levels of 5 and 10% along with PP as matrix and different ratios of hardwood and softwood.[4] The material properties and other details have already been mentioned in this chapter. For the sake of simplicity and interest of reader, only the results with hardwood content of 40% are presented here in Table 5.1.

A significant increase of about 37% in tensile strength is observed with the addition of MAPP. Other properties also showed some improvement but it is obvious that MAPP is more effective at lower loading. The decrease in the damping index, a measure of improved impact resistance, with the addition of coupling agent is most probably due to better chemical interaction between matrix and wood flour.

In yet another study,[17,18] MAPP (Hercoprime, Himont) was used in concentrations of 2 and 5% in combination with HDPE which was obtained from mixed recycled milk bottles. Aspen hardwood fibers (Canfor Canadian Forest Products), 30% by weight, were blended with the granulated mixed dairy bottles (20% unused bottles and 80% postconsumer bottles) through a corotating twin-screw extruder maintained at 150 rpm and 150 °C. WPC samples were prepared by compression molding at 150 °C and 30 000 psi for 10 min. Although no statistical significance was found in case of modulus for samples with and without MAPP, a positive improvement was observed compared with 100% recycled HDPE (Fig. 5.13). In case of tensile strength, statistical significance was found between samples containing no additive and 5% MAPP as shown in Fig. 5.14.

Table 5.1 Mechanical properties of WPC containing 40% hardwood[4]

Property	MAPP loading		
	0%	5%	10%
Tensile strength (MPa)	27	37	36.5
Elongation at break (%)	2.45	2.5	2.35
Stiffness (GPa)	3.8	4.2	4.3
Charpy impact strength (mJ/mm^2)	7	9	8.4
Damping index	1.6	0.9	0.8

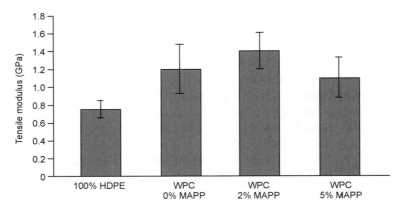

5.13 Effect of MAPP on tensile modulus of WPC (30% HW).[21,22]

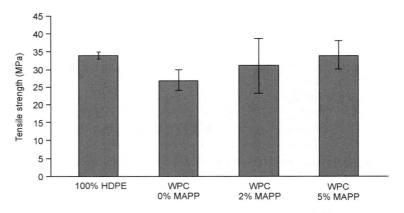

5.14 Effect of MAPP on tensile strength of WPC (30% HW).[21,22]

A well-established method to optimize the impact properties of WPC products is to enhance the ductility of matrix itself. The toughness of PP can be significantly increased through the use of block copolymers, elastomers, or ethylene polymerization. The most successful and frequently used impact modifiers for PP are reported as; ethylene/propylene copolymers (EPM) and ethylene/propylene/diene terpolymer (EPDM).[19,20]

In a related research project,[20] the effects of such impact modifiers in combination with MAPP were studied on the mechanical performance of WPC samples. The impact modifiers used in this study were; ethylene/propylene/diene terpolymer EPDM, maleated ethylene/propylene/diene terpolymer EPDM-MA, and maleated styrene–ethylene/butylene–styrene triblockcopolymer SEBS-MA. The filler used was pine wood flour of nominal 40 mesh and particle size of 420 mm. All formulations had wood flour ratio as 40% whereas MAPP and impact

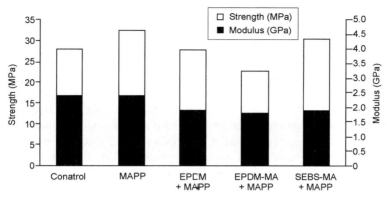

5.15 Effect of MAPP and impact modifiers on tensile strength and modulus of WPC samples.[20]

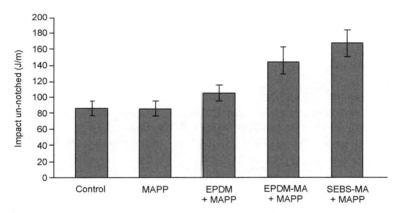

5.16 Izod impact (un-notched) values of WPC after addition of different impact modifiers.[20]

modifiers were 2 and 10% respectively. After preblending, the matrix, filler and impact modifiers were compounded in a corotating twin-screw extruder. The compounded pellets were then injection molded through a conventional reciprocating screw injection molder into representative WPC specimens.

The MAPP compatibilizer had a positive effect on strength and stiffness on WPC samples but did not show any improvement upon impact property as shown in Figs 5.15 and 5.16 respectively. On the other hand a significant increase in impact properties was observed after addition of maleated impact modifiers. This may be attributed to the fact that maleated elastomers are expected to form a flexible interface around the wood particles thereby enhancing impact strength of WPC products.

5.5 Conclusions

The development of a mechanically durable and maintenance-free WPC can help not only in the economic growth for both the wood and plastic-based industry but also ensures exciting new options for the end user. Although WPCs represent a relatively new era of materials development, the range of materials being explored for different applications is very wide, covering a variety of polymer matrix types, fillers, and innovative additives. The ultimate optimization of such a complex material as WPCs needs a thorough understanding of material performance and a wide-ranging evaluation of mechanical properties intended for desired applications. In spite of the quantity and complexity of the influencing parameters, most of the desired mechanical properties can be achieved by different manipulations during material and process selection. Apart from frequently used techniques to improve durability of WPCs, such as interfacial modification and surface treatment, the future research efforts are heading towards nano-scale modifications of materials which can impart many new possible combinations of physical as well as mechanical properties such as hardness and scratch resistance.

5.6 References

1. Wood–Plastic Composites, A technical review of materials, processes and applications. Tangram Technology Ltd, UK, 2002.
2. TechLine – Wood Plastic Composites. Forest Products Laboratory, Madison, MI, USA.
3. Bledzki, A.K., Reihmane, S., Gassan, J., Thermoplastics reinforced with wood fillers: a literature review. *Polym.–Plast. Technol. Eng.*, 1998, **37**, 451.
4. Bledzki, A.K., Faruk, O., Huque, M., Physico-mechanical studies of wood fiber reinforced composites. *Polym.–Plast. Technol. Eng.*, 2002, **41**(3), 435–451.
5. Mohanty, A.K., Misra, M., Drzal, L.T., *Natural Fibers, Biopolymers, and Biocomposites*, CRC Press, Taylor & Francis Group, 2005.
6. Callister, W.D. Jr., *Materials Science and Engineering – An Introduction*, 5th edn, John Wiley & Sons Inc., 2000.
7. Rowell, R.M., Young, R.A., Rowell, J.K., *Paper and Composites from Agro-based Resources*, CRC Press, 1996.
8. Dale Ellis, W., O'Dell, J.L., Wood–polymer composites made with acrylic monomers, isocyanate, and maleic anhydride. *J. Appl. Polym. Sci.*, 1999, **73**, 2493–2505.
9. Gramlich, W.M., Gardner, D.J., Neivandt, D.J., Surface treatments of wood–plastic composites (WPCs) to improve adhesion. *J. Adhesion Sci. Technol.*, 2006, **20**(16), 1873–1887.
10. Raj, R.G., Kokta, B.V., Daneault, C., A comparative study on the effect of aging on mechanical properties of LLDPE-glass fiber, mica, and wood fiber composites. *J. Appl. Polym. Sci.*, 1990, **40**, 645–655.
11. Woodhams, R.T., Thomas, G., Rodgers, D.K., Wood fibers as reinforcing fillers for polyolefins. *Polym. Eng. Sci.*, 1984, **24**(15), 1166–1171.
12. Gatenholm, P., Bertilsson, H., Mathiasson, A., The effect of chemical composition

of interphase on dispersion of cellulose fibers in polymers. I. PVC-coated cellulose in polystyrene. *J. Appl. Polym. Sci.*, 1993, **49**, 197–208.
13. Raukola, J. et al., *Wood Plastic Composites with Conical Conex® Wood Extruder*, VTT Processes, Finland, 2004.
14. Stark, N., Berger, M., Effect of species and particle size on properties of woodflour-filled polypropylene composites, In: *Conference Proceedings: Functional Fillers for Thermoplastics and Thermosets*, Madison, p. 119, 1997.
15. Gosselin, R. et al., Injection molding of postconsumer wood–plastic composites. II: Mechanical properties. *J. Thermoplastic Composite Mater.*, 2006, **19**.
16. Smook, G.A., *Handbook for Pulp and Paper Technologists*, 2nd edn. Angus Wilde Publications, 1992.
17. Selke, S.E. et al., Wood fiber/polyolefin composites. *Composites: Part A*, 2004, **35**, 321–326.
18. Nieman, K.A., Mechanical property enhancement of recycled high density polyethylene and wood fiber composites due to the inclusion of additives. MS Thesis, Michigan State University, 1989.
19. Oksman, K. et al., Mechanical properties and morphology of impact modified polypropylene–wood flour composites. *J. Appl. Polym. Sci.*, 1998, **67**, 1503–1513.
20. Inoue, T., Suzuki, T., Selective crosslinking reaction in polymer blends. III. The effects of the crosslinking of dispersed EPDM particles on the impact behavior of PP/EPDM blends. *J. Appl. Polym. Sci.*, 1995, **56**, 1113.

6
Micromechanical modelling of wood–polymer composites

R C NEAGU, Ecole Polytechnique Fédérale de Lausanne (EPFL), Switzerland and E K GAMSTEDT, Kungliga Tekniska Högskolan (KTH), Sweden

6.1 Introduction

When selecting a wood–polymer composite (WPC) material for certain applications, the cost and engineering properties are of primary concern. If the cost constraints are fulfilled, the goal is to produce a composite material with optimal properties to fulfil the desired performance of a component. The production method and choice of constituents will influence the microstructure of the material, which in turn will influence the material properties. The microstructure–property link is therefore of interest, not only for improved understanding, but also in materials development, such as identification of microstructural features that affect a certain property. With quantitative models of this link, it would also be possible to perform parametric studies of how microstructure could be optimised for best possible material properties. The chosen models should, however, be validated by experiments and more accurate models. Considering the variability and complexity of the microstructure of WPCs, viable models should be as simple as possible, and yet be sufficiently accurate to describe the physics of the phenomena associated with the property under study. Since most WPCs are used in structural applications, the mechanical properties are the most important ones from an engineering point of view. Some of these are stiffness, dimensional stability, strength and fracture toughness. The objective of this chapter is to give some examples of how these properties can be predicted by micromechanical models. References are made to useful sources in the literature, where detailed derivations and explanations of the models are found. Only the main features and comments of the applicability of the models are presented here. Hopefully, these resources will be useful for engineers and researchers working to develop WPCs with improved mechanical properties.

For particle composites and particularly laminated continuous fibre composites, a large framework of models has been developed to link the microstructure to mechanical properties. These composite materials are typically reinforced with carbon or glass fibres. Before considering any of the models developed for glass or carbon fibre composites, it is necessary to appreciate the difference between these fibres and wood reinforcement. For instance, glass fibres are isotropic, unlike carbon fibres and wood fibres which are anisotropic. Furthermore, wood fibres are hollow and show a large variability compared with the synthetic fibres. Other special features of wood fibres that may affect the validity of existing framework of models are the irregular shapes, helical structure, pits, dislocations, etc. Nevertheless, an engineering model should be as simple as possible with sufficient accuracy. The large scatter in properties in wood fibre composites does not warrant the use of the most exact and complex models developed for composite materials for aerospace applications. These have generally a much more regular microstructure than wood fibre composites originating from biological resources. The relatively simple models that describe the main physical features of the wood fibre composite are probably the most suitable for engineering purposes.

The most common type of WPC is based on thermoplastics, typically polypropylene, in extruded structural parts. To a much smaller extent, wood fibre reinforced thermosets are also used in, for example, interior panels. Essentially, the micromechanical modelling of mechanical properties does not differentiate between thermoplastic or thermoset-based wood fibre composites, since both are solid materials, and the input parameters of the two different types of matrix are of the same type. From a solid mechanics point of view, it is only the values of these parameters that may differ, e.g. elastic moduli, hygroexpansion coefficient, interfacial strength, etc. The models covered are still equally appropriate. Only higher-order effects pertain to differences between thermoplastics and thermosets, although these effects are deemed to be negligible compared with scatter in measurement and desired accuracy for engineering purposes. Such particulars include transcrystallinity of semicrystalline thermoplastics around the wood fibres, which may induce cylindrical orthotropy. Another feature is that most thermoplastics have melting points and glass transition temperature well above the curing temperature (ambient room temperature for larger structure) of most conventional thermoset systems, which leads to higher residual stresses at the fibre–matrix interface in the thermoplastic case. In any case, these effects are not relevant to the models discussed in the present chapter. Even if they are exemplified by thermoset matrix here, they are likewise applicable for thermoplastic matrix composites.

6.2 Elastic properties

The elastic properties of a wood fibre composite depend on the properties of the constituent fibres (phases), their relative volume fractions and on the microarchitecture, i.e. the fibre orientation distribution, the fibre aspect ratio (i.e. length to diameter ratio), the fibre shape, etc. Other important aspects are the ultrastructural features that govern the fibre properties (Neagu *et al.*, 2006a). It is important to highlight some materials aspects in order to choose a suitable micromechanical model for wood fibre reinforced plastics. Thereafter, a micromechanical approach to predict the elastic properties of wood fibre composites is presented. Suitable micromechanical models are discussed, with the intention of providing a short overview of the work conducted in order to direct further reading. Finally, it is illustrated how a micromechanical approach can be used for a quantitative determination of the contribution of various wood fibres to the elastic properties of a general composite with an arbitrary in-plane fibre arrangement.

6.2.1 Material microstructure

The bulk of wood composites used today is based on wood chips or flour rather than slender separated wood fibres (Morton et al., 2003), where the wood material is often used as filler instead of reinforcement. To take advantage of wood fibres in composites, they should be gently defibrated, have high aspect ratios and retain a relatively high stiffness and strength (Neagu et al., 2006b). Typical wood pulp fibres are about 1–3 mm in length and 20–40 μm in width (Sjöholm et al., 2002). Hence, the aspect ratio is about 100, which means that the stress transfer zone extends over a small part of the interface and perturbation effects related to fibre ends may be neglected (Cox, 1952; Nairn, 1997). The use of oriented separated fibres with a relatively large aspect ratio should ensure considerably improved reinforcement.

Two possible routes to produce wood fibre composites fulfilling these requirements are (i) manufacturing wood fibre mat preforms that can subsequently be impregnated with a thermosetting resin (Neagu et al., 2006b), and (ii) commingling wood fibres and thermoplastic fibres in preforms which can be consolidated into different shapes afterwards (Bogren et al., 2006). Figure 6.1(a) shows a layer from an oriented wood fibre mat manufactured with a dynamic sheet former (Neagu et al., 2005). The fibres are predominantly oriented in the machine direction (MD). For fibre mat preforms manufactured with conventional papermaking technologies the fibre orientation distribution can be considered planar (cf. Fig. 6.1(a)). The microstructure of a composite manufactured with the resin transfer moulding technique (RTM) using an epoxy vinyl ester and an oriented fibre mat of unbleached softwood kraft fibres is shown in Fig 6.1(b). Hot pressing preforms of commingled polylactic acid (PLA) and bleached softwood sulphite fibres resulted in the microstructure shown in Fig. 6.1(c).

It is clear that wood fibre composites exhibit a complex microstructure that is difficult to model precisely. Wood fibres exhibit variability in fibre cross-sectional dimensions, not to mention variability along the fibre length, curl of the fibres, etc. Figure 6.1(b–c) show that the cross-sectional shape varies from being thick-walled boxlike for most of the latewood fibres to a relatively slender

Micromechanical modelling of wood–polymer composites

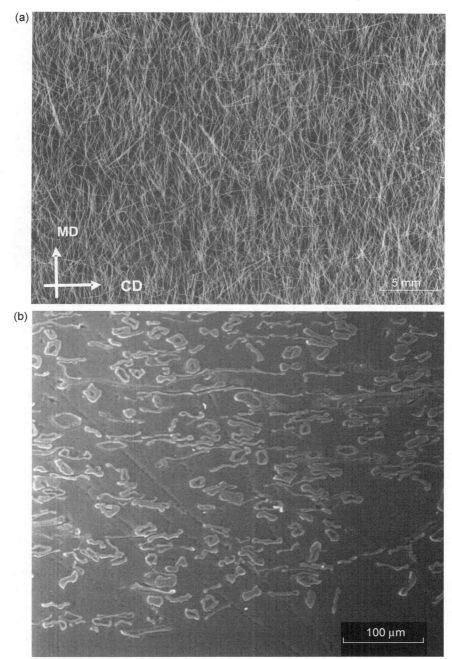

6.1 (a) Scanned photograph of an oriented wood fibre mat (Neagu *et al.*, 2005) and cross-section of (b) a wood fibre/epoxy vinyl ester composite (Neagu *et al.*, 2006b) and (c) wood fibre/PLA composite (Bogren *et al.*, 2006).

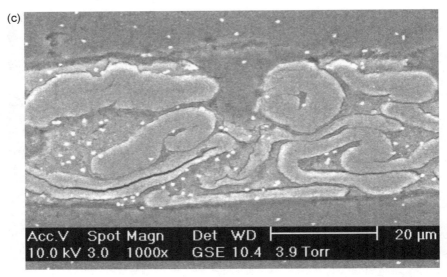

6.1 continued

rectangular form for the thin-walled earlywood fibres, some which are collapsed. During the composite manufacture with RTM the lumen is likely to be infiltrated by the resin as can be seen in Fig. 6.1(b), while lumen filling might not be attainable for thermoplastic-based wood fibre composite (cf. Fig. 6.1(c)). Wood fibres are themselves multiphase composite materials with a highly intricate ultrastructure and anisotropic material properties. The reinforcing elements are helically arranged and the cellulose microfibrils embedded in a stress-transferring matrix of the amorphous wood polymers, lignin and hemicelluloses (Mark, 1967). This structural domain, which is critical to the mechanical behaviour of wood fibres, has been reviewed in Neagu et al. (2006a). To facilitate modelling, simplifications and assumptions must be made with regard to the microstructure of the composite and wood fibres. For example the helical structure of the fibre means that axial deformation is coupled with torsion. Taking this coupling into account would require a detailed analysis with additional unknown parameters (Neagu et al., 2006a). The very complicated wood fibre structure can be simplified to a fibrous structure which has a high intrinsic degree of mechanical anisotropy and is stiff in the direction of the cellulose microfibrils and more compliant in the perpendicular direction.

Simplifications that are needed to make analytical treatment tractable are: (i) consider the wood fibres transversely isotropic (simplest anisotropic case) and (ii) consider the wood fibres as homogeneous and cylindrical. These assumptions imply that the extension–twist interaction, the presence of lumen and effect of the cross-sectional geometry are overlooked. These issues have been addressed by Marklund (2005) who showed that for fibres with a filled lumen a change of the cross-section geometry has an insignificant effect on the composite properties. Also that the error introduced by neglecting the orthotropic nature of the cell wall and replacing it by a transverse isotropic material was evaluated. It was found that its value can reach 10–20%.

6.2.2 The in-plane Young's modul

Laminate analogy is a method which can be applied to describe the elastic behaviour of non-aligned short fibre composites with planar fibre orientation distribution. It is based on modelling composites reinforced by non-aligned discontinuous fibres by using classical lamination theory for a stack of laminae, each of which accounts for one fibre orientation (e.g. Chow, 1992). In this way the in-plane Young's moduli of a composites can be estimated and their dependence on the loading direction can be studied (Fu and Lauke, 1998a,b). Other ways to determine the elastic moduli of short-fibre composites could be the continuum approach where the composite is regarded as a homogeneous body with an internal field of distributed fibres (Advani and Talreja, 1993), the method of cells by Aboudi (1991), direct numerical simulation (Gusev, 1997; Hine et al., 2002), etc.

For wood fibre composites with a planar fibre orientation distributions, the laminate analogy approach should be sufficient. They can thus be treated as a stack of laminae of different orientation and with a fixed layer volume fraction obtained from an experimentally determined fibre orientation distribution, $p(\theta)$ (Neagu et al., 2005). The stiffness matrix of the laminate in an off-axis coordinate system, defined by the loading direction angle φ, is related to the stiffness matrix of a lamina by:

$$[\bar{Q}] = \int_{-\pi/2}^{\pi/2} [T(\theta - \varphi)]^{-1} [Q][T(\theta - \varphi)] p(\theta) d\theta \qquad 6.1$$

where $[Q]$ is the stiffness matrix of a lamina and $[T(\theta - \varphi)]$ is the transformation matrix from an angle θ, relative to the global coordinate system, to the direction angle φ, respectively. The generalised Hooke's law gives the constitutive equation for a lamina

$$\begin{Bmatrix} \sigma_1 \\ \sigma_2 \\ \tau_{12} \end{Bmatrix} = \begin{bmatrix} Q_{11} & Q_{12} & Q_{16} \\ Q_{12} & Q_{22} & Q_{26} \\ Q_{16} & Q_{26} & Q_{66} \end{bmatrix} \begin{Bmatrix} \epsilon_1 \\ \epsilon_2 \\ \gamma_{12} \end{Bmatrix} \qquad 6.2$$

where the components of $[Q]$ are related to engineering constants as $Q_{11} = E_L/(1 - \nu_{LT}\nu_{TL})$, $Q_{22} = E_T/(1 - \nu_{LT}\nu_{TL})$, $Q_{12} = \nu_{LT}E_T/(1 - \nu_{LT}\nu_{TL})$, $Q_{16} = Q_{26} = 0$ and $Q_{66} = G_{LT}$. The transformation matrix that gives the proportionality between the stresses and strains in the off-axis coordinate system is defined as

124   Wood–polymer composites

$$[T] = \begin{bmatrix} c^2 & s^2 & 2sc \\ s^2 & c^2 & -2sc \\ -sc & sc & c^2 - s^2 \end{bmatrix} \qquad 6.3$$

where $c = \cos(\theta - \varphi)$ and $s = \sin(\theta - \varphi)$. To evaluate the integral in Equation (6.1) a fibre orientation distribution function $p(\theta)$ can be fitted to experimental data. The fibre orientation function can be represented as a normalised Fourier series expansion of the probability density function for symmetric distributions as:

$$p(\theta) = \frac{1}{\pi} \sum_{n=0}^{\infty} a_n \cos(2n\theta) \qquad 6.4$$

with $a_0 = 1$, and the rest of an arbitrary number of Fourier cosine coefficients a_n can be determined experimentally. Figure 6.2 shows the histogram of the fibre orientation with a fitted distribution function corresponding to the fibre mat shown in Fig. 6.1(a).

Insertion of the constitutive equation for a lamina and the transformation matrix, Equations (6.2) and (6.3), and $p(\theta)$, Equation (6.4), into Equation (6.1), and evaluation of the integral gives the relation for the laminate global stiffness matrix. The Young's modulus of the composite in the loading direction φ can then be obtained from the first component of the inverse of this matrix:

$$E_c(\varphi, E_\mathrm{L}, E_\mathrm{T}, G_\mathrm{LT}, \nu_\mathrm{LT}, a_1, a_2) = \left\{ [\bar{Q}]^{-1} \right\}_{11}^{-1} = \frac{\bar{Q}_{11} \bar{Q}_{22} - \bar{Q}_{12}^2}{\bar{Q}_{22}} \qquad 6.5$$

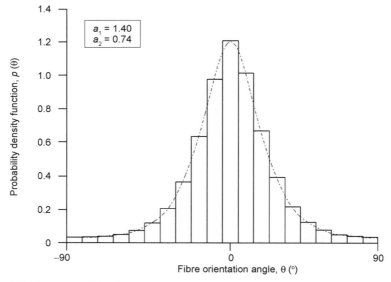

6.2 Histogram of the fibre orientation of the fibre mat shown in Fig. 6.1(a) with a fitted distribution function $p(\theta)$, Equation (6.4) (Neagu et al., 2005).

The longitudinal and transverse Young's modulus of the composite can be obtained from Equation (6.5) evaluated at $\varphi = 0°$ and $\varphi = 90°$, respectively. The composite moduli depend of course also of the properties of the lamina E_L, E_T, ν_{LT} and G_{LT}, and the orientation parameters a_1 and a_2 of $p(\theta)$ in Equation (6.4).

6.2.3 Micromechanical modelling for stiffness prediction

The key feature in the laminate analogy approach is determination of the elastic properties of a lamina. These can be obtained from different micromechanical models, some of which will be discussed in the following. The aim of a micromechanical model is the determination of the elastic constants of the lamina. This requires the knowledge of the complex stress and strain microfields developed around the fibre and matrix phase. This is usually achieved with the help of the mean field methods which assume that averaged values of the stress and strain are representative of the behaviour of each phase. It is also assumed that, under far field boundary conditions, the mean stress and strain in the phases are related to the effective stress and strain by certain mechanical influence functions called concentration tensors (Hill, 1963). These functions depend on the shape (i.e. particles, short or long fibres), spatial distribution and volume fraction of the reinforcement. Different micromechanical models provide different ways to approximate these concentration functions.

Orientation and aspect ratio of the fibres are among two key parameters that describe the microstructure of wood fibre composites. In the laminate analogy modelling approach adopted here it is assumed that the oriented material can be regarded as an aggregate of units of structure, i.e. the laminae. The orientation function describes the state of orientation and each lamina can be regarded either as an ideal perfectly aligned short fibre composite or as a unidirectional layer of continous fibres, depending on the aspect ratio of the fibres.

The predictive capability of the most commonly used micromechanical models for aligned short fibre composites have been compared in a review article by Tucker and Liang (1999). Many of the models make use of the fundamental Eshelby's equivalent inclusion result, which is the solution for the elastic stress field in and around an ellipsoidal particle in an infinite matrix (Eshelby, 1957). If the particle is considered as a prolate ellipsoid of revolution, this result can be used to model stress and strain fields around a cylindrical fibre. One can use Eshelby's result to find the stiffness of a composite, but the predictions are only accurate at very low volume fractions (about 1%) since interaction between fibres are not included in the model. To treat non-dilute composite materials, models have been developed based on the proposal made by Mori and Tanaka (1973) who made the assumption that within a composite the fibre 'sees' a far-field strain equal to the average strain in the matrix. This assumption was later used by Tandon and Weng (1984) and Qiu and Weng (1990) to develop a complete set of elastic constants for a composite with short isotropic and anisotropic fibres, respectively. The Mori–Tanaka approach has also been used when considering interactions between fibres at different orientations to predict the stiffness of misoriented short fibre composites (Takao *et al.*, 1982; Chen and Cheng, 1996). This is in contrast to the laminate analogy approach, where the interaction between differently oriented neighbouring fibres is neglected which might lead to errors for high volume fractions and high elastic contrasts between fibres and matrix.

Another approach to find the properties of aligned short fibre composites is the self-consistent scheme (Budiansky, 1965; Hill, 1965a). Here the properties are found by considering a composite in which a single fibre is embedded in an infinite matrix that has the average properties of the composite (Laws and McLaughlin, 1979; Chou *et al.*, 1980). A different, but very popular, way to predict the properties of aligned short fibre composistes is to use the so-called Halpin–Tsai equations (Halpin, 1969; Halpin and Kardos, 1976). The Halpin–Tsai equations are semi-empirical, since they are based on reducing the work of Hermans (1967) and Hill (1965b) on continuous fibre composites to a simple form and accounting for the fibre geometry through use of some empirical factors. The Halpin–Tsai equations are known to fit some data very well at low volume fractions, but to underestimate some elastic parameters at high volume fractions. Nevertheless, the Halpin–Tsai equations have been applied with success to predict the longitudinal Young's modulus of different wood fibre reinforced thermoplastic composites (Lundquist *et al.*, 2003; Bogren *et al.*, 2006; Facca *et al.*, 2006a).

Shear lag type of models (e.g. Cox, 1952; Nairn, 1997; Mendels *et al.*, 1999) can only predict the longitudinal Young's modulus, but have had quite widespread use. The shear lag analysis focuses on a single fibre of a given length and radius, which is encased in a concentric cylinder shell of matrix with a certain radius. The average fibre stress can be computed and combined with an average matrix stress to produce a modified rule of mixtures for the axial modulus. Facca *et al.* (2006a) found that shear-lag models (Nairn, 1997; Mendels *et al.*, 1999) consistently overestimated the longitudinal Young's modulus of different wood fibre reinforced thermoplastic composites. A source of discrepancy was thought to be the assumption of perfect adhesion. It is known that poor adhesion is a particular problem in composites with hydrophilic wood fibres and hydrophobic polymer matrices, such as thermoplastics, which could result in slip at the interface that reduces the average stress in the fibres. Nairn and Liu (1997) and Nairn (2004) have extended the shear lag analysis to include stress transfer from the matrix to a fibre through an imperfect interface, which could be used to predict the longitudinal Young's modulus in terms of the interface quality. In wood fibre reinforced thermoplastics poor interface quality is usually improved by addition of some coupling agents to improve the chemical compatibility between the matrix and the wood fibres. Many authors, e.g. Tandon and Weng (1984), have recognised that the other engineering constants are only weakly dependent on aspect ratio and orientation and so can be approximated by a continuous fibre model. Hence, one could obtain the remaining elastic parameters using some model for continuous fibres (e.g. Hashin, 1979), and for anisotropic fibres their axial modulus should be used in the shear lag equations. Comparison and validation with numerical models of the micromechanical models mentioned here (Tucker and Liang, 1999; Hine et al., 2002; Lusti et al., 2002), led to the conclusion that a Mori–Tanaka approach (i.e. Tandon and Weng, 1984; Qiu and Weng, 1990), gives the best all-round prediction of elastic properties for aligned short fibre composites.

If the aspect ratio is very large, micromechanical models for continuous fibre composites can be used. It is difficult to specify an exact limit when composites are relative insensitive to the aspect ratio, since it depends on the difference in elastic moduli of the constituents. An example is the work of Laws and McLaughlin (1979), which showed that the effective elastic moduli of aligned or completely random short-fibre composites are independent of the aspect ratio when it exceeds about 20. In other words, the elastic response of the composite differs little from the response it would exhibit if it were reinforced by continuous fibres with infinite aspect ratio. The simplest way to obtain the elastic properties of a unidirectional lamina is to use the classical isostrain and isostress aproaches, known as Voigt and Reuss averages (Hill, 1952). Facca et al. (2006b) investigated the predictive abilities of modified rules-of-mixtures for composites containing several different kinds of fibres and accounting for the effect of fibre ends. However, these models are semi-empirical and do not have a very high accuracy especially when the fibre and matrix have substantially different stiffness. Some of the early theories that have been developed to predict the elastic properties of unidirectional composites have been reviewed by Chamis and Sendeckyj (1968). More recently, a survey of the analysis of the elastic behaviour of continuous fibre composites has been compiled by Hashin (1983). The micromechanical models developed by Hashin and Rosen (1964), Hashin (1979) and Christensen and Lo (1979) consider the constituents as concentric circular cylinders, and are usually referred to as composite cylinder assemblage (CCA) models. Hashin and Rosen (1964) developed the stiffness expressions for isotropic hollow circular fibre composite, while Hashin (1979) considered transversely anisotropic fibre (not hollow). The limitation with the micromechanical model of Hashin (1979) is that it renders only upper and lower bounds for the transverse shear modulus. This problem was solved by Christensen and Lo (1979) using a generalised self-consistent scheme. Here, a cylindrical subdomain which has an equivalent fibre matrix microstructure is embedded in an infinite composite with unknown properties. Furthermore, the work of Hashin (1979) and Christensen and Lo (1979) has been generalised to the case of composites made of heterogeneous multilayered and transversely isotropic fibres (Hervé and Zaoui, 1995) and orthotropic properties (Marklund, 2005). These models could be used to include the effect of the properties of the different cell wall layers and lumen on the composite properties.

A suitable and handy micromechanical model to determine the properties of a lamina, i.e. Equation (6.2), is the CCA model of Hashin (1979). Four effective elastic moduli are exactly known: the transverse bulk modulus k, the longitudinal Young's modulus E_L, Poisson's ratio ν_{LT}, and the longitudinal shear modulus G_{LT}. The transverse stiffness modulus E_T is not determined exactly, but close bounds have been established based on the classical extremum principles of elasticity. The final formulae are presented here for the sake of completeness. They are given in terms of the fibre V_f and matrix volume fractions V_m, the matrix elastic properties (i.e. E_m, G_m, ν_m) and the transversely isotropic elastic properties of the fibre (i.e. E_{f1}, E_{f2}, G_{f12}, ν_{f12}, ν_{f23}). The engineering elastic constants entering the constitutive equation for a lamina, Equation (6.2), are:

$$E_L = E_{f1}V_f + E_m V_m + \frac{4(\nu_{f12} - \nu_m)^2 V_f V_m}{V_m/k_f + V_f/k_m + 1/G_m} \qquad 6.6$$

$$\nu_{LT} = \nu_{f12}V_f + \nu_m V_m + \frac{(\nu_{f12} - \nu_m)(1/k_m - 1/k_f)V_f V_m}{V_m/k_f + V_f/k_m + 1/G_m} \qquad 6.7$$

$$G_{LT} = G_m \frac{G_m V_m + G_{f12}(1 + V_f)}{G_m(1 + V_f) + G_{f12}V_m} \qquad 6.8$$

$$E_T^{(\pm)} = \frac{4G_T^{(\pm)}}{1 + \frac{G_T^{(\pm)}}{k}\left(1 + \frac{4k\nu_{LT}^2}{E_L}\right)} \qquad 6.9$$

where $E_T^{(-)} < E_T < E_T^{(+)}$. The effective transverse bulk modulus k is given as a function of the fibre and matrix transverse bulk moduli, k_f and k_m, respectively:

$$k = \frac{k_m(k_f + G_m)V_m + k_f(k_m + G_m)V_f}{(k_f + G_m)V_m + (k_m + G_m)V_f} \qquad 6.10$$

$$k_f = \frac{E_{f1}E_{f2}}{2E_{f1}(1 - \nu_{f23}) - 4\nu_{f12}^2 E_{f2}} \qquad 6.11$$

$$k_m = \frac{E_m}{2(1 - \nu_m - 2\nu_m^2)} \qquad 6.12$$

The bounds on the effective transverse shear modulus G_T applicable for composites with fibres stiffer than matrix are given in closed form as follows:

$$G_T^{(-)} = G_m + \frac{V_f}{\dfrac{1}{E_{f2}/(2+2\nu_{f23})_G m} + \dfrac{k_m + 2G_m}{2G_m(k_m + G_m)}V_m} \qquad 6.13$$

$$G_T^{(+)} = G_m\left(1 + \frac{(1+\beta_m)V_f}{\rho - V_f\left(1 + \dfrac{3\beta_m^2 V_m^2}{\alpha V_f^3 + 1}\right)}\right) \qquad 6.14$$

where

$$\alpha = \frac{\beta_m - \gamma\beta_f}{1 + \gamma\beta_f} \qquad 6.15$$

$$\rho = \frac{\gamma + \beta_m}{\gamma - 1} \qquad 6.16$$

$$\beta_m = \frac{k_m}{k_m + 2G_m} \qquad 6.17$$

$$\beta_f = \frac{k_f}{k_f + E_{f2}/(1 + \nu_{f23})} \qquad 6.18$$

and

$$\gamma = \frac{E_{f2}}{2G_m(1 + \nu_{f23})} \qquad 6.19$$

It is not known which of the bounds on the effective transverse shear modulus G_T is the most accurate for wood fibre composites. However, following the discussion on the nature of the bounds in Hashin (1979), the effective transverse moduli of a unidirectional wood fibre/epoxy composite E_T can be determined by omitting the $\pm$ superscripts and using the lower bound given in Equation (6.13) into Equation (6.9).

6.2.4 Approach to determine the reinforcement efficiency of wood fibres

The micromechanical approach that has been described can, among others, be applied to determine the contributing stiffness of the fibres to the composite. Representative stress–strain tensile curves of RTM produced laminates with an epoxy vinyl ester and softwood kraft fibres with different kappa number are presented in Fig. 6.3(a). Details of the material properties, composite manufacture and microstructural characterisation are found in Neagu et al. (2006b). The kappa number is used as a measure of the lignin content of a pulp. Lignin is responsible for the brown coloration of paper, and is removed by bleaching. Lignin is also known to act as the matrix between fibrils in the cell wall, and removal of lignin by bleaching could reduce stress transferability between the fibrils, which could lead to a reduction in the fibre stiffness (Åkerholm and Salmén, 2003; Salmén, 2004; Neagu et al., 2006a). Owing to difficulties in the *a priori* control of the fibre contents, the composites contained different relative amounts of fibres. Some of the composites had a uniform fibre orientation distribution, while others had an orientation distribution shown in Fig. 6.2. A direct comparison of the efficiency of the reinforcement can therefore not be made from the curves in Fig. 6.3(a). Hence, the incompatibility of fibre contents and fibre orientation distributions makes it difficult to interpret the effect of

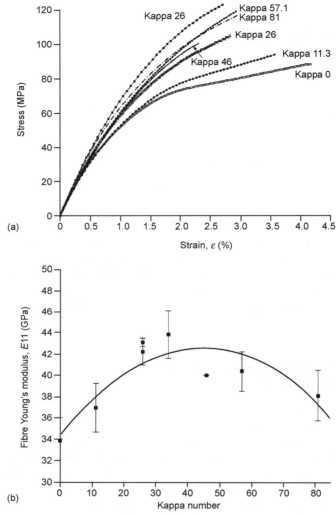

6.3 (a) Stress–strain curves of the composites with softwood kraft fibres with different kappa number (Neagu *et al.*, 2006b) and (b) the effect of kappa number on the longitudinal Young's modulus of the fibres.

lignin content on the reinforcing contribution of the fibres to composite stiffness.

A quantitative measure of the effective Young's modulus of the fibres can be determined by back-calculation. The stiffness is identified as the parameters that minimise the difference between the experimental data (slope of the linear part of the stress–strain curves in Fig. 6.3a) and the calculated values from the model, i.e. Equation (6.5). In Fig. 6.3(b) the estimated longitudinal Young's moduli of the softwood kraft fibres are plotted against the kappa number. A kappa number around zero represents fully bleached fibres while a kappa number around 80 should reflect the processing conditions for producing kraft liner fibres although at laboratory scale. From 6.3(a) it becomes clear that the lignin content of the fibres plays an important role in the effective reinforcement efficiency in composites, which could not be deduced directly from the gross stress–strain behaviour of the composites in Fig. 6.3(a). In Fig. 6.3(b), there seems to be an optimal kappa number that gives the highest fibre stiffness for the kraft fibres. The fitted trendline reveals that within an intermediate range of kappa number say 30–50 the stiffest fibres are to be expected. At lower and higher values of the kappa number, the effective stiffness of the fibres drops considerably for the softwood kraft fibres. During pulping and bleaching it is inevitable that the fibres will be degraded not only chemically but also mechanically. The removal of lignin matrix impairs the ability of stress transfer between fibrils and damage may develop even at low strains. The conclusion is that unbleached fibres are generally more suitable than bleached fibres for use as reinforcement in polymers in terms of stiffness for undamaged fibre of the same dimensions. Also, the micromechanical approach is a useful tool that can be used, e.g. to find out at what stage in the pulping process the most suitable reinforcing fibres can be retrieved. The fibre properties, fibre arrangement and fibre content to the mechanical properties of the composite can be quantitatively assessed and the most suitable reinforcing fibres can be selected. This in contrast to being limited to merely qualitatively rank different composite materials from their gross tensile curves.

6.3 Hygroexpansion

The key drawback of wood fibre composites is probably their propensity to swell on water uptake. This is due to the abundance of hydrophilic hydroxyl groups in the hemicelluloses and amorphous cellulose within the cell wall of the wood fibre (Neagu et al., 2006a). This hydrophilic nature can never be fully alleviated, but still mitigated to some degree by sensible chemical or enzymatic treatments. The hydrophilicity and swelling of the polymer matrix are usually limited for the matrix materials, and focus is usually placed on reducing the moisture swelling of the fibres. However, for certain biopolymers, such as thermoplastic starch and PLA, the hygroexpansion may be similar to or exceed that of the fibres. Since the matrix materials are usually monolithic and isotropic, they can easily be tested in neat form. How the swelling of the fibres affects the dimensional stability of the composite is more complicated owing to the fibrillar and random structure of the fibre assembly.

Since wood fibres are hygroscopic and show a complex interaction between the stress–strain behaviour and the moisture content, the discussion on elasticity here is confined to low moisture contents where the fibre is assumed to be linearly elastic with constant elastic parameters. Other simplifying assumptions are that the hygroexpansion is linear and that the moisture content is independent of residual stresses. It should be noted, however, that for large variations in moisture content, local stresses will develop due to the mismatch in hygroexpansion of the constituents. The corresponding changes in free volume can then affect the equilibrium moisture content (Neumann and Marom, 1986).

One way to isolate the hygroexpansion of the fibres, and quantify its contribution to the hygroexpansion of the composite, is to use a micromechanical model. These are similar to models primarily developed for thermal expansion and residual stresses in ceramic–matrix composites. Thermal expansion and hygroexpansion are governed by the same physical equations, where the hygrothermal strains are proportional to the temperature and moisture content, respectively. It has been concluded that the best approach for predicting the coefficient of thermal expansion of aligned short fibre composites (Hine et al., 2002) is to use the explicit treatment of Levin (1967) which is identical to that of Rosen and Hashin (1970) and Christensen (1991).

6.3.1 Free deformation due to change in moisture content

A particular nuisance in composite applications is when the structure curls as the relative humidity changes after manufacturing or in use. A micromechanical approach would be useful to quantify and predict the curl, and would suggest means to suppress this kind of buckling. It could also be used to back-calculate the contribution of the fibres to the curl, and thereby provide a method to select fibres that prevents this deformation mechanism. Analogous to the back-calculation of the stiffness described previously, the hygroexpansion properties of fibres may be determined from deformation measurements.

Hygroscopic strains in the laminate result from changes in the moisture content. These strains are assumed to depend linearly on the moisture content over the humidity range. This proportionality is given by the coefficients of moisture expansion. If the laminate is completely free to expand, bend and twist when subjected to a change in moisture content ΔC, the relationship between the changes in fictitious hygroscopic forces and moments $\{\Delta N^H\}$ and $\{\Delta M^H\}$ that produce the changes in the midplane strains $\{\Delta \epsilon^0\}$ and plate curvatures $\{\Delta \kappa\}$ can be expressed as

$$\left\{ \begin{array}{c} \Delta N^H \\ \Delta M^H \end{array} \right\} = \begin{bmatrix} A & B \\ B & D \end{bmatrix} \left\{ \begin{array}{c} \Delta \epsilon^0 \\ \Delta \kappa \end{array} \right\} \qquad 6.20$$

The $[A]$, $[B]$ and $[D]$ matrices are conventionally defined as extensional stiffness matrix, extension-bending coupling stiffness matrix and bending stiffness matrix, respectively. The hygroscopic forces and moments can be evaluated from the hygroscopic strains induced in the laminate by a uniform change in moisture content ΔC of the laminate. By integration of the lamina stresses induced by the constraints placed on its deformation by adjacent layers, the hygroscopic force and moment vectors can be calculated as:

$$\{\Delta N^H\} = \Delta C \sum_{i=1}^{n} [\bar{Q}]_i \{\bar{\beta}\}_i (z_i - z_{i-1}) \qquad 6.21$$

and

$$\{\Delta M^H\} = \frac{\Delta C}{2} \sum_{i=1}^{n} [\bar{Q}]_i \{\bar{\beta}\}_i (z_i^2 - z_{i-1}^2) \qquad 6.22$$

where $[\bar{Q}]_i$ is the effective global stiffness matrix of layer i, Equation (6.1), $\{\bar{\beta}\}_i$ is a vector with the coefficients of moisture expansion for layer i and z_i are layer coordinates with the origin of the z-coordinate placed in the midplane. The laminate mechanics model, as outlined in Equations (6.20)–(6.22), can be used to determine the average coefficient of transverse hygroexpansion by relating it to macroscopic measurements of changes in curvature and thickness with respect to change in moisture content. A micromechanical model is necessary to link hygroelastic behaviour of the fibres to the mesoscopic behaviour of each layer and macroscopic expansion behaviour of the laminate.

6.3.2 Approach to determine the hygroexpansion of wood fibres

Since hygroexpansion is intimately linked with the elastic properties of the constituents, the latter need to be characterised before any conclusions can be made on how the hygroexpansion coefficients of the constituents affect the resulting expansion of the composite material. The hygroexpansion strains and elastic properties are inversely related. Schulgasser (1987a) considered experimental data on materials such as wood, particle board and paper, and showed that for a material constituted at a microscopic scale of the fibre cell wall, the expansion caused by a moisture content change is determined by the microstructure of the material predominantly through the elastic compliance of the material in that direction, i.e. the aforementioned linear relationship. It was mentioned that although the compliance dominates the relation, the influence of Poisson's ratios need not necessarily be negligible. This findings suggest that no reduction or increase of the expansion in a given direction can be achieved through varying the microstructure without concomitantly reducing or increasing the compliance in the direction of interest in the same proportion.

Based on the fundamental work of Levin (1967) on the effective thermal expansion coefficients of polycrystalline materials in terms of the anisotropic thermoelastic properties of the constituting crystals, Schulgasser (1987b) obtained an exact relationship between hygro- (or thermal) expansion coefficients and elastic compliances of polycrystalline aggregates. For a composite constituted of many anisotropic phases (i.e. wood fibres) of the same kind with various orientation in space, having the effective compliance tensor $\bar{S}_{ijkl}$ and the constituent compliance tensor S^c_{ijkl}, the hygroexpansion coefficients of the constituent material β_1 and β_2 in the longitudinal and transverse direction, respectively, are related to the effective hygroexpansion tensor of the composite by:

$$\bar{\beta}_{ij} = \frac{(\bar{S}_{11ij} + \bar{S}_{22ij})\beta_2 \left(\frac{\beta_1}{\beta_2} - 1\right)}{S_2 \left(\frac{S_1}{S_2} - 1\right)} + \frac{\beta_2}{\left(\frac{S_1}{S_2} - 1\right)} \left(\frac{S_1}{S_2} - \frac{\beta_1}{\beta_2}\right) \delta_{ij} \qquad 6.23$$

where δ_{ij} is Kronecker's delta function, and S_1 and S_2 are the strain in the longitudinal and transverse directions due to a unit uniform pressure applied to the composite, respectively. Hence, Equation (6.23) is a linear relation between the macroscopic strain in an unrestricted composite in a certain direction due to a uniform change in moisture content and the strain in that direction due to a unit uniform pressure applied to the composite (Schulgasser, 1987a). The direct implication of the above approach is that the macroscopic hygroexpansion and elastic properties of the composite are sufficient to determine hygroexpansion of the fibres. From manipulation of Equations (6.20)–(6.23), the hygroexpansion coefficients of the wood fibre can be estimated.

Using this methodology, curl measurements of composite or fibre-mat strips, as shown in Fig. 6.4, can be used to determine the transverse hygroexpansion coefficient of wood fibres which turned out to be approximately 0.10 strain per relative moisture content (Neagu et al., 2005). This is in accordance with a few scarce data found in the literature on the hygroexpansion properties of wood fibres. More work along this line would be useful to assess physicochemical treatments of fibres to suppress moisture swelling, such as hornification and cell-wall crosslinking by, for example, butyl tetracarboxylic acid treatment.

6.4 Strength

When selecting a material for a load-carrying application, the perhaps most important property that the engineer considers is strength. Strength is, however,

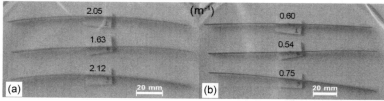

6.4 Measured curvature for three non-symmetric specimens at (a) 0% RH and (b) 50% RH. The specimens are curled around the longitudinal axis, i.e. the machine direction (Neagu et al., 2005).

not a unique property listed in materials handbooks. Strength depends on the stress multiaxiality, presence of stress concentrations, strain rates, temperature, etc. Most literature data, however, concern uniaxial tensile strength of dog-bone specimens of standardised dimensions and test conditions. This is suitable for a general rank of strength of materials, but for structural design purposes of composite components, more data are needed to formulate a criterion for more complex stress states. Nevertheless, only unixial tensile strength at ambient conditions will be considered in this section, since the topic concerns microstructure–property relations rather engineering design of components.

6.4.1 Empirical models

Strength is a difficult issue to address by micromechanics modelling, since strength is inherently controlled by local phenomena and depends on the largest or most severe microstructural heterogeneity. Strength is therefore not amenable to unit-cell modelling and similar appoaches, such as for homogenised properties such as elasticity and hygroexpansion. In other words, it is not fruitful to predict and simulate by extrapolation the strength of composites based on the strength of the fibres and the matrix. Nevertheless, interpolating composite strength from constituent stengths and microstructurally related fitting parameters can work in practice.

For wood fibre composites with higher aspect ratios of the reinforcing fibres, the influence of the interface is not the only dominating factor that affects the strength. From a mechanics point of view, there is no difference whether the fibres come from wood, glass or synthetic polymers. Essentially, it is only the values of the properties of constituent materials that change. Hence, models developed, e.g. for short glass fibre composites, can be used also for wood fibre composites for sufficiently high aspect ratios, i.e. not for wood–flour composites. A theory on strength of short-fibre composites has been developed by Fukuda and Chou (1982). The variation in fibre length and orientation is taken into account in this modified rule-of-mixtures. The coefficient of alignment stems from the work by Fukuda and Kawata (1974). This approach has been adopted by Toftegaard and Lilholt (2002), who found a good correlation with experimental results by using an empirical expression based on a combination of a modified rule-of-mixtures (Kelly and Macmillan, 1986) and a factor for the effect of porosity suggested (Davidge, 1979). For uniaxial tensile loading, composite strength is then expressed as:

$$\sigma_c = (\eta \sigma_f V_f + \sigma_m^* V_m)(1 - V_p)^n \qquad 6.24$$

where σ_f is the fibre strength, σ_f^* is the matrix stress at ultimate failure of the composite, V_f, V_m and V_p are the volume fractions of fibres, matrix and pores, respectively, η is an empirical factor related to the microstructure (e.g. fibre orientation and fibre length) and n is an empirical exponent describing the effect of porosity.

The empirical factor, η, can be related to a factor that depends on the fibre length η_L and a factor that depends on the fibre orientation distribution, η_O, such that

$$\eta = \eta_L \eta_O \qquad 6.25$$

where the multiplicative relation is an empirical assumption, that has shown to work relatively with experimental data (Fu and Lauke, 1996). The length factor can be expressed by the model of Kelly and Tyson (1965):

$$\eta_L = \begin{cases} 1 - \dfrac{l_c}{2l} & l \geq l_c \\ \dfrac{l_c}{2l} & l < l_c \end{cases} \qquad 6.26$$

where l is the fibre length, and the critical or ineffective fibre length is

$$l_c = \dfrac{\sigma_f r}{\tau} \qquad 6.27$$

where r is the fibre radius, and τ is the interfacial shear strength. The latter can be determined from, for example, pull-out tests or single-fibre fragmentation tests. For instance, Sanadi et al. (1993) have characterised the interfacial shear strength from pull-out through loading of partially embedded wood fibre specimens. The single-fibre fragmentation test has been shown to work for wood fibres embedded in a ductile thermoset matrix (Thuvander et al., 2001). Joffe et al. (2003) have shown that this method can also be used for interfacial characterisation, where treated and untreated flax fibres were fragmented until saturation.

The factor ascribed to the fibre orientation distribution, η_O, can be interpreted in the same way as for elastic analysis, where $\eta_O = 3/8$ for in-plane random orientation distribution (Fukuda and Kawata, 1974). However, the fibre orientation is likely to have a different impact on strength than on stiffness, since the underlying physical mechanisms pertaining to these properties are different in nature. Fitting the model in Equation (6.24) to measured composite strengths for short glass-fibre reinforced polypropylene, a value of $\eta_O = 0.20$ was obtained (Thomason et al., 1996).

6.4.2 Interfacial effects

For WPCs with wood particles used as fillers, the composite strength is frequently used as a measure of the interfacial strength (e.g. Oksman, 1996) since the wood particles act as stress raisers. For particle composites, it can be assumed that the strength improves with stronger fibre–matrix interfaces, provided that all other material parameters remain unchanged. For fibre composites, a stronger interface does not necessarily mean a stronger composite, since a too strong interface can result in a brittle and flaw-sensitive composite material. However, for most polyolefin thermoplastics, the interface should be as strong as possible if a strong composite is the aim, since the interaction between the polar natural fibre and the non-polar aliphatic polymer matrix is very small. A useful characteristic of natural cellulose fibres is that they can be more easily modified than the relatively inert carbon, aramid and glass fibres. This characteristic stems from the abundance of reactive functional groups, most notably hydroxyl groups, on the surface of cellulosic fibres, as well as in the interior bulk of the fibres. It opens up opportunities to modify cellulosic fibres for improved strength properties. Since the microstructure in wood fibre composites is far more complex than in, for example, laminated composites based on glass or carbon fibres, the influence of the fibre-matrix interface on strength properties is far from understood. Energy-based criteria, such as Tsai–Hill used for laminated prepreg composites, are not directly suitable for wood fibre composites. These composites usually have a distribution of fibre orientation, and the fibres are relatively short with varying properties. For instance, micrographs of fracture surfaces for wood fibre composites with various fibre orientations are shown in Fig. 6.5, where the role of the interface depend on the fibre orientation (Gamstedt *et al.*, 2002). In general, these kinds of complexity call for experimental work to quantify how the fibre–matrix interface controls strength, as well as other properties such as fracture toughness and moisture uptake/swelling in wood fibre composite. With the resulting understanding, it would be possible to chemically tailor the interface for optimal mechanical performance.

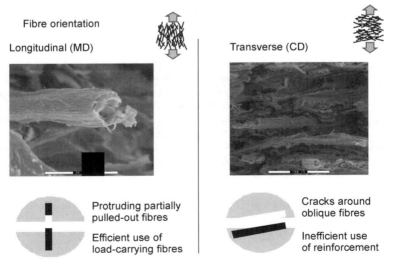

6.5 Effects of fibre orientation on damage mechanisms leading to ultimate fracture.

6.5 Conclusions

This chapter covers an overview of existing micromechanical models to predict the hygroelastic and strength properties of wood fibre composites based on properties of the fibre and matrix constituents, and the microstructure. The idea is to refer the reader to relevant resources to find a suitable model. Most models have previously been developed for composite materials based on synthetic fibres, such as glass, carbon or aramid fibres. These fibres generally have a simpler geometry and show less variability than wood fibres. If a model has not been validated thoroughly to be used for wood fibre composites, care must thus be taken when choosing models that have been developed for other materials. The particularities of the wood fibres that could affect the applicability of models developed for composites with synthetic fibres are their rather large variability in properties, helical structure, general anisotropy, pits, etc.

The micromechanical models can be used to back-calculate a certain fibre property, since it may prove to be easier to test a composite than individual wood fibres. The models may also be used for simulations to identify which microstructural feature has the largest impact on the given macroscopic property. This could give processing engineers and chemists an indication of where to spend most of their efforts in materials development.

6.6 References

Aboudi, J (1991), *Mechanics of composite materials. A unified micromechanical approach*, Amsterdam, The Netherlands, Elsevier Science Publishers B.V.

Advani, S G and Talreja, R (1993), 'A continuum approach to determination of elastic properties of short', *Mechanics of Composite Materials*, **29** (2), 171–183.

Åkerholm, M and Salmén, L (2003), 'The oriented structure of lignin and its viscoelastic properties studied by static and dynamic FT-IR spectroscopy', *Holzforschung*, **57** (5), 459–465.

Bogren, K M, Gamstedt, E K, Neagu, R C, Åkerholm, M and Lindström, M (2006), 'Dynamic-mechanical properties of wood-fiber reinforced polylactide: experimental characterization and micromechanical modeling', *Journal of Thermoplastic Composite Materials*, **19** (6), 613–637.

Budiansky, B (1965), 'On the elastic moduli of some heterogeneous materials', *Journal of the Mechanics and Physics of Solids*, **13** (4), 223–227.

Chamis, C C and Sendeckyj, G P (1968), 'Critique on theories predicting thermoelastic properties of fibrous composites', *Journal of Composite Materials*, **2** (3), 332–358.

Chen, C H and Cheng, C H (1996), 'Effective elastic moduli of misoriented short-fiber composites', *International Journal of Solids and Structures*, **33** (17), 2519–2539.

Chou, T W, Nomura, S and Taya, M (1980), 'A self-consistent approach to the elastic stiffness of short-fiber composites', *Journal of Composite Materials*, **14**, 178–188.

Chow, T-W (1992), *Microstructural design of fiber composites*, Cambridge, Cambridge University Press.

Christensen, R M (1991), *Mechanics of composites materials*, Malabar, FL Kreiger Publishing Company.

Christensen, R M and Lo, K H (1979), 'Solutions for effective shear properties in three phase sphere and cylinder models', *Journal of Mechanics and Physics of Solids*, **27** (4), 315–330.
Cox, H L (1952), 'The elasticity and strength of paper and other fibrous materials', *British Journal of Applied Physics*, **3** (3), 72–79.
Davidge, RW (1979), *Mechanical behaviour of ceramics*, Cambridge, Cambridge University Press.
Eshelby, J D (1957), 'The determination of the elastic field of an ellipsoidal inclusion and related problems', *Proceedings of the Royal Society A*, **241**, 379–396.
Facca, A G, Kortschot, M T and Yan, N (2006a), 'Predicting the elastic modulus of natural fibre reinforced thermoplastics', *Composites Part A: Applied Science and Manufacturing*, **37** (10), 1660–1671.
Facca, A G, Kortschot, M T and Yan N (2006b) 'Predicting the elastic modulus of hybrid fibre reinforced thermoplastics', *Polymers & Polymer Composites*, **14** (3), 239–249.
Fu, S-Y and Lauke B (1996), 'Effects of fiber length and fiber orientation distributions on the tensile strength of short-fiber-reinforced polymers', *Composites Science and Technology*, **56** (10), 1179–1190.
Fu, S-Y and Lauke, B (1998a), 'An analytical characterization of the anisotropy of the elastic modulus of misaligned short-fiber-reinforced polymers', *Composites Science and Technology*, **58** (12), 1961–1972.
Fu, S-Y and Lauke, B (1998b), 'The elastic modulus of misaligned short-fiber-reinforced polymers', *Composites Science and Technology*, **58** (3–4), 389–400.
Fukuda, H and Chou, T-W (1982), 'A probabilistic theory of the strength of short-fibre composites with variable fiber length and orientation', *Journal of Materials Science*, **17** (4), 1003–1011.
Fukuda, H and Kawata, K (1974), 'On Young's modulus of short-fiber composites', *Fibre Science and Technology*, **7**, 207–222.
Gamstedt, E K, Sjöholm, E, Neagu, R C, Berthold, F and Lindström, M (2002), 'Effects of fibre bleaching and earlywood-latewood fractions on tensile properties of wood fibre reinforced vinyl ester', in Lilholt, H, Madsen, B, Toftegaard, H L, Cendre, E, Megnis, M, Mikkelsen, L P & Sørensen, B F (Eds.) *Proceedings of the 23rd Risø International Symposium on Sustainable and Natural Polymeric Composites – Science and Technology*, Risø National Laboratory, Denmark, 185–196.
Gusev, A A (1997), 'Representative volume element size for elastic composites: a numerical study', *Journal of the Mechanics and Physics of Solids*, **45** (9), 1449–1459.
Halpin, J C (1969), 'Stiffness and expansion estimates for oriented short fiber composites', *Journal of Composite Materials*, **3** (4), 732–734.
Halpin, J C and Kardos, J L (1976), 'Halpin–Tsai equations – review', *Polymer Engineering and Science*, **16** (5), 344–352.
Hashin, Z (1979), 'Analysis of properties of fiber composites with anisotropic constituents', *Journal of Applied Mechanics*, **46** (3), 543–550.
Hashin, Z (1983), 'Analysis of composite materials – a survey', *Journal of Applied Mechanics*, **50** (3), 481–505.
Hashin, Z and Rosen, B W (1964), 'The elastic moduli of fiber-reinforced materials', *Journal of Applied Mechanics*, **31** (2), 223–232.
Hermans, J J (1967), 'The elastic properties of fiber reinforced materials when the fibers are aligned', *Konigl. Nederl. Akad. v Wetensch B*, **65**, 1–9.
Hervé, E and Zaoui, A (1995), 'Elastic behaviour of multiply coated fibre-reinforced composites', *International Journal of Engineering Science*, **33** (10), 1419–1433.

Hill, R (1952), 'The elastic behaviour of a crystalline aggregate', *Proceedings of the Physical Society of London Section B*, **65** (389), 396–396.

Hill, R (1963), 'Elastic properties of reinforced solids: some theoretical principles', *Journal of the Mechanics and Physics of Solids*, **11** (5), 357–372.

Hill, R (1965a), 'A self-consistent mechanics of composite materials', *Journal of the Mechanics and Physics of Solids*, **13** (4), 213–222.

Hill, R (1965b), 'Theory of mechanical properties of fibre-strengthened materials – III. self-consistent model', *Journal of the Mechanics and Physics of Solids*, **13** (4), 189–198.

Hine, P J, Lusti, H R and Gusev, A A (2002), 'Numerical simulation of the effects of volume fraction, aspect ratio and fibre length distribution on the elastic and thermoelastic properties of short fibre composites', *Composites Science and Technology*, **62** (10–11), 1445–1453.

Joffe, R, Andersons, J and Wallström, L (2003), 'Strength and adhesion characteristics of elementary flax fibres with different surface treatments', *Composites Part A*, **34** (7), 603–612.

Kelly, A and Macmillan NH (1986), *Strong solids*, Oxford, Clarendon Press.

Kelly, A and Tyson, W R (1965), 'Tensile properties of fibre-reinforced metals: copper/tungsten and copper/molybdenum', *Journal of the Mechanics and Physics of Solids*, **13** (6), 329–338.

Laws, N and McLaughlin, R (1979), 'The effect of fibre length on the overall moduli of composite materials', *Journal of the Mechanics and Physics of Solids*, **27** (1), 1–13.

Levin, V M (1967), 'Thermal expansion coefficients of heterogeneous materials', *Mekanika Tverdogo Tela*, **2** (1), 88–94.

Lundquist, L, Marque, B, Hagstrand, P O, Leterrier, Y and Manson, J-A E (2003), 'Novel pulp fibre reinforced thermoplastic composites', *Composites Science and Technology*, **63** (1), 137–152.

Lusti, H R, Hine, P J and Gusev, A A (2002), 'Direct numerical predictions for the elastic and thermoelastic properties of short fibre composites', *Composites Science and Technology*, **62** (15), 1927–1934.

Mark, R E (1967), *Cell Wall Mechanics of Tracheids*, New Haven, Yale University Press.

Marklund, E (2005), *Micromechanism Based Material Models for Natural Fiber Composites*, Department of Applied Physics and Mechanical Engineering, Division of Polymer Engineering, Luleå, Sweden, Luleå University of Technology.

Mendels, D A, Leterrier, Y and Manson, J-A E (1999), 'Stress transfer model for single fibre and platelet composites', *Journal of Composite Materials*, **33** (16), 1525–1543

Mori, T and Tanaka, K (1973), 'Average stress in matrix and average elastic energy of materials with misfitting inclusions', *Acta Metallurgica*, **21** (5), 571–574.

Morton, J, Quarmley, J and Rossi, L (2003), 'Current and emerging applications for natural and woodfiber-plastic composites', in *The 7th International Conference on Woodfiber-Plastic Composites*, Forest Products Society, Madison, WI, 3–6.

Nairn, J A (1997), 'On the use of shear-lag methods for analysis of stress transfer in unidirectional composites', *Mechanics of Materials*, **26** (2), 63–80.

Nairn, J A (2004), 'Generalized shear-lag analysis including imperfect interfaces', *Advanced Composites Letters*, **13** (6), 263–274.

Nairn, J A and Liu, Y C (1997), 'Stress transfer into a fragmented, anisotropic fiber through an imperfect interface', *International Journal of Solids and Structures*, **34** (10), 1255–1281.

Neagu, R C, Gamstedt, E K and Lindström, M (2005), 'Influence of wood fibre hygroexpansion on the dimensional instability of fibre mats and composites', *Composites Part A – Applied Science and Manufacturing*, **36** (6), 772–788.

Neagu, R C, Gamstedt, E K, Bardage, S L and Lindström, M (2006a), 'Ultrastructural features affecting mechanical properties of wood fibres', *Wood Material Science and Engineering*, **1** (3), 146–170.
Neagu, R C, Gamstedt, E K and Berthold, F (2006b) 'Stiffness contribution of various wood fibers to composite materials', *Journal of Composite Materials*, **40** (8), 663–699.
Neumann, S and Marom, G (1986), 'Free-volume dependent moisture diffusion under stress in composite materials', *Journal of Materials Science*, **21**, 26–30.
Oksman, K (1996). 'Improved interaction between wood and synthetic polymers in wood/polymer composites', *Wood Science and Technology*, **30**, 197–205.
Qiu, Y P and Weng, G J (1990), 'On the application of Mori–Tanaka's theory involving transversely isotropic spheroidal inclusions', *International Journal of Engineering Science*, **28** (11), 1121–1137.
Rosen, B W and Hashin, Z (1970), 'Effective thermal expansion coefficients and specific heats of composite materials', *International Journal of Engineering Science*, **8**, 157–173.
Salmén, L (2004), 'Micromechanical understanding of the cell-wall structure', *Comptes Rendus Biologies*, **327** (9–10), 873–880.
Sanadi, A R, Rowell, R M and Young, R A (1993), 'Evaluation of wood–thermoplastic–interphase shear strengths', *Journal of Materials Science*, **28** (23), 6347–6352.
Schulgasser, K (1987a), 'Moisture and thermal expansion of wood, particle board and paper', in *Proceedings of the International Paper Physics Conference*, Quebec, 53–63.
Schulgasser, K (1987b), 'Thermal-expansion of polycrystalline aggregates with texture', *Journal of the Mechanics and Physics of Solids*, **35** (1), 35–42.
Sjöholm, E, Berthold, F, Gamstedt, E K, Neagu, C and Lindström, M (2002), 'The use of conventional pulped wood fibres as reinforcement in composites', in Lilholt, H, Madsen, B, Toftegaard, H L, Cendre, E, Megnis, M, Mikkelsen, L P & Sørensen, B F (Eds.) *Proceedings of the 23rd Risø International Symposium on Sustainable and Natural Polymeric Composites – Science and Technology*, Risø National Laboratory, Denmark, 307–314.
Takao, Y, Chou, T W and Taya, M (1982), 'Effective longitudinal Young's modulus of misoriented short fiber composites', *Journal of Applied Mechanics*, **49** (3), 536–540.
Tandon, G P and Weng, G J (1984), 'The effect of aspect ratio of inculsions on the elastic properties of unidirectionally aligned composites', *Polymer Composites*, **5** (4), 327–333.
Thomason, J L, Vlug, M A, Schipper, G and Kriker, H G L T (1996), 'Influence of fibre length and concentration on the properties of glass fibre-reinforced polypropylene. 3. Strength and strain at failure', *Composites Part A*, **27** (11), 1075–1084.
Thuvander, F, Gamstedt, E K and Ahlgren, P (2001), 'Distribution of strain to failure of single wood pulp fibres', *Nordic Pulp and Paper Research Journal*, **16** (1), 46–56.
Toftegaard, H and Lilholt H (2002), 'Effective stiffness and strength of flax fibres derived from short fibre laminates', in Lilholt, H, Madsen, B, Toftegaard, H L, Cendre, E, Megnis, M, Mikkelsen, L P & Sørensen, B F (Eds.) *Proceedings of the 23rd Risø International Symposium on Sustainable and Natural Polymeric Composites – Science and Technology*, Risø National Laboratory, Denmark, 325–334.
Tucker III, C L and Liang, E (1999), 'Stiffness predictions for unidirectional short-fiber composites: review and evaluation', *Composites Science and Technology*, **59** (5), 655–671.

7
Outdoor durability of wood–polymer composites

N M STARK, USDA Forest Service, USA and
D J GARDNER, University of Maine, USA

7.1 Introduction

Wood–plastic composite (WPC) lumber is promoted as a low-maintenance, high-durability product (Clemons, 2002). However, after a decade of exterior use in the construction industry, questions have arisen regarding durability. These questions are based on documented evidence of failures in the field of WPC decking products due to such impacts as polymer degradation (Klyosov, 2005), wood decay (Morris and Cooper, 1998), and susceptibility to mold which negatively impact the aesthetic qualities of the product. The industry has responded to problems associated with first-generation products by improving WPC formulations. Manufacturers have also made great strides in making more reasonable claims and in educating consumers on the proper care and maintenance of WPC products to maintain the aesthetic quality of the surface finish. Research groups throughout the world are working toward a fundamental understanding of WPC durability that will help improve and/or identify new strategies for protecting WPCs. WPC durability will continue to be an important subject regarding the use of these products in building construction and other related applications in the field.

7.2 Characteristics of raw materials

This section will discuss the characteristics of wood and plastics that make them susceptible to degradation, and the degradation mechanisms of each component. A discussion of structure and composition of wood and polymers in WPCs can be found in Chapter 1.

7.2.1 Wood

Weathering

Wood is very susceptible to the effects of weathering including the primary impacts of ultraviolet (UV) light, oxidation, rain, and combinations of light, oxidation, and rain. Wood is hygroscopic, readily absorbing moisture due to the prevalence of hydroxyl groups. All wood components are also susceptible to degradation by UV radiation. However, lignin is primarily responsible for UV absorption. Of the total amount of UV light absorbed by wood, lignin absorbs 80–95% (Fengel and Wegener, 1983). Chromophoric functional groups present in lignin can include phenolics, hydroxyl groups, double bonds, and carbonyl groups. In addition, lignin can form free radicals as intermediates. Photodegradation of wood begins with an attack on the lignin-rich middle lamella. Longer exposure leads to a degradation of secondary walls (Fengel and Wegener, 1983). UV light degrades lignin into water-soluble compounds that are washed from the wood with rain, leaving a cellulose-rich surface with a fibrous appearance.

The effects of UV degradation are largely surface phenomena. UV light cannot penetrate deeper than 75 μm into wood. However, studies investigating depth of degradation show that degradation occurs deeper than this. The result is a proposed mechanism where wood components at the surface initially absorb UV light, and then an energy transfer process from molecule to molecule dissipates excess energy to create new free radicals. In this way, free radicals migrate deeper into wood and cause discoloration reactions (Hon and Minemura, 2001). After long exposure times, lignin content through the thickness of wood changes gradually even beyond the discolored surface layer, with less lignin at the surface and more in the center portion (Hon and Minemura, 2001).

Photodegradation leads to changes in wood's appearance such as discoloration, roughening and checking of surfaces, and destruction of mechanical and physical properties. All wood species eventually will fade from a brown, red, or yellow color to a light grey during weathering. Wood weathering is visually apparent in untreated wood and in WPCs that do not contain a colorant.

Biological attack

The natural origin of wood predisposes it to degradation by a variety of biological deterioration agents. Wood chemical components provide a food source for a variety of biological organisms including insects, fungi, bacteria, and marine borers.

Brown-, white-, and soft-rot fungi (see below) all contribute to the decay of wood. The conditions essential for fungal growth in wood are food, sufficient oxygen, suitable temperature, and adequate moisture. Wood itself provides the necessary food, and oxygen is readily available in the environment. A wood moisture content of approximately 20% is required for decay. Below this level degradation due to fungal attack will not occur, and fungi that may have already begun to grow will cease growing (Naghipour, 1996). As a general rule, if wood is kept dry, i.e. moisture content below 20% then fungi typically will not attack wood. However, in unprotected outdoor or marine exposures, wood can be exposed to high levels of moisture that provide the necessary conditions for biological attack.

Fungi that attack wood include the decay fungi and stain or mold fungi. Decay fungi have the most deleterious impact on wood in service because the organisms consume the primary wood chemical components. Decay fungi are grouped into three types including white-rot, brown-rot (both basidiomycetes), and soft-rot (ascomycetes). White-rot fungi primary attack lignin and leave behind a cellulose-rich residue, while brown-rot primarily attack carbohydrates and leave behind a lignin-rich residue. Brown-rot fungi attack mainly softwood while white-rot fungi attack primarily hardwoods. However, each fungal type can be found on both softwoods and hardwoods (Ibach, 2005). Soft-rot fungi typically produce chains of cavities with conical ends in the secondary wall of wood (Daniel, 2003). Stain or mold (mildew) fungi will use extractable materials from wood as a food source. Mildew does not impact the strength properties of wood but does have a negative impact on the aesthetic quality of a wood surface.

Mold requires similar conditions for growth as decay fungi. However the food sources for mold include stored sugars, starches, and other compounds. Similar to decay fungi, mold growth on wood also appears above the threshold of 20% moisture content. Although mold does not usually affect the strength of wood, it can increase the absorptivity, making the wood more susceptible to moisture (Ibach, 2005). Discoloration from the spores is usually confined to the surface.

Insects that attack wood include termites, wood destroying beetles, carpenter ants, and carpenter bees. Marine borers that attack wood include shipworms (teredineds) and piddocks (pholads).

7.2.2 Polymers

Polymers degrade by chemical, mechanical, photo(light)-induced, and/or thermal modes. The major polymer matrices used in WPCs include high-density polyethylene (HDPE), low-density polyethylene (LDPE), polypropylene (PP), and polyvinyl chloride (PVC). More recent work describes the use of nylon as a polymer matrix for WPCs (Chen and Gardner, 2008). Polymer additives used in WPC processing include lubricants, colorants, light stabilizers, anti-oxidants and coupling agents. Specific additives or combinations of additives can potentially contribute to degradation. For the most part, polymers and polymer additives are much less susceptible to biological attack than wood; but polymers, like wood, are susceptible to the weathering effects of UV light and oxidation.

PE and PP are linear molecules consisting of carbon and hydrogen. The energies of photons in the UV region (290–400 nm) are significantly higher than bond energies typically found in PE and PP (e.g. C–C and C–H bonds).

However, the excitation of single bonds requires an amount of energy that is also significantly higher than the bond energy (Gugumus, 1995). Therefore, photodegradation of polyolefins is caused mainly by the presence of chromophores, functional groups that readily absorb UV light, introduced during polymer manufacturing, processing, or storage. They include catalyst residues, hydroperoxide groups, carbonyl groups, and double bonds (Gugumus, 1995). Photodegradation of the chromophores yields free radicals, which then give rise to compounds containing hydroxyl groups, carbonyl groups, and vinyl groups (Wypych, 1995). As semicrystalline polymers, the packing of the crystalline phase is much tighter than that of the amorphous phase. Oxygen diffusion into crystalline segments is restricted (Wypych, 1995). Comparing PE and PP, the degree of branching determines the oxidation rate; more branching results in more labile hydrogen atoms attached to the tertiary carbon atoms.

PVC contains C–Cl, C–C, and C–H bonds. None of these absorb UV radiation. PVC is degraded by residual solvents, unsaturations, irregularities in polymer structure, and thermal history. Photodegradation proceeds via a free radical pathway and leads to dehydrochlorination, chain scission, and crosslinking. Incorporation of a plasticizer increases the rate of oxygen diffusion (Wypych, 1995).

Indications of photodegradation include oxidation of the polymer, changes in crystallinity, and structural changes such as crosslinking and chain scission. Photodegradation is a surface phenomenon and can lead to the formation of surface cracks, embrittlement, and loss of strength and modulus of elasticity (MOE).

7.3 Changes in composite properties with exposure

Environmental, biological, chemical, mechanical, photo(light)-induced, and/or thermal modes of degradation all contribute to the degradation of WPCs. Outdoor durability, by its very nature, exposes WPCs to degradation modes that can act synergistically.

7.3.1 Moisture effects

In WPCs, the hydroxyl groups on wood or other lignocellulosic materials are primarily responsible for the absorption of water, which causes the wood to swell. When WPCs are exposed to moisture, the swelling wood fiber can cause local yielding of the plastic due to swelling stress, fracture of wood particles due to restrained swelling, and interfacial breakdown (Joseph et al., 1995). Figure 7.1 illustrates this mechanism. Initially, there is adhesion between the wood particle and matrix in a dry WPC. As the wood particle absorbs moisture, it swells. This creates stress in the matrix, leading to the formation of microcracks. It also creates stress in the wood particle, causing damage. After drying the

146   Wood–polymer composites

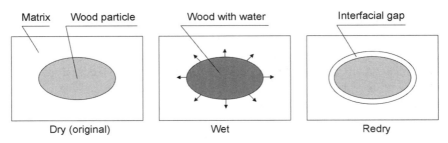

7.1 Schematic of moisture damage mechanism in WPCs.

composite, there is no longer adhesion at the matrix and wood particle interface. Cracks formed in the plastic and the interfacial gap contribute to penetration of water into the composite at a later exposure.

Damage that occurs during moisture exposure degrades mechanical properties. Microcracks in the matrix and damage to wood particles cause a loss in MOE and strength. Interfacial damage is primarily responsible for a loss in composite strength. The effects can be dramatic. For example, after soaking 40% wood flour filled PP composites in a water bath for 2000 hours, the water absorption of the composites was 9%. This corresponded with a 39% decrease in flexural MOE and a 22% decrease in flexural strength (Stark, 2001).

The amount of moisture absorbed by WPCs can be influenced by wood flour content, wood particle size, processing method, and additives. Steckel *et al.* (2007) examined the effects of wood flour content, particle size, coupling agent, and milling of the original surface on the moisture absorption of WPCs. After soaking the composites in distilled water, it was reported that increasing wood flour content, removing the original composite surface, and increasing the particle size increased the equilibrium moisture content of the composite while adding a coupling agent decreased it (Table 7.1; Steckel *et al.*, 2007). Lin *et al.* (2002) also found that increasing wood flour content and, to a smaller extent, increasing particle size increases the moisture absorption of WPCs. Lignocellulosic type also changes water absorption properties. Tajvidi *et al.* (2006) demonstrated that PP containing kenaf fiber and newsprint absorbed more moisture than WPCs, while composites containing rice hulls absorbed less. Because the diffusion of water into WPCs is slow, it can take some time to reach equilibrium moisture content. As the thickness of WPCs increases, the distribution of absorbed water in the material becomes non-uniform with higher moisture content in the outer surface layer than in the core (Wang and Morrell, 2004).

Moisture exposure that is cumulative can cause some irreversible damage to WPCs. Clemons and Ibach (2004) subjected 50% wood flour filled HDPE composites to five moisture exposure cycles. Each cycle consisted of 2 hours of boiling and 22 hours of drying. After each boiling cycle, the final moisture content of the composite was higher than for the previous cycle (Fig. 7.2).

Outdoor durability of wood–polymer composites 147

Table 7.1 Average moisture content of PP-based composites soaked in distilled water for 238 days (adapted from Steckel *et al.*, 2007)

Wood flour content (%)	Wood flour particle size	Coupling agent (%)	Surface treatment	Moisture content (%)
25	Coarse	0	None	5.08 (0.03)
25	Coarse	0	Milled	6.33 (0.21)
25	Coarse	3	None	4.42 (0.16)
25	Coarse	3	Milled	5.75 (0.05)
25	Fine	0	None	4.76 (0.17)
25	Fine	0	Milled	5.73 (0.04)
25	Fine	3	None	4.43 (0.06)
25	Fine	3	Milled	5.19 (0.20)
50	Coarse	0	None	13.33 (0.34)
50	Coarse	0	Milled	14.12 (0.43)
50	Coarse	3	None	10.92 (0.27)
50	Coarse	3	Milled	12.56 (0.16)
50	Fine	0	None	11.51 (0.16)
50	Fine	0	Milled	12.41 (0.07)
50	Fine	3	None	10.77 (0.17)
50	Fine	3	Milled	11.51 (0.10)

Values in parentheses are one standard deviation.

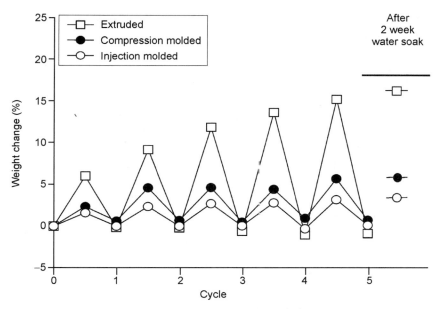

7.2 Moisture sorption of 50% wood flour filled HDPE during five moisture cycles consisting of 2 hours of boiling followed by 22 hours of drying (Clemons and Ibach, 2004).

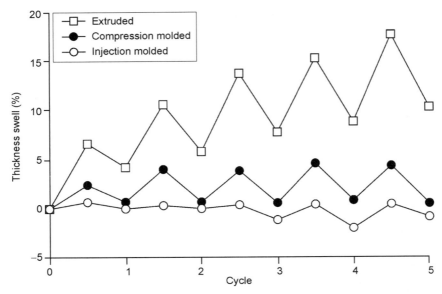

7.3 Thickness swell of 50% wood flour filled HDPE during five moisture cycles consisting of 2 hours of boiling followed by 22 hours of drying (Clemons and Ibach, 2004).

Thickness swell also occurred (Fig. 7.3). Drying composites that absorbed more water, i.e. extruded composites, did not result in the composite returning to its original thickness.

Moisture exposure also becomes important in cold climates during freeze–thaw cycles. Freeze–thaw cycling WPCs results in a loss of mechanical properties (Panthapulakkal *et al.*, 2006; Pilarski and Matuana, 2005, 2006). Standard freeze–thaw cyclical testing consists of three parts: (1) water soak until weight gain in a 24 hour period is not more than 1%; (2) freezing for 24 hours at −29 °C; and (3) thawing at room temperature for 24 hours (ASTM D6662). WPCs comprising 46% pine wood flour filled PVC were exposed to five freeze–thaw cycles. After exposure, the modulus of rupture (MOR) decreased 13% and the MOE decreased 30%. To better understand the effect moisture has on freeze–thaw cycling, Pilarski and Matuana (2005) performed two variations of the water soak/freeze–thaw test in addition to the standard test. In the first variation, the water soak portion of the test was removed. Therefore, the composites were exposed to only freeze–thaw cycles. For the second variation, the freezing and thawing portions were removed. Composites were exposed only to the water soak. Figure 7.4 shows the loss in MOR and MOE after the standard water soak/freeze–thaw cycle and the two variations. The authors concluded that the water soak portion of the cycling process had the greatest impact on flexural properties. This was attributed to moisture influencing interfacial adhesion between PVC and wood flour.

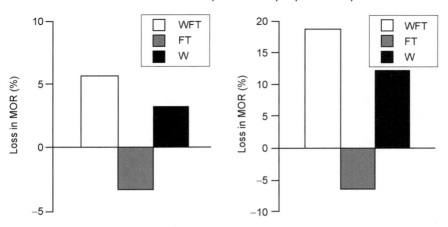

7.4 Loss in flexural properties after exposure of 46% pine wood flour filled PVC composites to water soak/freeze–thaw cycling (WFT), freeze–thaw cycling (FT), and water soak cycling (W) (adapted from Pilarski and Matuana, 2005).

7.3.2 Thermal changes

Thermal response of polymers can impact mechanical properties, especially creep, and physical properties, i.e. coefficient of thermal expansion which can impact in-service properties of WPCs. Depending on the particular climatic exposure, WPCs used in decking applications can experience temperatures ranging from −30 °C to 50 °C. The primary thermal changes impacting WPC properties include thermal expansion, mechanical creep, and thermal-oxidative degradation. Increased temperatures can also act synergistically with other chemical degradation mechanisms to increase reaction rates.

Thermal expansion

Thermal expansion in WPCs has been shown to be anisotropic in conventional extruded solid deck boards made from either HDPE or PP containing at least 50% wood flour and 50% polymer/additives (O'Neill and Gardner, 2004). As an example, for PP-based deck boards, the coefficient of thermal expansion is greatest in the thickness direction ($30.03 \times (1/°F) \times 10^{-6}$) followed by the width ($23.61 \times (1/°F) \times 10^{-6}$) and is the least in the length ($11.53 \times (1/°F) \times 10^{-6}$) or machine direction of the boards. Both the polymer chains and wood flour align with the flow of extrusion and this behavior is believed to contribute to the anisotropic thermal behavior of extruded deck boards. Thermal expansion behavior of WPC deck boards becomes important during installation because improper gapping can lead to warping.

Mechanical creep

Thermoplastic materials such as WPCs will experience changes in mechanical (stiffness) properties as a function of increased temperature. Materials under a mechanical load that might perform adequately at normal service temperatures may experience creep under loads for extended periods of time and at higher temperatures (Brandt and Fridley, 2003).

Thermal-oxidative degradation

Recently, a thermal-oxidative degradation issue in WPC deck boards was reported (Klyosov, 2005). WPC deck boards not containing sufficient levels of antioxidant exposed to extreme temperatures in the Arizona desert were experiencing crumbling and subsequent board failure. The failures were attributed to free radical oxidative degradation mechanisms that were exacerbated by high temperatures.

7.3.3 Weathering

Primary weathering variables include solar radiation, temperature, and water. Secondary variables include seasonal and annual variation, geographical differences, atmospheric gases, and pollution changes. The large number of variables accounts for the high variability in weathering studies, and does not readily allow for comparison among natural weathering studies. Results obtained are specific for the location and timeframe of the test.

Accelerated weathering is a technique used to compare performance by subjecting samples to cycles that are repeatable and reproducible. Primary weathering variables can all be measured during accelerated weathering. During accelerated weathering, test standards are typically followed that prescribe a schedule of radiation (ultraviolet, xenon-arc, etc.) and water spray (number and time of cycles).

WPCs exposed to weathering may experience color change, which affects their aesthetic appeal, as well as mechanical property loss, which limits their performance. Changes in mechanical properties after weathering can be due to changes such as composite surface oxidation, matrix crystallinity changes, and interfacial degradation. Although photodegradation of both polymers and wood has been extensively examined, the understanding of WPC weathering continues to evolve as several research groups work on characterizing and understanding changes that occur when WPCs weather.

Weathering results in the destruction of the WPC surface (Fig. 7.5). Wood particles exposed at the surface absorb water and swell. In addition, the plastic matrix cracks upon UV exposure. In combination, the result is a flaky, cracked surface. As WPCs weather, the surface chemistry changes. Oxidation at the surface is one measure of degradation; its increase can be followed throughout weathering. Fourier transform infra-red (FTIR) spectroscopy is one tool that can

Outdoor durability of wood–polymer composites 151

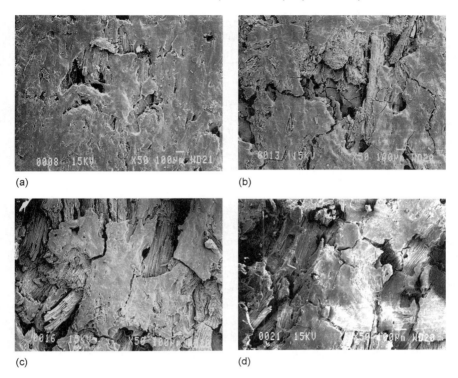

7.5 Micrographs of extruded 50% wood flour filled HDPE composites before weathering (a) and after 1000 hours (b), 2000 hours (c), and 3000 hours (d) of weathering (adapted from Stark et al., 2004).

be used to determine changes in surface chemistry during weathering. Peaks that appear on spectra are assigned to functional groups present at the composite surface. This method has been used to study surface oxidation and matrix crystallinity changes in weathered WPCs (Stark and Matuana, 2004a, 2004b; Muasher and Sain, 2006).

Weathering variables act independently and synergistically to degrade WPCs. In the following example, Stark (2006) exposed extruded WPCs (50% wood flour filled HDPE) to two accelerated weathering cycles for approximately 3000 hours. The first weathering cycle included water spray cycles (12 minutes of water spray every 2 hours); in the second weathering cycle there was no water spray. The change in composite properties is shown in Table 7.2 (Stark, 2006).

The color of the composite clearly lightened after weathering, i.e. increase in lightness (L^*). However, the increase in L^* was much less when the samples were exposed only to UV light, demonstrating that water spray had a large effect on color fade. Flexural MOE and strength decreased when the composites were exposed to UV radiation with water spray (Table 7.2). Exposing the WPCs to UV radiation resulted in only a small decrease in MOE and no significant change in

Table 7.2 Percentage change in extruded 50% wood flour filled HDPE composites after 3000 hours of accelerated weathering (adapted from Stark, 2006)

Property	Weathering cycle	
	UV + water spray	UV only
Lightness (L^*)	+46	+13
MOE	−52	−12
Strength	−34	+1^{NS}

NS: Change not significant at $\alpha = 0.05$.

strength. Exposure to UV radiation with water spray resulted in more destruction in mechanical properties than exposure to UV radiation only (Stark, 2006).

The color of WPCs primarily reflects the color of the wood during weathering, although some whitening is due to stress cracking of the matrix. Water and UV radiation jointly contribute to increasing composite L^*. Exposure to UV radiation degrades lignin leaving loose fibers at the composite surface. Water spray cycles wash away loose fibers, exposing more material for degradation. The result is a cyclical erosion of the surface (Williams *et al.*, 2001). Additionally, washing the surface can remove some water-soluble extractives that impart color (Stark, 2006). UV radiation and water also act synergistically to degrade WPCs in the following ways. Exposing WPCs to UV radiation degrades hydrophobic lignin, leaving hydrophilic cellulose at the surface and increasing surface wettability, causing the surface to become more sensitive to moisture (Kalnins and Feist, 1993). Swelling of the wood fiber also facilitates light penetration into wood and provides sites for further degradation (Hon, 2001).

Fiber swelling due to moisture absorption is primarily responsible for the loss in mechanical properties after weathering. Cracks form in the HDPE matrix due to swelling of the wood fiber and contribute to the loss of composite MOE and strength. The loss in strength is due to moisture penetration into the WPC, which degrades the wood–HDPE interface, decreasing the stress transfer efficiency from matrix to the fiber. Synergism between UV radiation and water also contribute to mechanical property losses by eroding the surface and increasing surface wettability as described above, causing exposure to UV radiation with water exposure to be much more damaging than exposure to UV radiation only (Stark, 2006).

7.3.4 Biological attack

Decay

Wood decays when its moisture content exceeds approximately 20%. In WPCs, it can be assumed that the plastic matrix does not absorb moisture. Therefore, a

50% wood filled WPC must reach a moisture content of about 10% for decay to occur (i.e. the wood component reaches a moisture content of 20%). However, the moisture content is typically measured for the bulk of the material. The slow diffusion of water through WPCs results in higher moisture contents at the surface than in the core (Wang and Morrell, 2004). In the field, the moisture content of the bulk WPC material may be significantly lower than the expected point at which decay begins.

The wood component in WPCs is responsible for its susceptibility to fungal attack. In the winter of 1996/97, brown-rot and white-rot fungi were observed on a commercial WPC after 4 years of service in Florida (Morris and Cooper, 1998). This observation led to the study of WPC decay. Because moisture is the critical factor, it is not surprising that the variables that influence moisture sorption, including fiber content and encapsulation, also influence decay. Mankowski and Morrell (2000) studied three commercial WPC products made from a combination of wood and HDPE. A sample from each product was soaked in water for 30 minutes, sterilized, and exposed to brown-rot or white rot fungi for 12 weeks. They showed that composites with more wood experienced a higher moisture content and also more decay. Verhey et al. (2001) manufactured WPCs with 30, 40, 50, 60, and 70% wood. After exposure to a white-rot or brown-rot fungus for 12 weeks, the composites with higher wood contents generally had more weight loss, and higher weight losses were observed after exposure to brown-rot fungi versus white-rot fungi (Table 7.3). Investigations into the effect of particle size on decay resistance of WPCs demonstrated that weight loss due to fungal decay decreased as wood particle size decreased (Verhey and Laks, 2002). This was attributed to better encapsulation of smaller wood particles by the plastic matrix.

Table 7.3 Percentage weight losses from wood flour/PP composites after 12 weeks exposure to fungi (adapted from Verhey et al., 2001)

Test fungus	Wood content (%)	Surface preparation	
		Unsanded	Sanded
White-rot	30	1	1
	40	1	1
	50	1	1
	60	2	5
	70	3	15
Brown-rot	30	8	8
	40	12	11
	50	14	16
	60	40	33
	70	54	58

WPCs decay only at the surface layer. Stakes cut from a commercial WPC consisting of approximately 56% wood and 44% PE were installed in the ground in Hilo, Hawaii and analyzed after 10 years of exposure. Evaluation of the in-ground portion of the stakes using SEM revealed surface pitting due to loss of wood particles. Only 5 mm in from the surface, there was no evidence of microbial attack (Schauwecker et al., 2006). This was attributed to limited moisture movement into the WPC.

Much of the loss in mechanical properties due to fungal attack can be attributed to moisture absorption. Following the procedure of Clemons and Ibach (2004), Schirp and Wolcott (2005) evaluated HDPE-based WPCs containing either 49% or 70% wood filler after exposure to modified agar-block tests for 3 months. Tests were conducted with a white-rot fungus, brown-rot fungus, or no innoculation to determine the contributions of wood decay and moisture exposure to changes in mechanical properties of WPCs. Loss of stiffness occurred for each of the formulations after exposure, while loss in strength occurred only for WPCs containing 70% wood fiber. The reported losses in mechanical properties were due primarily to moisture absorption.

Mold

Because mold has received attention as a potential health hazard, it has become more important to understand mold growth on WPCs. Much of the ongoing work has been proprietary. Dawson-Andoh et al. (2005/6) exposed wood flour filled HDPE composites to mold fungi. They were able to relate moisture exposure to mold growth and found more evidence of mold growth when composites were directly in contact with moisture versus exposed to an environment with a high relative humidity. Additionally, pre-conditioning WPCs through either exposure to UV weathering or freeze–thaw cycling was shown to have minimal effect on mold growth (Dawson-Andoh et al., 2005/6).

Laks et al. (2005) investigated the effect of several variables on mold growth on extruded wood flour filled HDPE composites. After 8 weeks of exposure, mold was hardly noticeable on composites containing 30% or 50% wood flour, very noticeable on WPCs containing 60% wood flour, and completely covered WPCs containing 70% wood flour. Lubricants are often used to aid in extrusion. Laks et al. (2005) evaluated two, ethylene bis-stearamide wax and zinc stearate. They suggested that the amine in each of these lubricants may provide sufficient nitrogen for more rapid colonization of mold fungi.

Insects and marine borers

Insect attack of wood can include attack by termites, beetles, and ants. There is little information in the area of insect attack on WPCs, but one on-going study reported termite activity after 3 years in-ground exposure in Mississippi (Ibach and Clemons, 2004). After the first and second year of in-ground exposure, there was no reported termite attack on 50% wood flour/HDPE composites. After the third year, nibbles of up to 3% cross-section were reported.

Marine borer attack includes attack of wood by shipworms and crustaceans. Although marine borers present a problem to untreated wood, they are not considered a threat to WPCs based on limited field studies. Pendleton and Hoffard (2000) exposed HDPE composites containing 50% or 70% wood flour to a natural marine environment in Pearl Harbor, Hawaii. After 1, 2, and 3 years the WPC specimens showed no visible marine borer attack (Pendleton and Hoffard, 2000).

7.4 Methods for protection

7.4.1 Moisture effects

Changes in processing conditions

Injection molding, compression molding, and extrusion are processing methods commonly used for manufacturing WPCs. Primary processing variables include temperature and pressure. Both processing methods and variables within a processing method greatly influence composite morphology and durability. Figures 7.2 and 7.3 clearly show that the weight change and dimensional change of extruded WPCs is higher than for compression molded and injection molded WPCs during cyclical water exposure. This is largely due to a polymer-rich surface layer and lower void content of compression molded and injection molded WPCs compared with extruded composites (Clemons and Ibach, 2004). As reported in Table 7.1, removing the polymer-rich surface layer from injection molded WPCs results in increased moisture contents (Steckel *et al.*, 2007).

Fiber modification

The hygroscopicity of wood can be reduced by replacing some of the hydroxyl groups with alternative chemical groups. Acetylation has been an active area of research in improving the moisture performance of wood and wood composites. Acetic anhydride reacts with hydroxyl groups in the wood cell wall to yield and acetylated fiber. For example, acetylating pine wood fiber reduced its equilibrium moisture content at 90% relative humidity and 27 °C from 22% to 8% (Rowell, 1997). This method has been investigated to a more limited extent to provide moisture resistance to WPCs. For example, a 50% wood fiber filled PP composite absorbed 5% moisture after soaking for 34 days while a 50% acetylated wood fiber filled PP composite absorbed only 2.5% (Abdul Khalil *et al.*, 2002).

Additives

Coupling agents are commonly used to improve bonding between the hydrophobic matrix and hydrophilic wood fiber. This has been the most active area of research in employing an additive to provide moisture resistance to WPCs. Coupling agents can reduce the amount of moisture WPCs absorb by reducing gaps at the wood–matrix interface and by reducing the number of hydroxyl groups available for hydrogen bonding with water. However, the decreases in water absorption can be minor. Steckel *et al.* (2007) examined the moisture sorption properties of several WPCs using a full factorial design and determined that wood flour content had the largest influence on equilibrium moisture content, followed by removal of the original polymer-rich composite surface. The inclusion of a coupling agent and reducing particle size had a much smaller effect on equilibrium moisture content (Table 7.1). Panthapulakkal *et al.* (2006) examined water absorption of 65% rice husk filled HDPE composites. After immersion in water for 1608 hours the control composites absorbed 14% water. Composites containing a coupling agent absorbed either 12% or 9% depending on the coupling agent used. However, the use of a lubricant as a processing aid negated any positive effect of the coupling agent. Lin *et al.* (2002) reported an 18% decrease in moisture absorption when 2% coupling agent was added to 15% wood flour filled PP composite. Coupling agents may be more effective at mitigating changes in strength and modulus after exposure to cyclical freeze–thaw testing. The increase in weight change of rice husk filled HDPE composites was 9.6% after 12 freeze–thaw cycles. Adding a coupling agent resulted in increases of 3.3–3.5% (Panthapulakkal *et al.*, 2006).

7.4.2 Thermal changes

Methods for protection of thermal changes in WPCs are primarily formulation based and need to consider protection of both the wood and the polymer components. Thermoplastic polymers can experience thermal oxidative effects during processing especially for PVC–wood composites. Both antioxidants and thermal stabilization aids can be added to WPC formulations to decrease thermal degradation during processing. In addition to protecting the polymer during processing, antioxidants can also help protect the polymer matrix during environmental exposure.

Using a plastic matrix that is not as susceptible to thermal changes may be an important way to provide thermal stability in the future. Chen and Gardner (2008) manufactured WPCs using a nylon matrix. The storage modulus of nylon–wood composite was found to be more temperature stable than pure nylon 66 and wood flour reduced the physical aging effects on nylon in the wood composites. Comparing the nylon–wood composite thermal mechanical properties with other, similar glass-filled nylon composites shows that nylon–wood composites are a promising low-cost material for industrial applications.

Thermal expansion in WPCs can be reduced by creating a cellular or foamed microstructure (Finley, 2000). Since wood is more thermally stable with temperature than plastic, WPCs with higher wood contents should exhibit lower coefficients of thermal expansion. Mechanical creep in WPCs can be reduced by crosslinking the polymer and this has been demonstrated for HDPE-based WPCs (Bengtsson *et al.*, 2006). Thermal-oxidative degradation can be reduced by the addition of antioxidants (Klyosov, 2005).

7.4.3 Weathering

Changes in processing conditions

Higher processing pressures and temperature during injection molding compared with extrusion result in more plastic at the surface. The presence of a hydrophobic surface delays some changes that occur during weathering by preventing some degradation due to water exposure.

As processing conditions change, the changes in WPC surface components and morphology that influence moisture performance also influence weathering performance. In the following example, Stark *et al.* (2004) characterized this difference by analyzing FTIR spectra. Figure 7.6 illustrates the changes in composite surface components when 50% wood flour filled HDPE composites were injection molded, extruded, or extruded with the surface removed (planed). The FTIR spectra of the planed surface had larger peaks associated with wood (a broad peak at $3318\,\text{cm}^{-1}$ and a strong peak at $1023\,\text{cm}^{-1}$) compared with the injection molded surface. This suggested more wood at the surface of the planed samples and a plastic-rich layer at the surface of the injection molded samples (Stark *et al.*, 2004).

The changes in the surface were related to weathering performance. Weathering resulted in composite lightening; WPCs lightened to a similar L^* after 3000 hours of weathering. Weathering resulted in a decrease in both flexural strength and MOE (Table 7.4). The relative flexural properties show that the retention of flexural properties was higher for injection molded composites, followed by extruded and planed composites (Stark *et al.*, 2004).

Additives

The first plan of attack to improve the weatherability of WPCs is often to add photostabilizers. Photostabilizers are compounds developed to protect polymers and combat UV degradation. They are generally classified according to the degradation mechanism they hinder. Ultraviolet absorbers (UVAs) and free radical scavengers are important photostabilizers for polyolefins. Commercial UVAs are readily available as benzophenones and benzotriazoles (Gugumus,

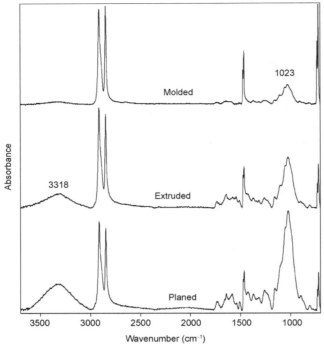

7.6 FTIR spectra of injection molded, extruded, and planed WF/HDPE composites. The broad peak at 3318 cm^{-1} and the sharp peak at 1023 cm^{-1} are due to the wood component (adapted from Stark *et al.*, 2004).

1995). A relatively new class of materials, hindered amine light stabilizers (HALS), has also been extensively examined for polyolefin protection as free radical scavengers (Gijsman *et al.*, 1993; Gugumus, 1993, 1995). Pigments physically block light, thereby protecting the composite from photodegradation.

Table 7.4 Relative flexural properties of 50% wood flour filled HDPE composites manufactured by injection molding, extrusion, or extrusion with the surface removed (planed) after accelerated weathering (adapted from Stark *et al.*, 2004)

Exposure time (hours)	Strength (MPa)			MOE (GPa)		
	Injection molded	Extruded	Planed	Injection molded	Extruded	Planed
0	1	1	1	1	1	1
1000	0.88	0.77	0.66	0.81	0.81	0.60
2000	0.82	0.65	0.63	0.67	0.67	0.47
3000	0.68	0.66	0.63	0.57	0.57	0.48

Many photostabilizers that were developed for use in unfilled polyolefins are being adapted for use in WPCs and this is an active area of research. Pigments, UV absorbers, and hindered amine light stabilizers have been used with some success in mitigating changes that occur during WPC weathering.

Pigments were shown to mitigate the increase in lightness and significantly increase the flexural property retention of WPCs after accelerated weathering (Falk et al., 2000). Lundin (2001) investigated the effect of hindered amine light stabilizer (HALS) content on the performance of WPCs. The author reported that the addition of HALS to the composites did not affect color change caused by accelerated weathering and slightly improved the mechanical property retention (Lundin, 2001). Stark and Matuana (2003) examined the effect of a low molecular weight HALS, a high molecular weight HALS, a benzotriazole ultraviolet absorber (UVA), and a pigment on the changes in lightness and mechanical properties of WPCs after weathering. Only the UVA and pigment significantly reduced composite lightening and loss in mechanical properties. Regardless of molecular weight, HALS was found to be ineffective in protecting the composite against surface discoloration and flexural property loss. Muasher and Sain (2006) also evaluated the performance of HALS and UVAs in stabilizing the color WPCs. They found that high molecular weight diester HALS exhibited synergism with a benzotriazole UVA.

FTIR identified functional groups present on the surface of unexposed and weathered WPCs. Following carbonyl growth indicated that both the pigment and UVA delayed the eventual increase in surface oxidation and decrease in HDPE crystallinity that would occur at later exposure times (Stark and Matuana, 2004a). Following this approach, Muasher and Sain (2006) also used FTIR to evaluate carbonyl growth in photostabilized WPCs. They identified a correlation between carbonyl growth and photostabilizer effectiveness at reducing color fade.

Table 7.5 illustrates the percentage change in property that occurs after injection molded, photostabilized WPCs weather (Stark and Matuana, 2006). The results clearly showed that composites with an ultraviolet absorber (UVA) or pigment (P) lightened less than unstabilized composites. The pigment (P) was more efficient at preventing composite lightening than UVA. Lightness (L^*) decreased with increase in pigment concentration. By contrast, increasing UVA content had little, if any, effect on L^*. Composites with the least amount of lightening had a combination of UVA and P. It was concluded that UVA reduced lightening by absorbing some UV radiation, resulting in less UV radiation available to bleach the wood component, while P physically blocks UV radiation, which also results in less available UV radiation to the wood component. In addition, P masked some lightening (Stark and Matuana, 2006).

Adding 0.5% UVA did not greatly influence the loss in MOE but did improve the loss in strength (Table 7.5). Increasing the UVA concentration to 1% resulted in further retention of MOE and strength. Adding P at 1% resulted in

160   Wood–polymer composites

Table 7.5 Percentage change in properties of photostabilized 50% wood flour filled HDPE composites after 3000 hours of accelerated weathering (adapted from Stark and Matuana, 2006)

Formulations	Change in property (%)		
	L^*	Strength	MOE
—	+115	−27	−33
0.5% UVA	+98	−20	−32
1% UVA	+107	−15	−21
1% P	+73	−13	−18
2% P	+61	−5	−18
0.5% UVA, 1% P	+59	−9	−15
1% UVA, 2% P	+50	-2^{NS}	−16

UVA: Hydroxyphenylbenzotriazole, Tinuvin 328, Ciba Specialty Chemicals
P: Zinc ferrite in carrier wax, CedarTI-8536, Holland Colors Americas.
NS: Change not significant at $\alpha = 0.05$.

smaller MOE and strength losses than did adding 1% UVA. Increasing the concentration of P did not change the loss in MOE but decreased the loss in strength. FTIR work suggested that UVA and P delay changes in HDPE crystallinity (Stark and Matuana, 2004a). UVA was likely consumed during weathering. Therefore, the higher concentration was required to protect against mechanical property loss for the full weathering period. The P consisted of zinc ferrite in a carrier wax. The wax may protect the WPC by creating a hydrophobic surface and resulting in less degradation of the interface (Stark and Matuana, 2006).

7.4.4 Biological attack

Decay of WPCs is a function of moisture content. Therefore, the first step in preventing decay is to prevent or limit moisture sorption. Limiting the access of nutrients, i.e. encapsulating the wood in the plastic matrix, would also provide fungal durability. Limiting oxygen availability by submersing in water decreases the opportunities for mold growth and fungal attack.

Verhey *et al.* (2001) demonstrated that sanding the surface of a compression molded WPC, i.e. removing some of the plastic film at the surface, increased the amount of decay at high wood contents (Table 7.3). Clemons and Ibach (2004) examined WPCs manufactured using extrusion, compression molding, or injection molding. Manufacturing method influenced moisture sorption (Fig. 7.2), with extruded composites absorbing more moisture than compression molded and injection molded composites. After cyclical moisture exposure followed by 12 week exposure to brown-rot fungi, the weight losses of extruded, injection molded, and injection molded samples were 22.8%, 2.4%, and 0.4%, respectively (Clemons and Ibach, 2004). The difference was attributed to surface characteristics of the WPCs; a plastic layer formed at the surface of the compression molded and injection molded composites. Acetylation and silane treatment of wood fibers decreases moisture sorption, and this has been shown to be beneficial in preventing decay (Hill and Abdul Khalil, 2000).

The preservative zinc borate is often used to prevent decay and can be very effective (Ibach *et al.*, 2003; Verhey *et al.*, 2001). For example, mass loss of 50% wood flour filled PP composites was 12.9% after 12 weeks exposure to brown-rot fungi. Adding zinc borate at 1% reduced the mass loss to less than 1%; adding zinc borate at 3 or 5% resulted in virtually no mass loss (Verhey *et al.*, 2001).

There is minimal literature regarding protecting WPCs against mold growth, and much of the work that has been done to date is proprietary. Because mold and mold spores are always present in the air, it is often recommended to clean WPCs with a dilute bleach solution to remove mold that is already growing. Strategies to prevent mold growth often coincide with reducing moisture content and moisture susceptibility.

Biocides also can control mold growth. Dylingowski (2003) evaluated an isothiazole and zinc borate moldicide in a model WPC material. It was reported that isothiazole was more effective at controlling mold growth than zinc borate. Laks *et al.* (2005) evaluated chlorothalonil and zinc borate as fungicides in 70% wood flour filled HDPE composites. After 8 weeks exposure to mold fungi, mold growth was slightly noticeable on WPCs containing 1% or 1.5% chlorothalonil versus mold completely covering the control WPCs. Zinc borate was ineffective at preventing mold growth at 1% loading but, at 3% or 5% loading, mold was hardly noticeable after 8 weeks.

7.5 Future trends

The work summarized here suggests that controlling moisture is the key to not only decreasing losses in performance due to moisture absorption but also increasing weathering performance and resistance to biological attack. Therefore, it is not surprising that in the future WPCs with improved durability will be less susceptible to moisture. New methods to modify the fiber or the composite will be developed. Modifying the composite can include surface coatings, such as co-extrusion or the development of paints and coatings. Methods for embossing a grain pattern onto the surface without disrupting the protective plastic-rich layer will also provide improved moisture resistance.

Until recently most additives used in WPCs were developed for unfilled plastics. As the market has grown, the opportunities for additives suppliers to manufacture additives specifically for WPCs has grown. Additives that address some of the unique needs of WPCs are being developed and incorporated into commercial products, improving the durability of WPCs. It can be expected that this trend will continue in the future. Improvements in lubricants, photostabilizers, and biocides will occur, along with a better fundamental understanding of how they work and interact with each other. New methods for coloring WPCs as well as new, more stable colorants will result in a more natural-looking WPC that exhibits less color fade.

Emerging fields of science such as nanotechnology will also benefit WPCs. Nanotechnology as a characterization method will allow for more fundamental understanding of how changes occur during exposure, and help to identify schemes for preventing those changes. Nanoparticles in WPCs may be able to provide increased resistance to moisture, photodegradation, and biological attack. Nanoparticles may also be used to improve thermal expansion and creep.

7.6 Sources of information and advice

To learn more about the durability of wood, readers can turn to the book chapters titled 'Biological Properties' (Ibach, 2005) or 'Weathering and Photochemistry of Wood' (Hon, 2001). Good reviews of polymer degradation can be found in the texts by Schnabel (1982) and Wypych (1995). A book chapter by Schirp *et al.* (2008) provides good discussion on the biological degradation of WPCs including a discussion of current test methods.

To keep informed on science and technology developments in wood–plastic composites including durability, readers are urged to attend the various conferences focused on this topic. Over the past several years, many international conferences per year have been focused on wood plastic composites. The International Conference on Wood & Biofiber Plastic Composites, the oldest and among the most important of these conferences, started in 1991 and provides forum on the science and technology for the processing and development of these materials.

7.7 References and further reading

Abdul Khalil H P S, Rozman H D, Ismail H, Rosfaizal and Ahmad M N (2002), 'Polypropylene (PP)–*Acacia mangium* composites: the effect of acetylation on mechanical and water absorption properties', *Polymer Plast Tech Eng*, **41** (3), 453–468.

Bengtsson M, Oksman K and Stark N (2006), 'Profile extrusion and mechanical properties of crosslinked wood–thermoplastic composites', *Polymer Compos*, **27** (2), 184–194.

Brandt C W and Fridley K J (2003), 'Load-duration behavior of wood–plastic composites', *J Mater Civ Eng*, **15** (6), 524–536.

Chen J and Gardner D J (2008), 'Dynamic mechanical properties of extruded nylon–wood composites', *Polymer Compos*. In press.

Clemons C (2002), 'Wood–plastic composites in the United States: the interfacing of two industries', *Forest Prod J*, **52** (6), 10–18.

Clemons C M and Ibach R E (2004), 'The effects of processing method and moisture history on the laboratory fungal resistance of wood–HDPE composites', *Forest Prod J*, **54** (4), 50–57.

Daniel G (2003), 'Microview of wood under degradation by bacteria and fungi', in Goodell B, Nicholas D D and Schultz T P, *ACS Symposium #845: Wood deterioration and preservation. Advances in our changing world*, Washington DC, American Chemical Society, 34–72.

Dawson-Andoh B, Matuana L M and Harrison J (2005/6), 'Susceptibility of high-density polyethylene/wood-flour composite to mold discoloration', *J Inst Wood Sci*, **17** (2), 114–119.

Dylingowski P (2003), 'Maintaining the aesthetic quality of wood-plastic composite decking with isothiazolone biocide', in *Proceedings, The Seventh International Conference on Woodfiber–Plastic Composites*, May 19–20, 2003, Madison, WI, 177–186.

Falk R H, Felton C and Lundin T (2000), 'Effects of weathering on color loss of natural fiber-thermoplastic composites', in *Proceedings, 3rd International Symposium on Natural Polymers and Composites*, University of São Paulo, 382–385.

Fengel D and Wegener W (1983), *Wood*, New York, Walter de Gruyter.

Finley M D (2000), 'Foamed thermoplastic polymer and wood fiber profile and member', US Patent 6054207, assigned to Andersen Corporation.

Gijsman P, Hennekens J and Tummers D (1993), 'The mechanism of hindered amine light stabilizers', *Polymer Degrad Stabil*, **39** (2), 225–233.

Gugumus F (1993), 'Current trends in mode of action of hindered amine light stabilizers', *Polym Degrad Stab*, **40** (2), 167–215.

Gugumus F (1995), 'Light stabilizers', in Gachter R and Muller H, *Plastics Additives Handbook*, New York, Hanser Publishers, 129–262.

Hill C A S and Abdul Khalil H P S (2000), 'The effect of environmental exposure upon the mechanical properties of coir or oil palm fiber reinforced composites', *J Appl Polymer Sci*, **77** (6), 1322–1330.

Hon D N S (2001), 'Weathering and photochemistry of wood', in Hon D N S and Shiraishi N, *Wood and Cellulosic Chemistry*, New York, Marcel Dekker Inc., 513–543.

Hon D N S and Minemura N (2001), 'Color and discoloration', in Hon D N S and Shiraishi N, *Wood and Cellulosic Chemistry*, New York, Marcel Dekker Inc., 513–543.

Ibach R E (2005), 'Biological properties', in Rowell R M, *Handbook of Wood Chemistry and Wood Composites*, Boca Raton, CRC Press, 99–120.

Ibach R and Clemons C M (2004), 'Field evaluation of extruded woodfiber–plastic composites', in *Proceedings, Progress in Woodfibre–Plastic Composites*, May 10–11, 2004, Toronto, Ontario.

Ibach R E, Clemons C M and Stark N M (2003), 'Combined ultraviolet and water exposure as a preconditioning method in laboratory fungal durability testing', in *Proceedings, The Seventh International Conference on Woodfiber–Plastic Composites*, May 19–23, 2003, Madison, WI, 61–67.

Joseph K, Thomas S and Pavithran C (1995), 'Effect of ageing on the physical and mechanical properties of sisal-fiber-reinforced polyethylene composites', *Compos Sci Tech*, **53** (1), 99–110.

Kalnins M A and Feist W C (1993), 'Increase in wettability of wood with weathering', *Forest Prod J*, **43** (2), 55–57.

Klyosov A A (2005), 'Durability of natural fiber and wood composites', in *Proceedings, The Global Outlook for Natural Fiber & Wood Composites 2005*, November 15–16,

2005, Intertech, Orlando, FL.

Laks P, Vehring K, Verhey S and Richter D (2005), 'Effect of manufacturing variables on mold susceptibility of wood-plastic composites', in *Proceedings, The Eighth International Conference on Woodfiber–Plastic Composites*, May 23–25, 2005, Madison, WI, 265–270.

Lin Q, Zhou X and Dai G (2002), 'Effect of hydrothermal environment on moisture absorption and mechanical properties of wood flour-filled polypropylene composites', *J Appl Polymer Sci*, **85** (14), 2824–2832.

Lundin T (2001), 'Effect of accelerated weathering on the physical and mechanical properties of natural fiber thermoplastic composites', MS Thesis, University of Wisconsin–Madison.

Mankowski M and Morrell J J (2000), 'Patterns of fungal attack in wood-plastic composites following exposure in a soil block test', *Wood Fiber Sci*, **32** (3), 340–345.

Morris P I and Cooper P (1998), 'Recycled plastic/wood composite lumber attacked by fungi', *Forest Prod J*, **48** (1), 86–88.

Muasher M and Sain M (2006), 'The efficacy of photostabilizers on the color change of wood filled plastic composites', *J Appl Polymer Sci*, **91** (5), 1156–1165.

Naghipour B (1996), 'Effects of extreme environmental conditions and fungal exposure on the properties of wood-plastic composites', MS Thesis, University of Toronto.

O'Neill S and Gardner D J (2004), 'Anisotropic thermal expansion of wood–thermoplastic composites', in *Proceedings, Progress in Woodfibre–Plastic Composites*, May 10–11, 2004, Toronto, Canada.

O'Neill S, Gardner D J and Shaler S M (2005), 'Manufacture of extruded wood-nylon composites: processing and properties', in *Proceedings, The Global Outlook for Natural Fiber & Wood Composites 2005*, November 15–16, 2005, Intertech, Orlando, FL.

Panthapulakkal S, Law S and Sain M (2006), 'Effect of water absorption, freezing and thawing, and photo-aging on flexural properties of extruded HDPE/rice husk composites', *J Appl Polymer Sci*, **100** (5), 3619–3625.

Pendleton D E and Hoffard T A (2000), *Phase II Engineered Wood Biodegradation Studies*, Special Publication SP-2078-SHR, Naval Facilities Engineering Service Center, Port Hueneme, CA.

Pilarski J M and Matuana L M (2005), 'Durability of wood flour–plastic composites exposed to accelerated freeze–thaw cycling. Part I. Rigid PVC matrix', *J Vinyl Additive Tech*, **11** (1), 1–8.

Pilarski J M and Matuana L M (2006), 'Durability of wood flour–plastic composites exposed to accelerated freeze–thaw cycling. Part II. High density polyethylene matrix', *J Appl Polymer Sci*, **100** (1), 35–39.

Rowell R M (1997), 'Chemical modification of agro-resources for property enhancement', in Rowell R M, Young R A and Rowell J K, *Paper and Composites from Agro-based Resources*, New York, Lewis Publishers, 351–375.

Schauwecker C, Morrell J J, McDonald A G and Fabiyi J S (2006), 'Degradation of a wood-plastic composite exposed under tropical conditions', *Forest Prod J*, **56** (11/12), 123–129.

Schirp A and Wolcott M P (2005), 'Influence of fungal decay and moisture absorption on mechanical properties of extruded wood–plastic composites', *Wood Fiber Sci*, **37** (4), 643–652.

Schirp A, Ibach R E, Pendleton, D E and Wolcott M P (2008), 'Biological degradation of wood-plastic composites (WPC) and strategies for improving the resistance of WPC against biological decay', in Schultz T P, Holger H, Freeman M H, Goodell B and Nicholas D D, *Development of Commercial Wood Preservatives*, San Diego, CA, American Chemical Society, 480–509.

Schnabel W (1982), *Polymer Degradation – Principles and Practical Applications*, Cincinnati, Hanser Gardner.

Stark N (2001), 'Influence of moisture absorption on mechanical properties of wood flour–polypropylene composites', *J Thermoplast Compos Mater*, **14** (5), 421–432.

Stark N M (2006), 'Effect of weathering cycle and manufacturing method on performance of wood flour and high-density polyethylene composites', *J Appl Polymer Sci*, **100** (4), 3131–3140.

Stark N M and Matuana L M (2003), 'Ultraviolet weathering of photostabilized HDPE/ wood flour composites', *J Appl Polym Sci*, **90** (10), 2609–2617.

Stark N M and Matuana L M (2004a), 'Surface chemistry and mechanical property changes of wood-flour/high-density-polyethylene composites after accelerated weathering', *J Appl Polymer Sci*, **94** (6), 2263–2273.

Stark N M and Matuana L M (2004b), 'Surface chemistry changes of weathered HDPE/ wood-flour composites studied by XPS and FTIR spectroscopy', *Polymer Degrad Stabil*, **86** (1), 1–9.

Stark N M and Matuana L M (2006), 'Influence of photostabilizers on wood flour–HDPE composites exposed to xenon-arc radiation with and without water spray', *Polymer Degrad Stabil*, **91** (12), 3048–3056.

Stark N M, Matuana L M and Clemons C M (2004), 'Effect of processing method on surface and weathering characteristics of wood-flour/HDPE composites', *J Appl Polymer Sci*, **93** (3), 1021–1030.

Steckel V, Clemons C M and Thoemen H (2007), 'Effects of material parameters on the diffusion and sorption properties of wood-flour/polypropylene composites', *J Appl Polymer Sci*, **103** (2), 752–763.

Tajvidi M, Najafi S K and Moteei N (2006), 'Long-term water uptake behavior of natural fiber/polypropylene composites', *J Appl Polymer Sci*, **99** (5), 2199–2203.

Tichy R and Englund K (2006), 'Product performance assessment and acceptance criteria', in *Proceedings, Durability in Wood Plastic & Natural Fiber Composites 2006*, Intertech Pira, December 4–5, 2006, San Antonio, TX.

Verhey S A and Laks P E (2002), 'Wood particle size affects the decay resistance of woodfiber/thermoplastic composites', *Forest Prod J*, **52** (11/12), 78–81.

Verhey S, Laks P and Richter D (2001), 'Laboratory decay resistance of woodfiber/ thermoplastic composites', *Forest Prod J*, **51** (9), 44–49.

Wang W and Morrell J J (2004), 'Water sorption characteristics of two wood–plastic composites', *Forest Prod J*, **54** (12), 209–212.

Williams R S, Knaebe M T and Feist W C (2001), 'Erosion rates of wood during natural weathering. Part II. Earlywood and latewood erosion rates', *Wood Fiber Sci*, **33** (1), 43–49.

Wypych G (1995), *Handbook of Material Weathering*, Toronto Scarborough, ChemTec Publishing.

8
Creep behavior and damage of wood–polymer composites

N E MARCOVICH and M I ARANGUREN,
Universidad Nacional de Mar del Planta, Argentina

8.1 Introduction

Polymers and polymer composites used in engineering applications are often designed to be in service for extended periods of time, during which they are subjected to various combinations of mechanical and environmental loads. These materials show a time-dependent response (viscoelasticity) that is influenced by the particular viscoelastic properties and history of the material, as well as the external conditions of testing or service. Frequently, engineering applications require that the polymer or composite maintains its structural rigidity during the envisioned service lifetime. This is the case for the wood–plastic composites (WPCs), which are frequently utilized in low-to-medium load-supporting structural applications, such as pallets, decks, window and door components, furniture, water front applications, fendering, outdoor trails and tiles. WPCs, much in the same way as other thermoplastic-based composites, are prone to deform over time (during periods that extend over hours, days or years) under the application of external loads.

'Creep' is the term used to describe the time-dependent deformation or strain that a material undergoes under load and constitutes its viscoelastic response to an externally applied stress. Laboratory testing of the plastic composites rarely can cover periods of time long enough to be comparable to their expected service lifetime. More usually, accelerated test methods together with calculations based on different proposed models are used to predict long-term response based on short-term data.

The creep manifested by WPCs is the result of the combined action of many variables: the thermoplastic polymer selected and its state (e.g. nature, molecular weight and crystallinity of the polymer), filler (type of wood utilized, shape: fiber or particle and size, concentration), chemistry of the wood–matrix interface (possible to be modified by addition of interfacial agents), plasticizers or lubricants added, operating temperature, stress level and state of stresses, prior history (thermal or mechanical), humidity content (Sternstein and Van Buskirk, 1988; Lin et al., 2004). In particular, increasing molecular weight and

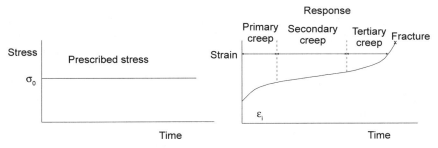

8.1 Creep response to a prescribed stress.

crystallinity of the polymer matrix lead to a decrease in the composite creep. Reduced deformation can also be achieved by increasing the filler concentration (Wolcott and Smith, 2004; Bengtsson et al., 2005). Additionally, the incorporation of compatibilizers usually improves interfacial adhesion and reduces the time-dependent deformation (Park and Balatinecz, 1998; Farid, 2000; Bledzki and Faruk, 2004; Lee et al., 2004). On the other hand, elevated temperature, high humidity environments or added plasticizers have a negative effect on the creep performance of the materials.

When a composite is subjected to a constant stress, progressive deformation takes place. The material suffers an instantaneous elastic deformation (ϵ_i) followed by a stage of decreasing creep rate (primary creep). Then, the creep rate reaches a steady-state value (secondary creep) and finally, the deformation rate accelerates, while the material suffers damage and its structure is irreversibly affected (tertiary creep). This behavior is illustrated in Fig. 8.1.

8.2 Viscoelasticity and creep

8.2.1 Theoretical background

Linear viscoelasticity

The simplest constitutive equation derived to explain the mechanical behavior of perfectly elastic solids is Hooke's law, which relates the applied force to the material deformation and is expressed as:

$$\sigma = E\epsilon \qquad 8.1$$

where σ is the stress (force per unit area) applied, E is the elastic modulus and ϵ is the strain or deformation of the material (the relative change of dimension suffered by the material). On the other hand, the simplest expression formulated for liquids is due to Newton and relates stress and rate of strain, as follows:

$$\sigma = \eta \frac{d\epsilon}{dt} \qquad 8.2$$

where $d\epsilon/dt$ is the strain rate and η is the viscosity of the liquid.

Although these simple equations are enough to represent the behavior of some materials, (e.g. metals and ceramics under small loads follow Hooke's law; small molecule liquids are usually Newtonian), there are a large number of cases when these equations cannot represent real materials. A paradigmatic example of this type of materials is provided by polymers, which show complex time-dependent behaviors that also depend on temperature, stress, and strain/stress rate. This complex behavior is said to be viscoelastic, indicating that it shares characteristics intermediate between those of a solid and a fluid. Correspondingly, the stress depends on both strain and strain rate.

In testing viscoelastic materials in creep, a constant stress, σ_0, is applied on the specimen and the resultant deformation, $\epsilon(\sigma, t)$, is measured as a function of time. The results are expressed as a time-dependent compliance (Findley et al., 1976):

$$\text{Compliance} = \frac{\epsilon(\sigma, t)}{\sigma_0} \qquad 8.3$$

The compliance is called $D(\sigma, t)$ when uniaxial tensile or compression stresses are applied, $J(\sigma, t)$ if the material is subjected to shear and $B(\sigma, t)$ for hydrostatically applied stresses. These functions are all different and not interchangeable. For the rest of the discussion, the compliance $D(\sigma, t)$ will be utilized, but the analysis and the resulting equations can be extrapolated to the other testing modes. Frequently, the term 'creep modulus' is invoked as the inverse of the compliance. Although it is related to the stiffness of the material, it should not be mistaken with the quasi-static Young modulus.

When small stresses are applied the curves of compliance obtained at different stress levels collapse in a single curve. In that case the material property is only a function of the time and (if tensile testing is considered) it is represented as $D(t)$, while the material response is said to be in the linear viscoelastic (LVE) region. Experimentally, the usual method to determine if the selected stress will produce a LVE response is to generate isochronous plots from the creep data. The strain results obtained at a fixed time and for different stress levels are used to produce an x–y plot, which should show a linear stress–strain functionality in the LVE region. Usually, the range of stresses that lead to a LVE response are limited to values below 40–50% of the composite ultimate strength, but they can be lower depending on the material (Farid, 2000; Kobbe, 2005).

Simple modeling of the linear viscoelastic creep has been achieved by using analogous mechanical models consisting in the combination of spring and dashpot elements. A spring represents a perfectly elastic component (Hookean) of the material response, and a dashpot represents a perfectly viscous (Newtonian) behavior. In particular, the Bürgers model is a combination of a Maxwell element (a series combination of an elastic and a viscous components) and Kelvin–Voigt elements (parallel combinations of an elastic and a viscous components), as shown in Fig. 8.2.

Creep behavior and damage of wood–polymer composites 169

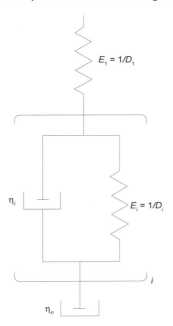

8.2 Scheme of the Bürgers model.

In Fig. 8.2, i corresponds to the 'ith' Kelvin–Voigt element in the arrangement. If i is equal to 1, the four parameters model is obtained (ASTM 2990). The spring connected in series is responsible for the instantaneous response of the material, the Kelvin–Voigt element is responsible for the following primary creep, an early stage in which creep occurs at a decreasing rate. The dashpot connected in series is responsible for the secondary creep, a viscous flow of the material that occurs at an essentially constant rate. The model does not account for the experimentally observed third stage in which creep occurs at an increasing rate to finally terminate with the material rupture.

In particular, the response to an applied constant stress can be easily solved analytically, giving a relatively good description of the polymer behavior. If the study is limited to the linear viscoelastic (LVE) response of the material, the compliance is given by:

$$D_{(t)} = \frac{\epsilon_{(t)}}{\sigma_0} = D_0 + \sum_{i=1}^{n-1} D_i \cdot \left[1 - \exp\left(-\frac{t}{\tau_i}\right)\right] + \frac{t}{\eta_n} \qquad 8.4$$

where σ_0 is the applied stress, t is the time, D_0 is the instantaneous compliance and it is related to the elastic immediate response of the material, D_i $(= 1/E_i)$ is the compliance of the elastic component in the ith Voigt element and τ_i is the retardation time of the ith element ($\tau_i = \eta_i \cdot D_i$). τ_i is a measure of the time required for the extension of the spring to its equilibrium length, while retarded by the viscous friction in the dashpot. Thus, the summation represents the retarded response of the material (primary creep). Finally, η_n is related to the creep rate in the steady state stage, where the viscous flow is the predominant behavior. Sometimes equation 8.4, modified by not including the last term (the viscous element connected in series), has also been used to represent creep in composites and is referred to as the Prony series (Lai and Bakker, 1995).

Although a completely different approach, similar expressions have been derived from molecular dynamic considerations (Ferry, 1980). On that ground, the retardation times are related to characteristic times for molecular relaxation mechanisms, and thus τ_i values acquire a fundamental meaning related to molecular structure. In practice, however, they are usually selected in an arbitrary manner, becoming fitting parameters of the model.

Other authors have preferred to use a simpler phenomenological function for $D(t)$, as follows (Schapery, 1969; Findley et al., 1976; ASCE, 1984; Rangaraj and Smith, 1999; Sain et al., 2000)

$$D(t) = D_0 + D_1 \cdot t^n \qquad 8.5$$

where D_0, D_1 and n are empirically determined parameters. D_0 is the instantaneous compliance, D_1 is a time-dependent compliance.

This expression, known as the power-law model, is very simple and has fewer fitting parameters than the Bürgers model. However, it can only fit primary creep (decreasing deformation over time), and it is unable, a priori, to model secondary (steady state creep) or tertiary creep stages. However, correction factors and terms have been added to this expression that extend its use, even to model damage effects on the composites.

The Boltzmann and the time–temperature superposition principles

Two important principles are frequently invoked in modeling the viscoelastic response of a material: the Boltzmann and the time–temperature superposition (TTS) principles. The Boltzmann superposition principle establishes that the response of the material to stresses applied at different times is the sum of the responses to each applied stress. For a given material, which has been subjected to a constant stress, σ_0, from time $= 0$ and for which the stress is step-increased to σ_1 at t_1 and so on, the total deformation at any given time can be mathematically written as:

$$\epsilon(t) = D(t)\sigma_0 + D(t - t_1)(\sigma_1 - \sigma_0) + \ldots D(t - t_i)(\sigma_i - \sigma_{i-1}) + \ldots \qquad 8.6$$

where $\epsilon(t)$ is the deformation of the material at time t, with t always larger than t_i.

If the material is subjected to a stress σ_0 at $t = 0$ and at time t_r the stress is removed (what can be interpreted as the application of a negative stress of magnitude equivalent to σ_0), the material response during unloading is called 'recovery'. During the recovery stage (after t_r), the material shows a decreasing

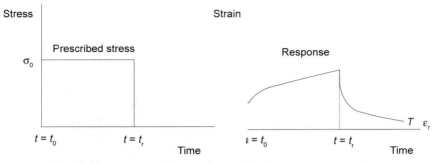

8.3 Creep-recovery response to a prescribed stress.

deformation to finally achieve zero creep, if there was no viscous flow during the deformation (the response expected for crosslinked thermosets at low stresses) or a non-zero permanent deformation (ϵ_r), which is the typical case in thermoplastics. The last case is illustrated in Fig. 8.3. Using the Bürgers model, the representative equation for the compliance during recovery would be:

$$D_{r(t)} = D(t) - \left\{ D_0 + \sum_{i=1}^{n-1} D_1 \cdot \left[1 - \exp\left(-\frac{t-t_r}{\tau_i}\right)\right] + \frac{(t-t_r)}{\eta_n} \right\} \quad 8.7$$

The other superposition principle is the time–temperature superposition (TTS). It is based on the observed fact that by elevating the temperature, the viscoelastic response resembles that one exhibited by the material at longer times. Thus, the creep compliance (or any other viscoelastic property) measured at a given temperature and plotted as a function of time can be superposed onto similar curves obtained at different temperatures, if they are horizontally shifted along a logarithmic time scale, as shown in Fig. 8.4. This simple procedure allows using short-time test data obtained at elevated temperatures to predict the material behavior over more extended periods of time than is experimentally feasible in the laboratory, owing to the expensive and time-consuming set-up and testing involved.

The TTS method was initially utilized for pure polymers and since then it has proved to be very useful, although presenting some limitations. The structure of the material should remain the same in the range of temperatures observed, so that the same molecular relaxation (or retardation) mechanisms are active at the different temperatures. Data for thermorheological complex materials can prove to be difficult to superpose, because the different relaxation mechanisms may have a different response to temperature changes, which is the case in most two-phase materials (from blends to semicrystalline polymers to polymer composites). However, different authors have successfully applied this principle to the behavior of different composite materials. Frequently, an additional vertical shift has been introduced to account for structural or stress-derived changes in the material (Penn, 1966; Onogi et al., 1970; Fukui et al., 1970; Crissman, 1986; Pooler, 2001).

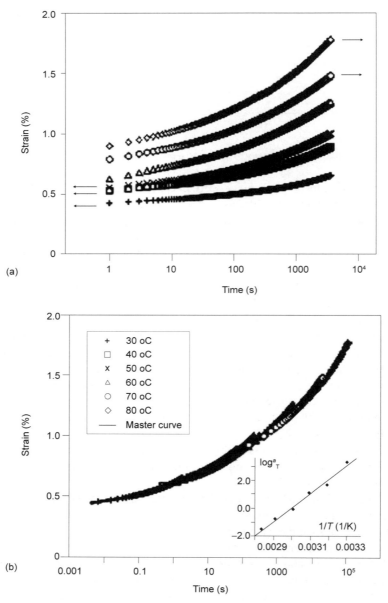

8.4 Time–temperature superposition of short-term experimental curves for a wood flour–polypropylene (WF–PP) composite (40 wt% WF) with 10wt% maleic anhydride polypropylene (MAH–PP) coupling agent (respect of filler content). Continuous curve was calculated using Bürgers model (nine parameters). Data taken from Nuñez *et al.* (2004).

According to the above, the first step consists in shifting the curves from the short-term tests performed at different temperatures and to observe if a single, smooth 'master curve' is obtained. This alone is not sufficient to ensure that the curve is a good prediction of the material response at long times and a validation is required.

If the shifts in the TTS are thought to be a process of matching retardation times, the horizontal shift factors, log a_T, are given by:

$$\log a_T = \log \frac{\tau(T)}{\tau(T_0)} \qquad 8.8$$

where $\tau_{(T)}$ are the corresponding material relaxation/retardation times at the given absolute temperatures, T and T_0, where T_0 is the reference temperature.

The shift factors can be correlated with temperature using the Williams–Landel–Ferry (WLF) equation or the Arrhenius equation. The former equation is given by:

$$\log(a_T) = \frac{-C_1(T - T_0)}{C_2 + (T - T_0)} \qquad 8.9$$

where C_1 and C_2 are constants and T_0 is the reference temperature.

On the other hand, the Arrhenius equation is:

$$\log(a_T) = 2.3 \frac{E_a}{R} \left(\frac{1}{T} - \frac{1}{T_0} \right) \qquad 8.10$$

where E_a is an activation energy for viscous flow and R is the universal gas constant.

In general, it is found that the WLF equation fits the horizontal shifts well if the range of working temperatures is in the region of T_g to $T_g + 100$ K, where T_g is the glass transition temperature of the polymer considered. At elevated temperatures, the Arrhenius equations provide a better fit (Tirrell, 1994).

Non-linear creep

If high stress levels are applied, the response of the material is no longer a single function of time, but it becomes also stress dependent (Papanicolaou et al., 1999; Jazouli et al., 2005). The deformations may be large due to viscous flow and the occurrence of permanent changes in the structure (e.g. enhanced crystallinity due to stretching under a tensile load, microcracking, aging).

Schapery (1969) presents a general form for the time-dependent non-linear compliance obtained from thermodynamic considerations, as follows (Lin et al., 2004):

$$D_{(\sigma,t)} = g_0 D_0 + g_1 g_2 \Delta D_{(\sigma,t)} \qquad 8.11$$

where D_0 is the instantaneous compliance, and $\Delta D_{(\sigma,t)}$ represents the transient component of the compliance, which is also dependent on the intensity of the stress applied. The g factors represent high-order terms of the Gibb's free energy dependence on the applied stress; g_0 introduces the effects of temperature or stress on the instantaneous component, correcting for the changes that these conditions produced in the stiffness of the material; g_1 introduces analogous corrections on the transient component of the compliance and g_2, which is related to the loading rate, also depends on temperature and stress.

If a power law expression is selected to account for the time-dependent component of the compliance, equation 8.11 can be written as:

$$D_{(\sigma,t)} = g_0 D_0 + g_1 g_2 D_1 \left(\frac{t}{a_\sigma}\right)^n \qquad 8.12$$

where D_1 is the transient compliance component. The stress dependence of the transient component appears through the contraction or expansion of the time introduced by the a_σ factor, a timescale shift factor that depends on the applied stress. When the applied load is small the g and the a_σ factors can be assumed to be unity and equation 8.5 is recovered.

The study of the non-linear response of materials is still the subject of ongoing research and different models are available (Findley, 1960; Schapery, 1969; Lou and Schapery, 1971; Rand, 1995; Zaoutsos et al., 1998). A full validation of the model should demonstrate that it performs correctly in conditions different from those originally studied (Kobbe, 2005).

8.2.2 Experimental methods

Creep testing is done in a test frame where a constant stress is applied, while the material deformation is determined with a strain-measuring device (which can be an optical or a strain gage device). The test is carried out in an environmental chamber to keep temperature and/or humidity in the desired ranges. The measurements can be done in tensile, flexural or compression modes, and, on some occasions, multi-axial testing has been reported. Most frequently, WPC creep is measured on flexural testing probes with geometries of three- or four-point bending, shown in Fig. 8.5 (flexural: ASTM D 790; creep: ASTM D 2990). Since WPCs are usually relatively ductile, they may reach the maximum strain allowed by norm (3%) under three-point bending without failing, and thus, four-point bending has also been used.

The surfaces of the supports and loading beams that are in contact with the specimen can have rounded shapes to avoid excessive indentation of the specimens. Besides, at least 10% of the support span should be maintained at each test specimen end as overhanging, to assure good support during testing. The deflection of the specimen is measured at the midpoint of the load span at the bottom face of the specimen. The deformation of the bars is calculated according to the norms for each geometry as follows (ASTM D 790):

Creep behavior and damage of wood–polymer composites

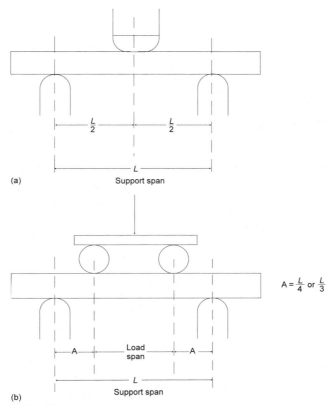

8.5 Geometries of (a) three- and (b) four-point bending.

3-point bending: $\epsilon = \dfrac{6D_{\text{ef}}d}{L^2}$ 8.13

4-point bending (one-half of the support span): $\epsilon = \dfrac{4.36 D_{\text{ef}}d}{L^2}$ 8.14

4-point bending (one-third of the support span): $\epsilon = \dfrac{4.7 D_{\text{ef}}d}{L^2}$ 8.15

where ϵ is the maximum strain in the outer fibers, L is the support span, D_{ef} is the maximum deflection of the center of the beam and b and d are the width and the depth of the beam, respectively. Additionally, measurements in these geometries have been done on small coupons or on life-size samples, similar to the long beams that can be used in decking, as shown in Fig. 8.6.

Accelerated creep testing (which is analytically handled through the use of the TTS) requires the measurement of short-term time-dependent deformations performed at different temperatures. Controlling the conditions of testing by

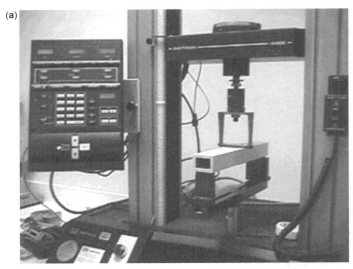

8.6 Life-size testing geometries: (a) three- and (b) four-point bending. From Tichy (2007).

maintaining the chamber at different constant temperatures is a requirement. Although not frequently reported, constant humidity conditions are also required to obtained reliable results, especially in long-term measurements.

8.3 Creep in wood–plastic composites

The WPC in service may be subjected to static and time-dependent loads, which motivates the studies on creep and damage to assess the durability of the materials and to evaluate the performance of future candidate composites. The addition of wood fibers to a polymeric matrix introduces several new variables that can affect the material response. These variables include filler/fiber volume fraction, aspect ratio, orientation, mechanical properties of the filler/fibers, interfacial adhesion, and the environmental conditions that will affect in a complex way the behavior of the two-phase material.

8.3.1 Filler concentration and temperature effects

Effect of the wood filler/fiber concentration

Essentially, all reports in the literature indicate (Park and Balatinecz, 1998; Wolcott and Smith, 2004; Bengtsson et al., 2005) that the time-dependent deformation of the composites is reduced with the addition of woody reinforcements (flour, fibers). Park and Balatinecz (1998) showed that creep in polypropylene (PP) composites is strongly dependent on the wood flour/fiber concentration. They reported a decrease of almost three times in the creep deformation of PP by addition of 40 wt% of wood fibers after 200 min testing. They fitted their data to the simple power law model and found that D_1 and n (eqn. 8.5) decrease with increasing fiber concentration, that is creep resistance is improved by wood fiber addition.

Also working with PP-based composites, Bledzki and Faruk (2004) corroborated that adding wood fibers increased creep resistance. Moreover, they noticed that long wood fibers perform better (reduced creep deformation) than short ones. Lee et al. (2004) tested samples prepared from wood flour (WF) and polypropylene, with different filler levels (0–40 wt%), and found significant improved creep resistance. They observed that as the wood flour content increases, the stiffness and also the brittleness of the composites increase. Thus, they inferred that the interface becomes increasingly weak with increasing filler loading by the formation of micro-voids (deficient wetting) between WF and matrix, which interferes with the stress distribution during testing.

As an example, Fig. 8.7 shows the linear creep response of injected WF–PP composites (Nuñez et al., 2003). The 30 min creep was reduced to about one-half with respect to the PP if 40 wt% of wood flour was added. The continuous curves were calculated with equation 8.4, using only the data from the creep stage. The recovery was predicted (not fitted) using equation 8.7 and the parameters calculated from creep, showing very good agreement with the experimental results. Creep was also fitted with the power law equation with equally good success. Table 8.1 shows the values of the fitting parameters calculated for both models. Clearly, the compliance parameters in both models decrease with the percentage of WF as a result of the increased rigidity of the composite as more WF is added. The steady state viscosity of the Bürgers model also increases as WF content increases as a result of the increased difficulty for viscous flow to take place due to the presence of the woody particles.

178 Wood–polymer composites

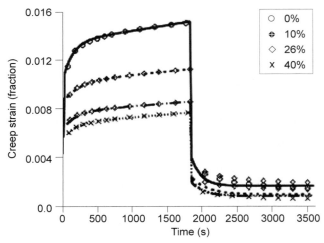

8.7 Linear creep response of injected WF–PP composites tested at 20 °C and $\sigma_0 = 10^7$ Pa. Experimental data were taken from Nuñez *et al.* (2003). Continuous curves were calculated with Bürgers model (four parameters).

Table 8.1 Linear creep fitting of injected WF–PP composites tested at 20 °C and $\sigma_0 = 10^7$ Pa, as a function of the WF concentration. Data from Nuñez *et al.* (2003)

WF (wt%)	Bürgers model (four parameters)				Time-dependent component (three-parameter model)	
	D_0 (Pa^{-1})	$D_{i=1}$ (Pa^{-1})	$\tau_{i=1}$ (s)	η_n (Pa s)	D_1 (Pa^{-1})	n
0	1.03×10^{-9}	3.24×10^{-10}	140.3	1.10×10^{13}	8.48×10^{-10}	0.07784
10	8.55×10^{-10}	1.74×10^{-10}	229.7	1.87×10^{13}	2.89×10^{-10}	0.11662
26	6.33×10^{-10}	1.45×10^{-10}	181.3	2.20×10^{13}	1.80×10^{-10}	0.12545
40	5.43×10^{-10}	1.39×10^{-10}	159.9	2.08×10^{13}	1.59×10^{-10}	0.13743

Effect of the temperature

As it occurs with unfilled thermoplastics, the creep deformation of WPCs increases with temperature (Park and Balatinecz, 1998; Marcovich and Villar, 2003; Nuñez *et al.*, 2003, 2004; Bledzki and Faruk, 2004; Reboredo *et al.*, 2007). Polymers are typically softened by elevated temperatures and this effect on the matrix is the controlling factor in reducing the resistance to creep of derived composites.

In an amorphous polymer the softening is the result of the increased mobility of the molecules as the temperature increases and get closer to the glass transition temperature (T_g) of the material. In a partially crystalline polymer

Table 8.2 Fitting parameters of the power law equation for the deformation (eqn. 8.16) as a function of the temperature. Data from Park and Balatinecz (1998)

T (°C)	Instant deformation (ϵ_0)	B/ϵ_0	n
23	9.513×10^{-3}	0.967	0.471
30	10.828×10^{-3}	1.110	0.471
40	13.907×10^{-3}	1.465	0.463
60	29.489×10^{-3}	1.907	0.457

(above T_g, below the melting point), the mobility of the amorphous 'tie molecules' (molecules that belong to more than one crystal and have intermediate amorphous segments) is greatly hindered by the reinforcing effect of the crystalline phase. Below the melting temperature, no changes in crystallinity are expected, but changes in the mobility of the amorphous phase can occur. As the temperature increases, the mobility of these chains (including molecules with previously hindered mobility) increases, which results in a larger creep deformation for the same applied stress.

Park and Balatinecz (1998) showed this effect on a 30 wt% WF–PP composite, using a stress level 35% of the ultimate stress measured at room temperature. The relative creep (relative to the instantaneous room temperature creep) at 60 °C almost doubles the room temperature value. The temperature affected the instantaneous creep as well as the retarded creep by increasing the compliance components. Table 8.2 shows the variation of the fitting parameters calculated by the authors, using the power law equation for the deformation:

$$\epsilon(t) = \epsilon_0 + Bt^n \qquad 8.16$$

The effect of temperature on the 'n' exponent was not appreciable. Similar results have been reported by Bledzki and Faruk (2004), who studied creep and impact properties of WF–PP composites.

Long-term creep and the time–temperature superposition (TTS) principle

The TTS principle is generally invoked as a first step to predict long-term creep. An example is offered by Nuñez *et al.* (2004), who generated a master curve for WF–PP composites, by shifting the individual creep curves according to the TTS principle. A smooth master curve was obtained and the shift factors were fitted as a function of temperature, using equations 8.9 and 8.10. In this particular case, the Arrhenius plot seemed to better adjust the temperature shift factors. The authors chose the Bürgers model to represent the response, using several retardation times. Figure 8.4 illustrates the excellent agreement obtained between the experimental data and the calculated curve. In order to validate the proposed LVE model, the authors also used the same parameters to predict the response of the material as a function of the frequency of a dynamically applied stress. A good agreement with experimental measurements was obtained.

In the particular example followed, long-term creep was also measured during about a week on a composite containing 40 wt% wood flour and 10 wt% of a compatibilizer. The time shifts calculated from short-term creep were used to shift the long-term data, with no success. As shown in Fig. 8.8, the predicted curve overestimated the creep deformation of the composite in the low-temperature experimental range. On the other hand, the high-temperature long-term creep was adequately predicted from short-term data.

In general, differences found between experimental data and master curves can be due to changes in the structure (aging, crystallinity), environmental effects (humidity, UV, microorganisms) or damage that can be provoked by one or more of the factors just mentioned or their combinations (Matsumoto, 1988; Brinson and Gates, 1995; Dean *et al.*, 1998; Marais and Villoutreix, 1998).

In the example of Fig. 8.8 the difference between prediction and long-term experimental data was related to aging of the PP matrix. Tie molecules and amorphous chains close to crystalline domains, showing hindered mobilities, require a significantly higher activation energy for relaxation than unrestricted amorphous chains. Thus, their relaxation occurs at higher temperatures. This transition occurs around 80 °C in PP, and aging of the polymer can take place at room temperature. Heating of the samples near this transition rejuvenates the

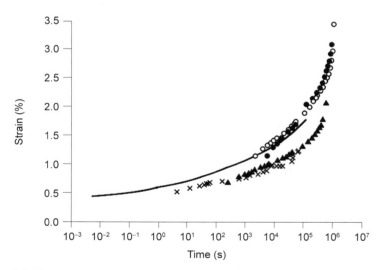

8.8 Time–temperature superposition of long-term experimental curves for a WF–PP composite (40 wt% WF) with 10 wt% MAH–PP coupling agent (respect of filler content). Continuous curve was calculated using Bürgers model (nine parameters) derived from short-term data. From Nuñez *et al.* (2004). With permission from John Wiley and Sons.

material, by erasing its previous history. For this reason, it was possible to predict long-term creep at high temperatures from the short-term creep data, but the method failed in predicting the long-term creep at lower temperature (Nuñez et al., 2004).

Previous work reported on aging of polypropylene at room temperature showed that there were no changes in the crystallinity of the samples, but a steadily increasing density. Gas sorption and diffusion experiments allowed the amorphous phase to be identified as the responsible for these changes (Schael, 1966; Agarwal and Schultz, 1981; Vittoria, 1988; Fiebig et al., 1999).

An important conclusion from the above is that the 'master curve' obtained from short-term creep data (even if a smooth curve is obtained) may not represent the experimental behavior of the composite for long-term creep and a validation of the method by comparison with experimental data at longer times is still required.

8.3.2 Interfacial adhesion

One of the pioneer works on creep behavior of WPCs was that published by Park and Balatinecz (1998), which also discusses the effect of improved interfacial adhesion on the creep of WPCs. They conducted short-term flexural creep tests to investigate the effect of the addition of a wetting agent (Epolene G 3002) and applied load (35% and 50% of the room temperature ultimate flexural strength) on the creep behavior of wood fiber–polypropylene composites prepared by injection molding. They found that the composites treated with the wetting agent crept less than those that were untreated. Moreover, at the highest stress level, the untreated specimens ruptured within 90 min after load application while treated specimens did not rupture after 200 min of testing. These observations were, *a posteriori*, corroborated in other works, for example, the one carried out by Bledzki and Faruk (2004) who investigated the effect of the addition of a commercial MAH–PP copolymer on short-term flexural creep. They explained the observed reduction of the deformation by the improved fiber dispersion in the matrix and the enhanced interfacial adhesion between the fibers and the polymer achieved by adding a compatibilizer.

In the same way, Lee et al. (2004) conducted short-term tensile creep tests to study the behavior of WF–PP composites modified with another commercial MAH-PP (Epolene G 3003). They found that using 3.0 wt% of MAH–PP based on the total weight, the creep deflection was approximately half that measured without the coupling agent, as a result of the improved compatibility between wood flour and polymeric matrix. Nuñez et al. (2004) also confirmed the reduction in creep deformation when MAH–PP compatibilizer is added to PP–WF composites, as shown in Fig. 8.9. They also noticed that increasing the coupling agent concentration from 5 wt% (respect to WF content) to 10% MAH–PP, a further decrease of creep strain is observed, although the difference

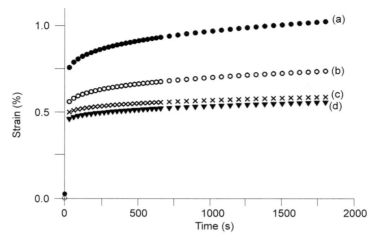

8.9 Effect of the interfacial interactions on the creep behavior of PP composites prepared with 40% WF. Creep behavior of neat PP is included for comparison (a); untreated WF (b); untreated WF and 5% MAH–PP (c); untreated WF and 10% MAH–PP (d). From Nuñez *et al.* (2004). With permission from John Wiley and Sons.

with the 5% MAH–PP addition is slight and probably the economy of the process does not justify the use of higher amounts of compatibilizer.

The properties of composites made from a linear low-density polyethylene (LLDPE) reinforced with WF were studied by Marcovich and Villar (2003). PE matrix was reacted with MAH (using an organic peroxide as initiator) and simultaneously compounded with untreated wood flour in a twin-screw extruder. Although the grafting of MAH onto polyethylene (PE) reduced the matrix crystallinity, it improved the dispersion of WF in the matrix and enhanced filler–matrix interactions. As a result, the creep deformation of modified composites was reduced. However, it has to be taken into account that in this example *in situ* formation of the coupling agent occurs, and also some crosslinking of the PE. The two changes contribute to reduce creep.

Other authors have also worked on the crosslinking of the polyolefins during processing in order to improve creep resistance, in the long-term performance. As indicated previously, adding wood filler to the plastic matrix decreases the creep response during loading, but does not completely alleviate the problem. Thus, crosslinking of the polymer matrix has been proposed as one additional way of reducing the creep during long-term loading (Bengtsson *et al.*, 2005). This is the way that Bengtsson *et al.* (2005) amd Bengtsson and Oksman (2006) selected to enhance the performance of HDPE–WF composites. They used wood flour from softwood (spruce and pine, average size 200–400 μm) in the form of fractured fiber bundles as reinforcement and produced silane crosslinked composites in one-step process. The short-term creep analysis of their samples showed that not only the total amount of creep during loading was lower in the crosslinked composites than in the non-crosslinked counterparts, but the unrecoverable part of the deformation also decreased during creep cycling. The lower creep in the crosslinked composites was also related to a reduced viscous flow due to crosslinking, as well as to improved adhesion between the PE matrix and WF. Moreover, they also prepared silane crosslinked composites with different amounts of silane (i.e. varying crosslinking density) and found that there was an inverse correlation between the degree of crosslinking and the creep response of the composites: the higher the degree of crosslinking, the lower the creep deformation.

8.4 Creep failure and material damage

When a composite is subjected to creep at relatively large stresses or under aggressive environments, irreversible changes in the structure can take place, so that if re-testing of the material is performed a different material response is obtained. Different mechanisms may be involved, such as crazing, crystallinity changes, formation of shear bands (localized deformation), microcracking, plasticization (or humidity effects), dewetting, or cracking at the interface. Since all these factors affect permanently the response of the composite, they can be considered responsible for creep damage in the material (Rangaraj and Smith, 1999; Pooler, 2001; Kobbe, 2005).

There have been attempts to incorporate damage into the creep models with varied success and different levels of complexity. Chia *et al.* (1987) presented a model that requires rigorous statistical analysis to model damage. Jerina *et al.* (1982) incorporated damage-correcting factors into the power law expression.

Damage models that do not identify individual cracks, crack propagation or crack growth (localized damage) have been proposed, which are based on the overall behavior of the composites. In that case, damage was defined as resulting in a degradation of the materials ability to carry a load, which could be measured as a reduction in stiffness. This is consistent with the work of Reifsnider (1982), who observed that stiffness changes are related to the damage and also to changes in residual strength. Andersons *et al.* (1991) also proposed that changes in modulus can be used to account for the amount of damage occurred in the composite structure.

8.4.1 Effect of the applied load

Creep deformation increases with stress level and the effect is more important at high stresses, usually above 40–50% of the ultimate strength of the material, where non-linearities appear. These effects are illustrated in Fig. 8.10, where the material shows tertiary creep (untreated sample) at stresses of 50% of the ultimate strength, while it only reaches primary creep if the stress is only 35% of

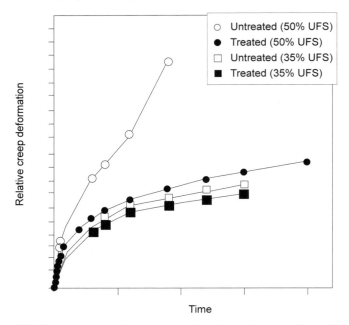

8.10 Relative creep deformation as a function of the stress level; UFS being the ultimate flexural strength. From Park and Balatinecz (1998). Reprinted with permission of Wiley-Liss, Inc., a subsidiary of John Wiley & Sons, Inc. Copyright (1998) Society for Plastics Engineers.

that value. It has been reported that the critical stress level for appearance of non-linearities decreases with increasing temperature, but the functionality is still a matter of study. A vertical shift (related to structure changes) as well as an horizontal shift (related to changes in the distribution of retardation times) may be required to superimpose compliance curves obtained at different levels of stress, even if measurements are made at constant temperature. The permanent deformation also increases with stress level as shown by Lin *et al.* (2004). These authors offer a good reference for the determination of non-linear material constants (equation 8.11) from experimental data fitting. Permanent damage is highly probably to occur at these high stress levels.

Rangaraj and Smith (1999) tested HDPE–WF composites at different stress levels and observed that creep was only satisfactorily modeled using linear viscoelasticity at stresses 45% below its ultimate strength. At higher stresses, the non-linear viscoelastic response obtained was attributed to damage and it was observed that this effect increases with creep stress and its duration. They incorporated non-linear effects into the simple viscoelastic model by considering an effective stress, $\sigma_{e(\sigma_0,t)}$, that increased with the measured damage:

$$\sigma_{e(\sigma_0,t)} = K_{(\sigma_0,t)} \sigma_0 \qquad 8.17$$

where $K_{(\sigma_0,t)}$ is a damage function that contains an instantaneous component, which depends on stress level and a time-dependent component that depends on stress and duration of the creep loading, as follows:

$$K_{(\sigma_0,t)} = K_{1(\sigma_0)} + K_{2(\sigma_0,t)} \qquad 8.18$$

At stress levels above the threshold value (which decreases with increasing temperature) time-dependent damage takes place and a transition from secondary to tertiary creep can be observed. On the other hand, the permanent deformation after creep recovery was found to correspond with the maximum strain reached during creep and, thus, it was related to stress and duration of the creep test.

8.4.2 Humidity effects

WPCs absorb less humidity than wood, but the effect of water on the composites is not negligible and deserves attention. Water is a plasticizer for woody fillers and additionally, owing to the mismatch between the hygrothermal expansions of the matrix and reinforcement, fracture of the fiber matrix interfaces can occur, as well as cracking through the matrix. Moisture transport on WPCs may occur through load-induced cracks (Smith and Weitsman, 1996). The water trapped inside cracks can act synergistically with the applied stress, accelerating damage. The damage of the composite results in the reduction of strength and stiffness of the material and, thus, deformation under creep is favored (Rangaraj, 1999). Besides, the effect will be more marked in cyclically applied loads, which reduce the creep resistance of the material, leading finally to rupture.

Maksimov *et al.* (1972) modeled the influence of moisture on the creep response of a polyester resin by using a time–moisture superposition method (analogous to the TTS) in which creep curves obtained at different humidity contents were superposed by shifting them in logarithmic timescale. The effect of moisture was modeled by considering the contraction of the timescales through a moisture shift factor. The basis for this method is the increased mobility of the plasticized chains (softened material) that accelerates the polymer relaxation mechanisms, reducing the time intervals in which they occur. An example of its use in long-term creep of wood lumber can be found in the work of Tissaoui (1996).

8.5 Conclusions and future trends

As the market for WPCs grows, so does the competition between different manufacturers. In particular, those focused on the construction applications are devoting their efforts to warrant durability for their products. Materials are marketed with 5–25 year limited warranties depending on the material and application and covering defect, splintering, splitting, rot decay, and termite damage. Cases have been documented on decking products that were reclaimed and replaced due to low performance and damage under high temperatures and sunlight (US CPSC, 2006).

Companies are working on improving the materials and studying the durability of their product under extreme conditions. On this venue, accelerated age testing studies are ongoing in industrial and academic grounds to give the best answer to the need of predicting performance (Stewart, 2007).

Considering the number of publications focused on the processing and performance of WPCs, the fraction of them devoted to the study of creep and damage on these materials is still relatively small. However, the general consensus is that durability of the materials is one of the key points to resolve in order to improve these materials' predictability and to sustain the growing market of WPCs.

At this point, it is clear that wood fillers/fibers added to polymeric matrices reduce the creep deflection, that the use of coupling agents also improves the creep resistance and that the more recent solution of crosslinking the polyolefin during wood–plastic processing is a valid convenient alternative to achieve creep reduction. The theoretical models (Bürger models/Prony series, or power law models) do good jobs in representing low stress-derived creep. More work has been developed in adjusting the power law models to include non-linear and damage effects. The TTS is a valuable tool, but experimental studies indicate that is not always trustworthy in the prediction of long-term performance and that validation over longer periods is still a requirement in making confident assessments of durability. In this sense, not only are more careful experimental data still needed (well-characterized materials and defined conditions of testing), but also versatile models (validated under complex and different conditions from those originally used in their formulation) that can predict creep performance in service and assess the effects of changing loadings and harsh environmental conditions (high temperature, humidity, cyclic loadings, etc).

8.6 References

Agarwal M K, Schultz J M (1981), 'Physical aging of isotactic polypropylene', *Polym Eng Sci*, **21** (12), 776–781.

Andersons J, Limonov V, Tamuzs V (1991), 'Fatigue of polymer composite materials', In: Cardon A H and Verchery G (eds), *Durability of polymer based composite systems for structural applications*, Elsevier Applied Science, London.

ASCE (1984), *Structural Plastic Design Manual*, ASCE Publications, American Society of Civil Engineers.

Bengtsson M, Oksman K (2006), 'The use of silane technology in crosslinking polyethylene/wood flour composites', *Composites: Part A*, **37**, 752–765.

Bengtsson M, Gatenholm P, Oksman K (2005), 'The effect of crosslinking on the properties of polyethylene/wood flour composites', *Compos Sci Technol*, **65**(10), 1468–79.

Bledzki A K, Faruk O (2004), 'Creep and impact properties of wood fibre–polypropylene composites: influence of temperature and moisture content', *Compos Sci Technol*, **64**, 693–700.

Brinson L C, Gates T S (1995), 'Effects of physical aging on long term creep of polymers and polymer matrix composites', *Int J Solid Structures*, **32**, 827–846.

Chia L H L, Teoh S H, Boey F Y C (1987), 'Creep characteristics of a tropical wood–polymer composite', *Radiation Phys Chem*, **29**(1), 25–30.

Crissman J M (1986), 'Creep and recovery behavior of a linear high density polyethylene and an ethylene–hexene copolymer in the region of small uniaxial deformations', *Polym Eng Sci*, **26**(15) 1050–1059.

Dean D, Husband M, Trimmer M (1998), 'Time–temperature-dependent behavior of a substituted poly(paraphenylene): tensile, creep, and dynamic mechanical properties in the glassy state', *J Polym Sci B: Polym Phys*, **70**, 2971–2079.

Farid S I (2000), 'Viscoelastic properties of wood-fiber reinforced polyethylene: stress relaxation, creep and threaded joints', MSc Thesis, University of Toronto.

Ferry J D (1980), *Viscoelastic properties of polymers*, 3rd edn, John Wiley & Sons, New York.

Fiebig J, Gahleitner M, Paulik C, Wolfschwenger J (1999), 'Ageing of polypropylene: processes and consequences', *Polym Testing*, **18**, 257–266.

Findley W N (1960), 'Mechanisms and mechanics of creep of plastics', *SPEJ*, **16**, 57–65.

Findley W N, Lai J S, Onaran K (1976), *Creep and relaxation of nonlinear viscoelastic materials*, North Holland Publishing Company, Amsterdam.

Fukui Y, Sato T, Ushirokawa M, Asada T, Onogi S (1970), 'Rheo-optical studies of high polymers. XVII. Time–temperature superposition of time-dependent birefringence for high-density polyethylene', *J Polym Sci Part A-2*, 1195–1209.

Jazouli S, Luo W, Bremand F, Vu-Khanh T (2005), 'Application of time–stress equivalence to nonlinear creep of polycarbonate', *Polym Testing*, **24**, 463–467.

Jerina K L, Schapery R A, Tung R W, Sanders B A (1982), 'Viscoelastic characterization of a random fiber composite material using micromechanics', in Sanders B A (ed.), *Short fiber reinforced composite materials*, ASTM STP 772, ASTM, Philadelphia, 225–250.

Kobbe R G (2005), 'Creep behavior of a wood–polypropylene composite', MSc. Thesis, Washington State University.

Lai J, Bakker A (1995), 'Analysis of the non-linear creep of high density polyethylene', *Polymer*, **36**(1), 93–99.

Lee S-Y, Yang H-S, Kim H-J, Jeong C-S, Lim B-S, Lee J-N (2004), 'Creep behavior and manufacturing parameters of wood flour filled polypropylene composites', *Composite Structures*, **65**, 459–469.

Lin W S, Pramanik, A K, Sain M (2004), 'Determination of material constants for non linear viscoelastic predictive model', *J Composite Mater*, **38**(1), 19–29.

Lou Y C, Schapery R A (1971), 'Viscoelastic characterization of a nonlinear fiber-reinforced plastic', *J Composite Mater*, **5**, 208–234.

Maksimov R D, Mochalov V P, Urzhumtsev Y S (1972), 'Time-moisture superposition', *Polym Mech*, **8**(5), 685–689.

Marais C, Villoutreix G (1998), 'Analysis and modeling of the creep behavior of the thermostable PMR-15 polyimide', *J Appl Polym Sci*, **69**, 1983–1991.

Marcovich N E, Villar M A (2003), 'Thermal and mechanical characterization of linear low density poly(ethylene)–woodflour composites', *J Appl Polym Sci*, **90**(10), 2775–2784.

Matsumoto D S (1988), 'Time–temperature superposition and physical aging in

amorphous polymers', *Polym Eng Sci*, **28**, 1313–1317.
Nuñez A J, Sturm P C, Kenny J M, Aranguren M I, Marcovich N E, Reboredo M M (2003), 'Mechanical characterization of polypropylene–wood flour composites', *J Appl Polym Sci*, **88**, 1420–1428.
Nuñez A J, Marcovich N E, Aranguren M I (2004), 'Analysis of the creep behavior of polypropylene–woodflour composites', *Polym Eng. Sci*, **44**(8), 1594–1603.
Onogi S, Sato T, Asada T, Fukui Y (1970), 'Rheo-optical studies of high polymers, XVIII. Significance of the vertical shift in the time temperature superposition of rheooptical and viscoelastic properties', *J Polym Sci: Part A-2*, 1211–1255.
Papanicolaou G C, Zaoutsos S P, Cardon A H (1999), 'Further development of a data reduction method for the nonlinear viscoelastic characterization of FRPs', *Composites Part A*, **30**, 839–848.
Park B D, Balatinecz J J (1998), 'Short term flexural creep behavior of wood-fiber/polypropylene composites', *Polym Compos*, **19**(4), 377–382.
Penn R W (1966), 'Dynamic mechanical properties of crystalline, linear polyethylene', *J Polym Sci: Part A-2*, **4**, 545–557.
Pooler D J (2001), 'The temperature dependent non-linear response of a wood plastic composite', MSc. Thesis, Washington State University.
Rand J L (1995), 'A nonlinear viscoelastic creep model', *Tappi J*, 178–182.
Rangaraj S (1999), 'Durability assessment and modeling of wood–thermoplastic composites', MSc. Thesis, Washington State University.
Rangaraj S V, Smith L V (1999), 'The nonlinear viscoelastic response of a wood-thermoplastic composite', *Mechanics Time Dependent Mater*, **3**, 125–139.
Reboredo M M, Aranguren M I, Marcovich N E (2007), 'Selected topics on polypropylene/wood flour composites: thermal, mechanical and time dependent response', in Nwabunma D, Kyu T, *Polyolefin composites*, Wiley Interscience, New Jersey, Chapter 6, 150–177.
Reifsnider K L (1982), 'Damage mechanisms in fatigue of composite materials', in *Fatigue and Creep of Composite Materials, Proceedings of Symposium*, Riso National Laboratory, 1982.
Sain M M, Balatinecz J J, Law S (2000), 'Creep fatigue in engineered wood fiber and plastic compositions', *J Appl Polym Sci*, **77**, 260–268.
Schael G W (1966), 'A study of the morphology and physical properties of polypropylene films', *J Appl Polym Sci*, **10**(6), 901–915.
Schapery R A (1969), 'On the characterization of nonlinear viscoelastic materials', *Polym Eng Sci*, **9**(4), 295–310.
Smith L V, Weitsman Y J (1996), 'The immersed fatigue response of polymer composites', *Int J Fracture*, **82**, 31–42.
Sternstein S S, Van Buskirk C S (1988), 'Polymer creep', in Kroschwitz J I, *Encyclopedia of Polymer Science and Engineering*, John Wiley & Sons, New York, Vol. 12, 470–486.
Stewart R (2007), 'Fierce competition drives advances in equipment, materials and processes', *Plastics Eng*, 20–27 February (www.4spe.org/pub/pe/articles/2007/february/20.pdf).
Tichy R J (2007), 'Wood–plastic composites. Emerging products and markets', (http://www.composites.wsu.edu/navy/Navy1/Presentations/TichyGermany99.pdf), accessed 1 June 2007.
Tirrell M (1994), 'Rheology of polymeric liquids', in Macosko C W, *Rheology, principles, measurements and applications*, VCH Publishers Inc., New York, 510.
Tissaoui J (1996), 'Effects of long-term creep on the integrity of modern wood

structures', PhD Thesis, Virginia Polytechnic Institute and State University.
US CPSC (Consumer Product Safety Commission) (2006), http://www.cpsc.gov/CPSCPUB/PREREL/PRHTML05/05247.html, accessed 10 June 2007.
Vittoria V (1988), Investigation of the ageing of isotactic polypropylene via transport properties', *Polymer*, **29**(6), 1118–1123.
Wolcott M P, Smith M S (2004), 'Opportunities and challenges for woodplastic composites in structural applications', in *Proceedings of Progress in Woodfibre–Plastic Composites Conference*. Toronto, May 2004, 1–10.
Zaoutsos S P, Papanicolaou G C, Cardon A H (1998), 'On the non-linear viscoelastic behaviour of polymer–matrix composites', *Composites Sci Technol*, **58**, 883–889.

9
Processing performance of extruded wood–polymer composites

K ENGLUND and M WOLCOTT,
Washington State University, USA

9.1 Introduction

The extrusion process of natural fiber–thermoplastic composites (NFTCs) deviates substantially from traditional polymer extrusion in a variety of different ways. Changes in the rheological behavior with highly filled systems, the thermally unstable nature of wood and other natural fibers (lignocellulosic material), the chemical and morphological differences within fiber types and species all play a major role in how these composites are produced. Extrusion processing has captured most of the wood–polymer composite (WPC) market, primarily with decking, guardrail, and fencing materials, and therefore has been the main interest for most of the present manufacturers and will be the main emphasis of this chapter.

One of the key principles for any type of NFTC processing method is the distribution and dispersion of the natural fiber into the polymer matrix. Natural fibers and thermoplastic resins have very little affinity for each other. Non-polar hydrophobic thermoplastics and polar hydrophilic natural fibers require added energy to homogeneously mix the two components. Dispersion can be enhanced through coupling and dispersing agents; however, melt blending through mechanical mixing is the primary method of creating a homogeneous composite.

Notwithstanding the frailties of the natural fiber/thermoplastic interface, many benefits can also be obtained with the blended composite. Increased fiber loading can alter the melt rheology to achieve higher output rates, improve some mechanical properties, create a more natural appearance than that of solid thermoplastics, and, most importantly to producers, the relatively inexpensive natural fibers extend the resin, which in turn, dramatically lowers the raw material costs. These attributes have been the impetus for the rapid growth of NFTCs and continue to open new markets.

9.2 Current extrusion processing methods for natural fiber–thermoplastic composites

Extrusion processing dominates the present NFTC manufacturing, due in part to the current market application requirements for continuous lineal elements. Extruders can either be single-or twin-screw designs. The twin-screw extruders offer a variety of options within the screw placement, directional rotation, and barrel shape. The two screws either can be placed to intermesh their flights or they can be independent of each other; however, intermeshing screws will provide a more dispersed composite. The barrel and screws in a twin-screw extruder can also either taper to a conical shape or remain parallel throughout the barrel. Screw design for all extruders can vary substantially and is usually very specific for the material processed.

Extrusion methods vary among producers. However, commercial processes generally follow two manufacturing platforms: direct and pre-compounded extrusion. In direct extrusion, the raw materials are dry blended and fed into the extruder, whereas pre-compounded extrusion requires the components to be melt-blended prior to profile extrusion. These extrusion methods have their pros and cons, and are both used extensively in the NFTC industry.

9.2.1 Dry blending and mixing

Dry blending develops a homogeneous mixture of the fiber, thermoplastic, and minors without imparting any thermal transition to the polymer. Having a uniform mix of the dry components prior to any melt-blending procedure aids in the final dispersion of all the components within the NFTC. No matter the extrusion platform some dry blending needs to be incorporated to efficiently disperse the natural fibers in the thermoplastic matrix. The only exception to this is extruders equipped with side stuffing capabilities, where different components of the formulation can be added to the mixing extruder throughout the length of the barrel.

The dry mixing procedure can be done either through batch or continuous mixers that are fed directly into the extruder or melt-blending operation. Batch dry blenders can be of a variety of types and sizes. Rotary drum and ribbon blenders are some of the most common and oldest technologies in the wood and plastics industries. Batch mixers blend a finite amount of material at one time; however, the feed into the melt blender or extruder is a continuous procedure. Continuous feeders are generally screw or auger feed mechanisms with blending capabilities. What makes them continuous is that the feed of the formulation components is not interrupted and supplies a constant stream into the mixing apparatus.

Metering of the individual components into a dry mixer is usually done with loss-in-weight or gain-in-weight feeders. With wood and other natural fibers, the density variation is too great to utilize volumetric feeding mechanisms. Gain-in-weight feeders are commonly used for batch mixing where the components are added independently to a main hopper that is balanced on load cells. The raw materials are fed into the hopper one at a time until the desired weight for each is obtained. The weighed material is then dropped or transferred into a blender. Most continuous blenders utilize loss-in-weight feeders, where the hoppers for the individual components are continuously weighed and monitored to provide the proper amount of material flow based upon the mixed output rate. Metering the correct amount of each component is imperative to maintaining output, product quality, and consistency for both feeding procedures.

9.2.2 Profile extrusion of wood–polymer composites

Direct extrusion

Direct extrusion utilizes one extruder to melt-blend and pump the composite through the profile die, therefore a mixing extruder is required to ensure a homogeneous product. An intermeshing counter rotating extruder is common for direct extrusion operations. To optimize direct extrusion, the wood fiber should be dried to a 1–5% moisture content prior to dry blending to minimize the amount of volatiles. Venting is also needed to evacuate residual moisture and volatile components resulting from thermal degradation.

Pre-compounded extrusion

Pre-compounding extrusion offers a wide degree of processing flexibility. The first step entails the melt-blending of the natural fiber and thermoplastic, which can be done in a variety of ways: twin-screw extruder, shear mixers, melt-blending pellet mills. The atmospheric exit of the material from the melt-blending operation can act as a venting zone for volatiles. The blended material is either pelletized or fed directly into another extruder that pumps the composite through a die and into the final profile. Single screw extruders are commonly used as pumps in pre-compounding extrusion for maintenance ease and lower costs than twin-screw extruders. Another advantage to pre-compounded extrusion is the potential elimination of drying. Natural fiber moisture contents can be higher than with direct extrusion in ranges above 6–12%.

Alternative melt blenders and pelletizers

Much of the melt-blending discussed thus far has been within a twin-screw extruder, however, alternative methods can also be applied to generate a homogeneous mixture of wood and thermoplastic. High-intensity mixers such as kinetic mixers (K-mixers) or melt granulators such as the Palltruder® can be incorporated to generate a pellet or granulated particle for use in a profile extruder and injection molder. These blending techniques often rely on the heat of friction to melt or plasticize the thermoplastic during blending.

9.3 Rheology of a wood fiber filled thermoplastic

One important factor controlling the processing of WPCs is the flow or rheological behavior of the melt. Understanding the melt rheology of these materials can not only aid in the design of dies and processing profiles, but can also provide information on the interface between the natural fiber, thermoplastic, and processing and performance additives.[1] Li and Wolcott[1–4] provided a series of studies on the rheology of wood-filled polymer systems. Their findings indicate that many facets of the wood component influence the melt flow and interfacial aspect of the WPC.

As with many polymer systems, WPC melts are represented as non-Newtonian fluids. In these systems the melt viscosity decreases with increasing shear rate and temperature. Although parallel plate techniques are commonly used by researchers to determine rheological properties of fluids, capillary rheometers produce shear rates that are more realistic of the actual die flow scenario. In addition, capillary rheometers facilitate the study of the two different components controlling die flow of WPC melts in extrusion: shear flow and wall slip. Most commonly, the velocity of a moving fluid is assumed to be zero at the wall of the opening. This assumption is quite reasonable for extrusion of many neat polymers. However, the wall slip phenomenon can be quite important for understanding the flow behavior for stiff melts like WPCs.

As non-Newtonian fluids, the measured viscosity of the melt is dependent on the shear rate of the test. To facilitate comparison among different melts, it is imperative to use a common shear rate, which is easily accomplished. Both the melt viscosity and the wall slip phenomena may be studied using capillary rheometry using a Mooney analysis. This analysis assumes that the total apparent shear rate in the die is a combined effect of the true corrected shear rate and that attributable to the wall slip phenomenon ($\dot{\gamma}_a = \dot{\gamma}_c + \dot{\gamma}_s$). This flow behavior can be represented for a capillary die by the Mooney equation:

$$\frac{8u}{D} = \left(\frac{4}{\tau^3}\int_0^\tau \tau^2 \dot{\gamma}\,d\tau\right) + \left(\frac{8u_s}{D}\right)$$

where:
$\frac{8u}{D}$ = total apparent shear flow in a capillary
u_s = melt velocity attributed to wall slip
τ = shear stress at the capillary wall
D = capillary diameter
$\dot{\gamma}$ = shear strain rate.

A detailed outline for determining these components is beyond the scope of this chapter but is available elsewhere.[1]

9.3.1 Influence of formulation components on the rheology of wood–polymer composites

With highly filled systems, such as WPCs, the magnitude and relative contribution of the various flow mechanisms are influenced by the various components of the formulation. A partial list of relevant component attributes that contribute to the melt rheology include (1) wood content, species, and particle size; (2) lubricant type and loading; and (3) coupling agents.

The addition of wood increases the viscosity of the polymer melt, thereby decreasing the contribution of shear flow in the overall output. While studying composite formulations composed of maple flour,[2] it was found that the addition of 70% wood increased the apparent shear viscosity by an entire order of magnitude over the base high-density polyethylene (HDPE) polymer (Fig. 9.1). The increased viscosity with wood loading was more pronounced with small wood particles. Shear viscosity continually increased with mesh sizes ranging from 20 to 140 at wood loadings of 60% by mass, although the magnitude of this increase was much smaller than that of wood loading.

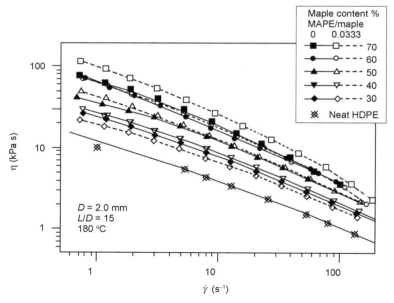

9.1 Apparent shear viscosity of HDPE/maple formulations. The influence of increasing wood content and the addition of a maleated polyethylene (MAPE) coupling agent are illustrated.[1]

When comparing formulations containing maple and pine flour,[2] similar trends were found. However the apparent shear viscosity of pine formulations did not seem to be as sensitive to changes with increased loading compared with maple. For instance, little difference was evident in shear viscosity between formulations containing 40 and 60% pine flour. The difference noted in the flow behavior of maple and pine formulations are also evident in the contribution of the wall slip phenomenon to overall flow. In maple formulations, the contribution of wall slip velocity averaged roughly 50% of the overall flow with formulations containing 40% wood. However, 60% maple formulations were dominated by wall slip at low stress levels. Pine showed exactly the opposite behavior. Where this difference might appear confusing at first consideration, it is important to realize that pine species contain a large amount of fatty-acid compounds that, at high levels, may act as lubricants in the system.

Polymeric lubricants are added to WPC formulations to control melt flow behavior, improve surface quality, and increase extruder output. Lubricants are classified differently depending on the type. Classically, internal lubricants act to decrease melt viscosity while external lubricants facilitate wall slip between the melt and the die. In reality, most lubricant packages perform both functions to different degrees.[4] Different commercial lubricant packages produce markedly different behaviors. For instance, the classical combination of zinc stearate and ethylene bisstearamide (EBS) was found to be very effective at decreasing the melt viscosity, especially at low shear rates. In contrast, a commercial polyester-based stearate was found to be most effective at increasing the contribution of wall slip in the overall flow (Fig. 9.2).

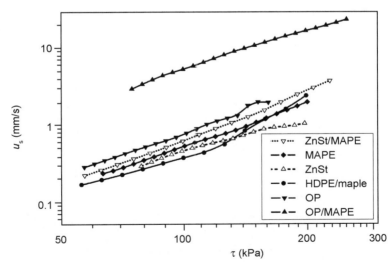

9.2 Wall slip velocity (u_s) of several lubricated formulations of HDPE/maple formulations. Higher levels of u_s are indicative of greater external lubrication.[3]

Coupling agents (CAs) such as maleated polyolefins have been widely studied for the influence on mechanical and physical properties of WPCs. These copolymers are believed to chemically react with the wood fiber and interact with the thermoplastic matrix through chain entanglement. As seen in Fig. 9.1, such a coupling phenomenon can also be realized in the melt form through an increased shear viscosity.[3] In addition, the low molecular weight nature of these additives tends to decrease the overall melt viscosity at lower shear rates and thereby act as an internal lubricant.[4]

9.3.2 Importance of rheology on product quality of wood–polymer composites

The flow behavior of the composite during manufacture, as controlled by the melt rheology, can have a strong influence on overall product quality (Fig. 9.3). In practice, surface tearing is often the limiting factor for increased production rates. Therefore, effective additive packages must not only control the shear viscosity but also enhance the surface lubrication.

The importance of rheology on product quality has recently been studied by Hristov *et al.*[5] who found that formulations at high levels of wood loading displayed behavior indicative of solids. Moreover, formulations that displayed high degrees of wall slip also displayed an enhanced surface quality. Enhanced wall slip velocity and surface quality can be imparted by increased shear rate.[5]

The role of wall slip in effective extrusion of NFTC materials highlights the pronounced differences from extrusion of neat polymers.[5] For filled systems, the change in processability appears to occur somewhere between 30 and 50% wood content. At low wood contents, the material will process similar to a neat polymer with increased viscosity. But wood content near 60% facilitates the production of products at high rates with smooth surfaces.

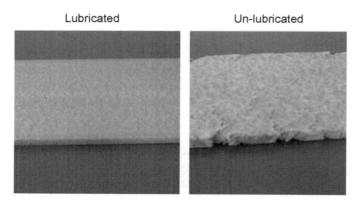

9.3 NFTC extruded with and without lubricant.

9.4 Commercial wood–polymer composites

Although there are a variety of manufacturers of WPCs, there is little information in the public domain that addresses the extrusion performance on a commercial production platform. Work by Englund and Olson[6] and Wolcott[7] have evaluated some of the processing and formulation influences on commercial extrusion of WPCs. Englund and Olson monitored processing parameters and composite physico-mechanical properties of HDPE and polypropylene (PP)-based WPCs extruded at varying screw speeds, temperature profile and formulations. In the work by Wolcott, an evaluation of CA performance at commercial rates was performed on PP-based WPCs.

Within both of these bodies of work, a direct extrusion platform was utilized. The dry-blended 60 mesh pine (*Pinus strobes*) wood flour, thermoplastic and minors were fed directly into a crammer-feed unit mounted on a Milaron® 86 mm counter-rotating conical twin-screw extruder. The final profile had a nominal $1 \times 5.5''$ cross-section and was cooled with chilled water-spray. Formulations were deviated to accommodate CA influences. Different lubricant types and amounts were used to minimize the detrimental attributes of some lubricants on coupling agent efficacy.[8] A polyester-based lubricant by Honeywell® called OptiPak100™ (OP100) was used with the coupled formulations and a common industrial lubricant package of 2:1 zinc stearate (ZnSt) and EBS wax was utilized for the uncoupled WPCs.

9.4.1 Ouput rate

The extrusion output rate of WPC is a process variable of much interest for all manufacturers. Maintaining product quality and surface appearance at higher production rates is often the goal of many extrusion manufacturers. Within the work by Englund and Olson,[6] production rates were examined with coupled and uncoupled formulations of HDPE and PF-based composites within different temperature profiles and screw speeds. An overview of the processing variables and formulation design can be seen in Tables 9.1 and 9.2, respectively. Two temperature profiles were used labeled T_1 and T_2, with the CA formulation being run with the T_2 profile (Table 9.2). With HDPE, the temperature profile had little influence on the production rate; however, the addition of the maleated polypropylene (MAPP) CA to the formulation caused a drop off of output rate at the higher screw speeds (Fig. 9.4). This reduction at higher rates can be attributed to the lower lubricant levels and/or the effect CA has on the rheology of the melt blend. Based upon the work of Li and Wolcott[2,4] the maleated coupling agent promoted internal lubrication in both shear and extensional flow with HDPE–wood composites, while the OP100 lubricant showed a significantly higher wall slip velocity.

With PP-based wood composites, the output rate was influenced more with

Table 9.1 Formulation design for the coupled and uncoupled extruded composites

	Uncoupled			Coupled		
Component	Component type	Formulation (%)	Component	Component type	Formulation (%)
Wood	60-mesh pine	58	Wood	60-mesh pine	58
Polymer	PP or HDPE	32	Polymer	PP or HDPE	32
Talc	Nicron 403	7	Talc	Nicron 403	6
Lubricant 1	ZnSt	2	Lubricant	OP 100	2
Lubricant 2	EBS	1	Coupling agent	950p	2

Table 9.2 Temperature profiles for the extruder and die zones for HDPE and PP formulations

	86 mm extruder barrel, screw and die temperatures (°F)			
	HDPE		PP	
	T_1	T_2 and CA	T_1	T_2 and CA
Barrel zone 1	340	380	415	450
Barrel zone 2	340	375	400	435
Barrel zone 3	340	365	385	400
Barrel zone 4	340	360	380	385
Screw	340	365	380	385
Die zone 1	340	355	370	370
Die zone 2	340	355	370	360
Die zone 3	340	355	370	360

temperature than with the formulation design (Fig. 9.5). The coupled and uncoupled formulations show the same output rate due primarily as a result of the same temperature profile. As the temperature profile was decreased, the melt stiffened and the output rate increased. The role of the coupling agent did not appear to alter the production rate for the PP-based system.[5]

9.4.2 Extrusion melt pressure

The melt pressure was significantly altered by the addition of coupling agents with the OP100 lubricant to the formulation of both HDPE and PP-based wood composites (Figs 9.6 and 9.7). A decrease in the melt pressure in both polymer systems was observed when the coupled/OP100 system was extruded. The inherent contribution of OP100 to increase the wall slip contribution likely lowered the melt pressure, while the internal lubrication of the CA aided in the reduction.[5]

Processing performance of extruded wcod–polymer composites 199

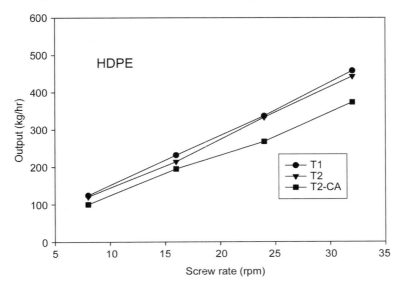

9.4 Output rate for the HDPE-based NFTCs at varying screw speeds and temperature profiles.[5]

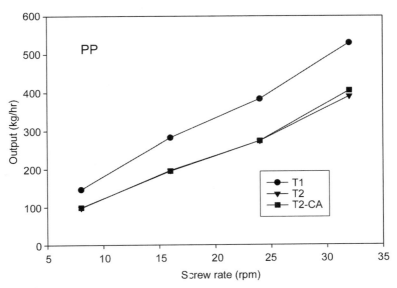

9.5 Output rate for the PP-based NFTCs at varying screw speeds and temperature profiles.[5]

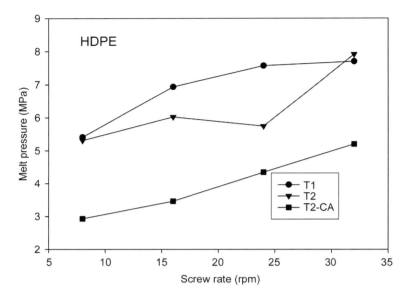

9.6 Melt pressure data for the HDPE NFTCs at varying screw speeds.[5]

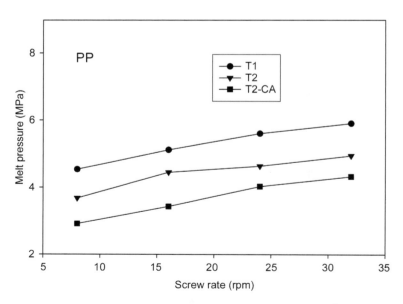

9.7 Melt pressure data for the PP NFTCs at varying screw speeds.[5]

Processing performance of extruded wood–polymer composites 201

The extruder profile temperature also imparted a reduction in melt pressure for the PP-based wood composites, which correlated to the output data. A higher output rate and melt pressure for the lower temperature profile indicates again increased melt stiffness. Melt stiffness can also be seen to increase with higher wood loadings, where contribution of wall slip becomes more prevalent in the volumetric output.[1,9]

9.4.3 Extrusion parameter influences on composite properties

Maintaining product properties is always a concern as production rates are increased with predominately all composite products and WPCs are no exception to this rule. As rates increase, dwell time of the material in the extruder reduces, causing manufacturers to accommodate with enhanced screw design and temperature profiles. However, care must be taken when temperatures from barrel, screws and dies are increased along with increased shear heating, to prevent or minimize thermal degradation and off-gassing which can result in poor surface and mechanical properties. In Fig. 9.8, a cross-sectional view of thermally degraded wheat straw NFTCs can be seen. The degradation of this composite was because the wheat straw material had been aged and its thermal properties reduced. The thermal gravimetric analysis (TGA) curves show a

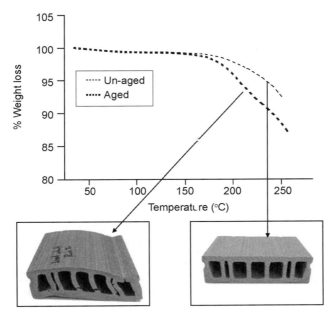

9.8 Thermal degradation of extruded wheat straw NFTCs with thermal gravimetric analysis.

weight loss reduction occurring at a much lower temperature. The degradative off-gasses resulted in poor extrudability and appearance.

Density

Within the work by Englund and Olson,[6] the composite density was assessed at the varying extrusion processing parameters discussed previously. With all natural fiber composites, density of the final product plays a significant role on the final properties. Within the NFTCs studied there was a general reduction of density with an increase in screw speed and with the addition of the CA/OP100 formulation (Figs 9.9 and 9.10). A reduction in density with NFTCs is associated with the insufficient dispersion of the natural fiber and a lack of penetration of the thermoplastic in the cellular anatomy.

Incorporating the thermoplastic resin into the natural fiber lumens and other orifices requires minimized resin viscosity, pressure, and fiber dispersion. The thermoplastic viscosity must be low enough to flow and penetrate the open cavities coupled with a sufficient pressure. With low cell wall mechanics and high pressures, cellular collapse can also be observed during extrusion processing. The images in Fig. 9.11 show the effects of extrusion processing on the hybrid-polar (*Populus* sp.) and maple (*Acer* sp.) filled PP composites. The fast-growth hybrid-poplar contains wood cells with thinner walls than the maple. The combination of pressure and heat develops an elastic collapse of the thin cell wall, increasing the composite density.

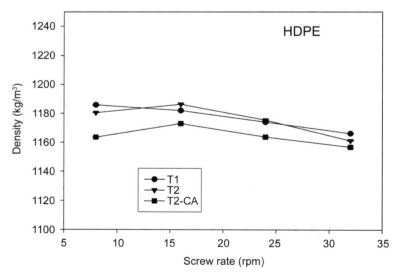

9.9 Density of the HDPE NFTCs with increasing screw speeds.[5]

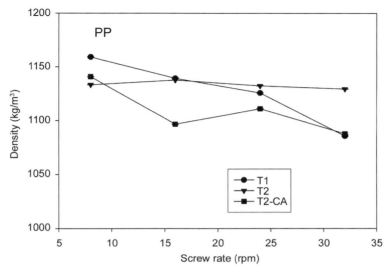

9.10 Density of the PP NFTCs with increasing screw speeds.[5]

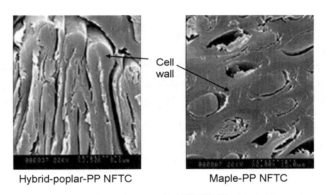

Hybrid-poplar-PP NFTC   Maple-PP NFTC

9.11 Cellular collapse in extruded HDPE filled hybrid-poplar and maple wood flour NFTCs.

Flexural properties

The static bending properties of HDPE and PP-based NFTCs vary based upon process variations and formulation design. The flexural strength behavior of HDPE remained unchanged with temperature and screw speed, with the coupled system at an expectantly higher load capacity (Fig. 9.12). The modulus of elasticity (MOE) (Fig. 9.13) for values show a slight decline as the screw speed was increased; however, temperature and coupling played little or no role in governing the stiffness of the composite. With PP designs, the lower temperature profile exhibited a sharp reduction in modulus of rupture (MOR) and MOE (Figs

204 Wood–polymer composites

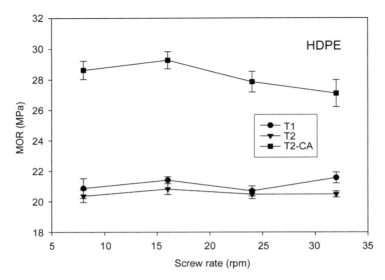

9.12 Flexural strength (MOR) of the HDPE NFTCs with increasing screw speeds.[5]

9.14 and 9.15) at the highest screw rate (32 rpm), indicating an insufficient thermal profile for PP-NFTC extrusion.[6]

The coupled PP NFTCs are notoriously difficult to run at production rates, while maintaining the full mechanical potential. The MOE data in Fig. 9.15 show a substantial drop of the coupled formulation at speeds above the 8 rpm

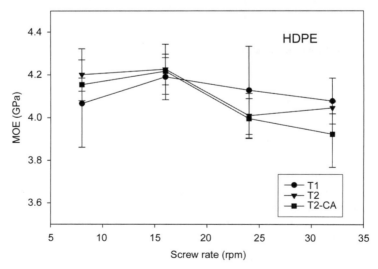

9.13 Flexural stiffness (MOE) of the HDPE NFTCs with increasing screw speeds.[5]

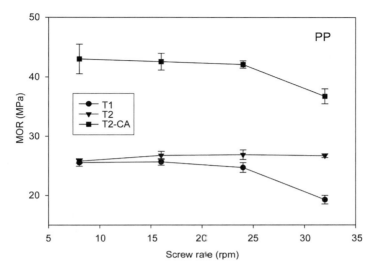

9.14 Flexural strength (MOR) of the PP NFTCs with increasing screw speeds.[5]

rate. Similar work by Wolcott[7] has shown the same behavior occurring with increased rate and with the addition of more OP100 lubricant.

Water sorption

One of the major drawbacks of incorporating natural fibers into any composite matrix is the sorption of water. The hydrophilic behavior of natural fibers and

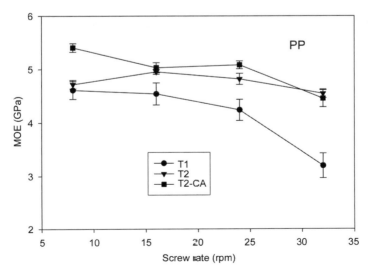

9.15 Flexural stiffness (MOE) of the PP NFTCs with increasing screw speeds.[5]

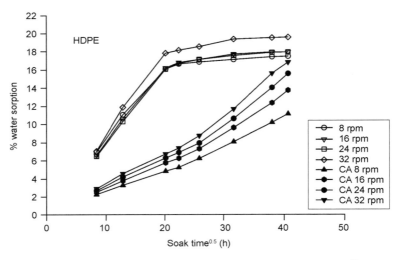

9.16 Water sorption of the coupled and uncoupled HDPE composites.[5]

the voids and capillaries developed in many composite structures, make NFTCs susceptible to damage and decrease the products' structural integrity when exposed to many exterior conditions. The sorption characteristics found in the work by Englund and Olson,[6] showed little difference in the screw speed with uncoupled formulations; however, MAPP did influence the water sorption of HDPE and to some extent the PP composites.

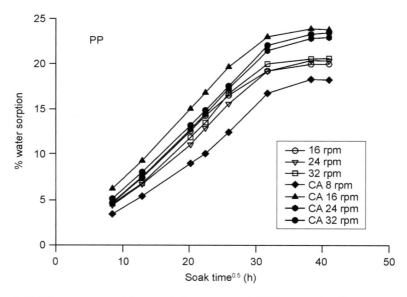

9.17 Water sorption of the coupled and uncoupled PP composites.[5]

With the HDPE NFTCs, a consistent increase was seen in the rate of diffusion for the coupled formulations (Fig. 9.16). In composite Fickian diffusion behavior, much of the initial linear sorption is promoted through the inherent reinforcement or filler and through the existing voids present in the composite after manufacture. The non-linear region of the sorption curve can often be associated with stress-induced cracks and voids developed from the initial sorption.[10] The data for the coupled HDPE and PP composites differ quite substantially in their sorption behavior (Figs 9.16 and 9.17). With the PP resin system, all formulations show similar trends, whereas with HDPE a long linear region is associated with the coupled material and the uncoupled systems exhibit a short but increased initial diffusion followed by a non-linear crack-propagated region. From these data, there would appear to be a stronger coupling interaction with HDPE composites that is subject to screw speeds. PP did show a reduced sorption behavior with the slow screw speed and coupled formulation.

9.5 References

1. Li, T.Q. and M.P. Wolcott, Rheology of HDPE-wood composites. I. Steady state shear and extensional flow. *Composites Part A – Applied Science and Manufacturing*, 2004. **35**(3): 303–311.
2. Li, T.Q. and M.P. Wolcott, Rheology of wood plastics melt. Part 1. Capillary rheometry of HDPE filled with maple. *Polymer Engineering and Science*, 2005. **45**(4): 549–559.
3. Li, T.Q. and M.P. Wolcott, Rheology of wood plastics melt, part 3: Nonlinear nature of the flow. *Polymer Engineering and Science*, 2006. **46**(1): 114–121.
4. Li, T.Q. and M.P. Wolcott, Rheology of wood plastics melt, part 2: Effects of lubricating systems in HDPE/maple composites. *Polymer Engineering and Science*, 2006. **46**(4): 464–473.
5. Hristov, V., E. Takacs, and J. Vlachopoulos, Surface tearing and wall slip phenomena in extrusion of highly filled HDPE/wood flour composites. *Polymer Engineering and Science*, 2006. **46**(9): 1204–1214.
6. Englund, K.R. and B.D. Olson. Processing Influences on Wood Plastic Composites (WPC's) Properties – Extrusion Screw Speed. In *9th International Conference on Wood & Biofiber Plastic Composites*. 2007. Madison, WI.
7. Wolcott, M.P., Foundation elements for naval low-rise buildings, in *Progress Report for the Office of Naval Research Grant N00014-06-1-0874*. 2007. p. 7.
8. Harper, D.P. and M.P. Wolcott, Chemical imaging of wood-polypropylene composites. *Applied Spectroscopy*, 2006. **60**(8): 898–905.
9. Hristov, V. and J. Vlachopoulos, A study of viscoelasticity and extrudate distortions of wood polymer composites. *Rheologica Acta*, 2007. **46**(5): 773–783.
10. Chateauminois, A., *et al.*, Study of the interfacial degradation of a glass epoxy composite during hygrothermal aging using water diffusion measurements and dynamic-mechanical thermal-analysis. *Polymer*, 1994. **35**(22): 4766–4779.

10
Oriented wood–polymer composites and related materials

F W MAINE, Frank Maine Consulting Ltd., Canada

10.1 Introduction

Ward and coworkers,[1-8] Woodhams and coworkers,[9-12] and Porter and coworkers[13] have shown in their pioneering work in the United Kingdom, Canada, and the United States, that elongating the polymer molecule so that an applied stress works on the carbon–carbon bond in the polymer backbone rather than uncoiling the polymer, results in significant strength and stiffness increases. The primary tools used to elongate the polymer molecules so as to orient them are the metal-forming processes. These processes include rolling, drawing, forging, and extrusion. In the applications development work that has been done, the process that gives the optimum in properties and economics is selected for commercialization of the product being developed.

10.2 Orientation of polymers

The term 'oriented polymers' is used to refer to thermoplastic polymers that have been processed in such a way as to straighten the polymer chain backbone so that when the polymer molecule experiences an applied stress, this stress is applied to the carbon–carbon bond in the chain backbone and not just to uncoiling the polymer helix. The result of this molecular orientation is an increase in strength, stiffness, and impact strength of the polymer. This is a general phenomenon that is applicable to all thermoplastic polymers to some degree. A common way to apply the forces that result in orientation is to use the processes used for metal processing such as rolling, drawing, and forging. If the orientation is all in one direction, the result is uniaxial orientation with the physical properties in the oriented direction being very much higher than the physical properties in the other two orthogonal directions. This type of material is similar to wood in which the properties along the grain are much higher than in the other two transverse directions. This type of orientation is fine for rod and profile products but is not adequate for sheet where reinforced properties in the

Oriented wood–polymer composites and related materials

plane of the sheet are desired. Here biaxial reinforcement is more suitable. The increase in physical properties using biaxial orientation can never be as high as with uniaxial orientation as some of the polymer molecules have to be oriented in two directions as compared with only one direction with uniaxial orientation. Different processes or variations of processes are used to obtain uniaxial or biaxial orientation.

The next question to be addressed is how much orientation to impart: although more is better since the increase in physical properties is directly related to how much the polymer molecule is oriented, the ultimate in uniaxial orientation results in a product with very weak transverse properties and the processing costs are very expensive. The draw ratio is used to define the amount of orientation imparted to the polymer molecules. In the case of rod and uniaxial orientation, the draw ratio (DR) is the ratio of the final length of the rod to the initial length of the rod.

$$DR = \frac{l_f}{l_i} \qquad 10.1$$

Since the volume of the initial cylindrical billet and the final oriented cylindrical rod are the same, equation [10.1] can be rewritten in terms of initial diameter of the billet and final diameter of the oriented rod:

$$DR = \frac{d_i^2}{d_f^2} \qquad 10.2$$

Therefore, as example, to obtain a draw ratio of 9 in producing an oriented polypropylene rod of 1 inch (2.5 cm) in diameter, in initial billet with a diameter of 3 inches (7.5 cm) is needed.

10.2.1 Processes

While there are several processes that can be used to impart orientation to the thermoplastic polymer, only ones that have been used and reported on will be discussed in this chapter.

Ram extrusion

Alcan[13] worked with the process used to make aluminum extrusions to make oriented profiles of polyethylene and polypropylene. They patented this work,[14–16] but the collapse of the aluminum industry on the dissolution of the Soviet empire, and Russia's dumping aluminum on the world market, forced Alcan to drastically cut back on their activities and instead of commercializing this work, they sold it to Symplastics, who also failed to commercialize it. The major limitations of this process are that it is a batch process and also that production speeds are relatively slow.

Die drawing

In an attempt to speed up the ram extrusion process and also to make it into a continuous process, die drawing was the natural evolution where the pressure used to push the billet through a die, as in ram extrusion, is replaced by a pulling force that pulls the billet through the orientation die. Ward and coworkers[6] had originally patented this work but that work was limited to pure polymer systems and not filled polymers. More recently, Ward has published his extensive work on die drawing.[17]

Radial compression

To make biaxially oriented polymer sheet, radial compression, or forging, which had been pioneered at the University of Toronto, was used to make sheet for ballistic protection applications. Initial evaluation[18–20] of sheet made was encouraging but financial support to continue the development work needed was not forthcoming and the work was discontinued.

10.2.2 Materials

Whereas most, if not all, thermoplastic polymers can be oriented, there are limitations and most work has been done with the commodity thermoplastics polyethylene and polypropylene. Because of the tendency of polyolefins to creep, where creep resistance is wanted or needed, a thermoplastic polyester, most commonly polyethylene terephthalate (PET) is used. From Gould's work,[21] Fig. 10.1 shows the relationship between strength and draw ratio for some thermoplastics. Figure 10.2 shows the relationship between strength and temperature for the same thermoplastics. Generally speaking, the higher the molecular weight of the polymer, the higher the physical property achieved upon orientation. The penalty paid, though, is that the higher the molecular weight of the polymer, the harder it is to process. Ultra-high molecular weight polyethylene (UHMWPE) is a good example of a high molecular weight member of the polyethylene family that is very difficult to process.

Polyethylene

The polyethylene family is the easiest to process because of the low melting temperature which is in the range of 125–137 °C. The disadvantage of this low melting temperature is that the oriented polymer has a low service temperature above which the orientation is lost.

Polypropylene

The polypropylene family is a little harder to process but has a higher melting temperature (in the range of 160–175 °C). This gives the oriented polymer a

Oriented wood–polymer composites and related materials 211

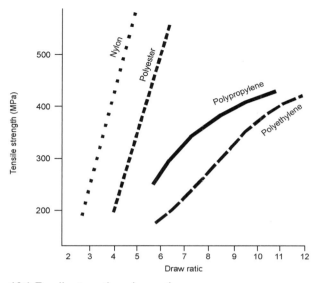

10.1 Tensile strength vs draw ratio.

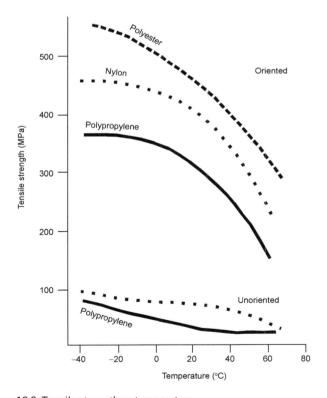

10.2 Tensile strength vs temperature.

higher service or use temperature. The temperature used for solid state orientation is 10 degrees below its melt temperature.

Polyethylene terephthalate (PET)

The thermoplastic polyester family, typified by PET, has a much higher melting temperature (in the range of 220–267 °C) and is very susceptible to hydrolysis. The major advantage of PET is the large increase in physical properties obtained upon orientation, making this a structural material.

10.3 Applications

The commercialization of oriented polymer technology was the primary intention of the work being reported. The challenge was to find applications for this technology that would transform the situation from technology-push to market-pull. Ward has been attempting this task in the United Kingdom with some success.

10.3.1 Historical development

To help the reader understand the genesis of oriented wood–polymer composites (WPCs), a chronological review of the development work leading to the discovery of producing low-density WPCs by orientation is reported.

Drumsticks

In the attempts to commercialize the oriented polymer technology pioneered at the University of Leeds, the University of Toronto, and the University of Massachusetts at Amherst, the challenge has been to find or develop applications that were economically viable. This has been a classic case of technology-push and not market-pull. This attempt to find viable applications, started in 1989, identified as a first product, drumsticks, from the music world. This application of the oriented polymer technology was attractive because the product was fairly simple to make, had no specifications to meet, had no regulations to pass, the market size was 12 million pairs per year in North America and the current many manufacturers (21) were primarily small, family-owned operations. The key test that had to be passed for customer acceptance was that the drumsticks had to feel like wood and sound like wood. The oriented polypropylene drumsticks did this and after two years of marketing them using the trademarked name EMMite™, sales increased to a respectable quantity.

The drumstick market historically is 92% made from wood, primarily hickory but also oak and maple. These are the strongest woods and have the highest density of the hardwoods at 0.8 g/cm^3. The different wood species have different

Table 10.1 Comparison of flexural strength and modulus

Material	Flexural strength (in psi)*	Flexural modulus (in psi)*
Polypropylene	7 000	270 000
Wood	14 000	1 300 000
Oriented polypropylene	40 000	1 100 000

*1 psi = 6.89 × 10^3 Pa.

densities covering a whole range but for generalities, softwoods are typically characterized with a density of 0.5 and hardwoods with a density of 0.8. The physical properties of strength and stiffness are directly related to density – the higher the density, the higher the strength and stiffness. Four pairs of our drumsticks made from oriented polypropylene were submitted to MacMillan Bloedel's research laboratory for testing and to be compared with wood. Table 10.1 summarizes the testing. The polypropylene numbers are from the polypropylene manufacturer. The oriented polypropylene had a draw ratio of 12. It can be readily seen that the orientation process increases the flexural strength of the polypropylene six-fold from one-half of that for wood to about three times that of wood. The flexural modulus is increased roughly four-fold from one-quarter of that of wood to approximately the same as wood.

In developing the oriented polymer drumsticks, one comment received from drummers was that the transparent 100% oriented polypropylene drumsticks did not *look* like wood. To address this criticism, wood fiber was added to the polypropylene which when oriented looked very much like wood with what seemed to be a grain much like wood. This was due to the orientation of the wood fiber particles in the orientation process. Because the use of wood fiber in this application was purely cosmetic, just enough was added to give a wood-like appearance. This amounted to about 8% wood fiber being added. This was the first example of an oriented wood fiber–plastic composite (OWPC) material.

One unexpected benefit that was observed when using the oriented polymer drumsticks was that there was less vibration reaching the drummer's wrists and that those drummers with carpal tunnel syndrome found relief in using the OWPC drumsticks. A preliminary study was undertaken to document this finding.[22]

Unfortunately, the company sold only 1000 pairs of drumsticks a month whereas it needed to sell 4000 pairs a month to break even. This market development attempt was finally terminated as unsustainable.

Hardwood flooring

In exploring the possible markets for this new material, different shapes of the wood fiber/plastic composite and different concentrations of wood fiber were made. The three dies that had been acquired when the technology was purchased from Symplastics were $\frac{7}{8}''$ diameter rod, a rectangular profile ($1\frac{1}{4}'' \times \frac{3}{4}''$), and a flat bar profile ($2'' \times \frac{3}{8}''$). When the flat bar profile was used in the ram extrusion process, the oriented polypropylene/wood fiber composite produced looked like hardwood flooring. A preliminary market research study on hardwood flooring indicated that this was a billion dollar per year market in North America and that Canada is primarily a softwood producer (95%) and a net importer of hardwood. Attempts were then made to make material to install as a test flooring to see if the market opportunity was worth exploring in detail. In December 1999, a test floor of 150 square feet (14 m^2) was laid by professional hardwood floor installers using flat bar material that had a tongue and groove machined on it similar to hardwood flooring. From this first trial installation, it was learned that coating in the field was not practical and that a snap-lock joint instead of tongue and groove was highly desirable.

Some preliminary testing was done to compare the properties of the synthetic hardwood with traditional hardwood flooring currently used and also laminate flooring that had recently been introduced into the flooring market. Commercially available material was obtained and the water immersion test was the first test conducted on commercial samples. These are shown in Fig. 10.3. The synthetic hardwood (SHW) had no water uptake whereas the laminate material has almost as high a water uptake as wood.[23] The polypropylene carpet manufacturers were approached to commercialize this new polypropylene synthetic hardwood flooring but there were no companies interested or able to undertake the work involved at that time.

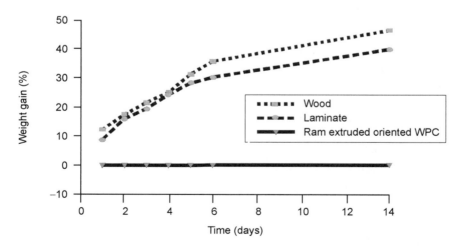

10.3 Water absorption test.

Low-density wood–polymer composites

Initial work with orientation involved ram extrusion, similar to the process used to extrude aluminum profiles. Whereas excellent physical properties were obtained,[23] the rate of producing the profile was very slow – typically 1″ to 6″/min (2.5–15 cm/min). To obtain a faster production speed, die drawing was investigated. The first material that was die drawn was 100% polypropylene. The results were excellent with similar or better physical properties being obtained compared with ram extrusion and with much faster production speeds – in the order of several feet/minute. Then, wood fiber–plastic composites were tried and a new phenomenon was observed. The die drawing process produced a low-density product. Whereas with ram extrusion with 30% wood fiber content, the OW/PP (oriented wood fiber–polypropylene) profile had a relative density of 0.98, using the die drawing process on the same starting billet, the EOW/PP (expanded oriented wood fiber–polypropylene) produced had a relative density of 0.52 – roughly half that obtained using the ram extrusion process. A totally new product had been formed – one never seen before and one that looked very much like wood. Physical testing was conducted on the new low-density material as well as the higher density OW/PP to see how the strength and stiffness compared with both wood and first generation WPCs. The first low-density material tested was the $1\frac{1}{2}″ \times \frac{5}{8}″$ (3.8 cm × 1.6 cm) rectangular profile.[24] The stress-strain graph is shown in Fig. 10.4. It shows the high-density ram extruded oriented WPC and the low-density die drawn oriented WPC, both containing 30% wood fiber as well as pine and commercially available WPC decking material (that contains 50% wood fiber). The oriented WPCs both have higher physical properties than the commercially available WPC.

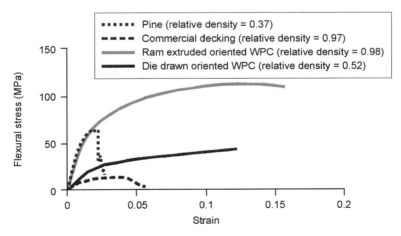

10.4 Comparison of stress–strain curves of four materials (pine, commercial decking, OW/PP and EOW/PP).

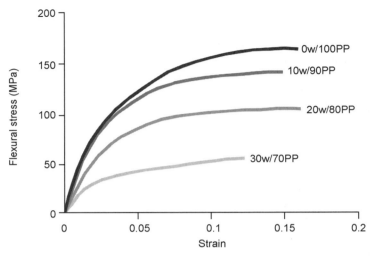

10.5 Stress–strain curves of polypropylene with various wood fiber concentrations made by the die drawing orientation process.

The stress–strain graph in Fig. 10.5 illustrates the effect of wood fiber concentration. The four curves represent wood fiber concentrations from 0 to 30%. It can be readily seen that as the wood fiber concentration increases, the strength and stiffness decreases. The relative density decreases from 0.91 for 100% PP (0% wood fiber), to 0.80 for 10% wood fiber–PP, to 0.68 for 20% wood fiber–PP, to 0.51 for 30% wood fiber–PP.

To determine what was happening, scanning electron microscope (SEM) micrographs were taken. Figures 10.6, 10.7 and 10.8 were taken by the Materials Group, Engineering Department, University of Cambridge and permission to use

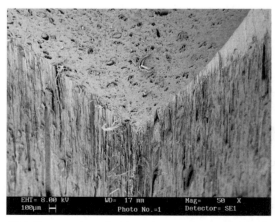

10.6 Corner view (end and two sides) of die-drawn 30% wood fiber–polypropylene with a nominal draw ratio of 8.

Oriented wood–polymer composites and related materials 217

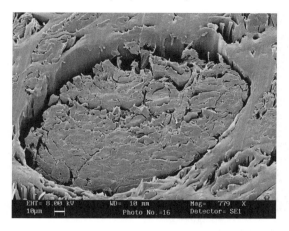

10.7 Enlargement of part of the end view of Fig. 10.6.

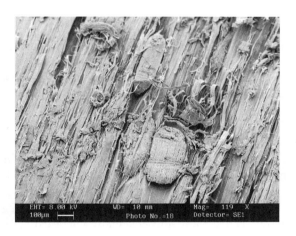

10.8 Enlargement of part of the side view of Fig. 10.6.

these pictures is acknowledged. Figure 10.6 shows the end and two sides of a rectangular section of die-drawn 30% wood fiber–polypropylene with a draw ratio of about 8. The wood fibers are all aligned and show in the end face as the cross-section of the fiber bundle. The two sides show the fibrous nature of the oriented polypropylene. Figure 10.7 is a magnified view of the end of the sample in Fig. 10.6. It shows a wood fiber bundle that has the lumens of the wood collapsed and the space between the wood fiber and the polypropylene shows the lack of adhesion between the two. Figure 10.8 is a magnified view of the side of the sample in Fig. 10.6. It shows clearly two wood fiber particles that are sitting in an elongated channel. These particles are not attached and can be readily removed from a cut surface by abrasion. The fibrils are the oriented

10.9 SEM of pine.

polypropylene and are all unidirectional. It can be seen by comparison with a SEM picture taken at the University of Toronto of the pine material used in the stress–strain testing (Fig. 10.9), the die drawn low-density oriented WPC has a structure somewhat similar to that of wood.

The next challenge that was addressed was to scale up to larger cross-sections to see if there were any changes in the physical properties obtained. From the $1\frac{1}{2}'' \times \frac{5}{8}''$ (3.8 cm × 1.6 cm) rectangular profile, the next size chosen was $1'' \times 4''$ (2.5 cm × 10 cm). To produce a low-density die drawn of this dimension with a draw ratio of 8, an initial billet with dimensions of $2'' \times 8''$ (5 cm × 20 cm) had to be extruded. This is a large solid extrusion especially with a $1\frac{3}{4}''$ (4.4 cm) diameter single screw extruder which was the R&D extruder being used. Material was made, tested, and submitted to potentially interested parties who indicated that to be interested, they would like to see the oriented WPC made on a commercially sized machine at commercial production rates. In November 2002, a demonstration was arranged in Onaga, Kansas at Onaga Composites, the suppliers of wood fiber–polypropylene concentrate. A Berstorff 84 mm single screw extruder was used to extrude the 2 × 8 billet at 1 foot/minute (0.3 m/min) which was die drawn to the 1 × 4 profile at 8 feet/minute (2 m/min) – a commercial machine operating at a commercial production rate. The response was that what was required was 5/4 board not 1 × 4. 5/4 Board is a scaled down 2 × 6. Lumber comes in nominal sizes which reflect the size of the board when initially cut. When it is dressed or finished, the size is somewhat smaller. For example a 2 × 4 is not $2'' \times 4''$ (5 cm × 10 cm) but dressed is $1\frac{3}{4}'' \times 3''$ (4.4 cm × 9.5 cm). It follows that 5/4 board which was supposed to be $1\frac{1}{4}'' \times 6''$ is really $1'' \times 5''$ (2.5 cm × 14.6 cm). The initial billet to make this profile is $2'' \times 11\frac{1}{2}''$ (5 cm × 29.2 cm) – no small challenge.

A pilot manufacturing facility, with American money, was built in Kent, Washington when Canadian money could not be found to finance this scale-up venture. This company was named PSAC LLC and was the American version of PSAC Inc. whose equipment was all moved to the United States. In September 2005, 5/4 board was successfully manufactured at 8 feet/minute (2 m/min) in this new pilot facility. This material is twice as strong and half the weight of the commercially available WPCs which have been dubbed 'first generation WPCs'. This 'second generation WPC', which is called oriented WPC, is now in the final stages of commercialization. Patents have been issued.[25–27] The first licensee is Green Forest Engineered Products of Nevada, Missouri. In October 2006, Weyerhaeuser purchased all the patents and assets of this technology from PSAC LLC and started plans for manufacturing trim, decking, railing, fencing, and siding. These products should be commercially available starting in 2009.

The Burgess Bridge Co. is a new company that has been formed to manufacture and market bridges made from oriented WPCs. The first bridge, a pedestrian bridge 31 feet (7.8 m) long, made completely from 5/4 board, was installed in the Arboretum at the University of Guelph in September 2006. The second bridge, another pedestrian bridge 28 feet (7.1 m) long, was made entirely of 1 × 4 and was installed in October 2006 in Millbrook, Ontario.

10.4 Current developments

While the scale-up work was proceeding, further work was done on developing the technology from a materials point of view.[28] From the SEM pictures (Figs 10.6–10.8), an explanation of the low-density or porosity of the WPC was postulated. There is a polarity mismatch between the wood fiber and the polypropylene. The wood fiber is polar and the polypropylene is non-polar. This is like oil and water where the two do not mix or stay together. When the composite material is pushed through the die as is the case with ram extrusion, the composite has no choice but to stay together. But when the composite is pulled through the die, as is the case in die drawing, the polypropylene does not adhere to the wood fiber and it pulls apart, creating a void. The wood fiber bundle just acts as a nucleus for this void formation. There is no reason to believe that this phenomenon is limited to wood fiber. Any polarity mismatch should do the same thing. Conversely, a polarity match should not pull apart and thereby not create voids on die drawing. This hypothesis was investigated by using other fillers.

Key variables for low density include the concentration of the filler, particle size of the filler, temperature of the composite billet prior to die drawing, uniformity of temperature of the composite billet, the pulling force applied in the die drawing process, and the rate of cooling of the composite after die drawing.

10.4.1 Materials

Inorganic fillers

The first class of fillers investigated was that of inorganic fillers. Instead of using coupling agents with the wood fiber composites, it was decided to better illustrate the adhesion between matrix and filler using a different filler. The first filler chosen was calcium carbonate, the filler with the highest usage in thermoplastics. To evaluate the value of adhesion between the filler and the matrix, a series of orientation runs were conducted using coated calcium carbonate with polypropylene. The calcium carbonate is coated with calcium stearate which acts as a coupling agent with the calcium part of the calcium stearate having an affinity for the calcium of the calcium carbonate filler and the stearate part of the calcium stearate having a compatibility with the polypropylene matrix. The first series varied the level of calcium carbonate – 10, 20 and 30%. The second series used both calcium carbonate and wood fiber – 10CC/10WF, 20CC/10WF and 10CC/20WF. The physical testing results are shown in Table 10.2. The strength and stiffness of the oriented calcium carbonate/polypropylene composites are not that much different from the values obtained for pure polypropylene. With the relative density of calcium carbonate being 2.65, the relative densities measured for the composites was lower than that calculated from the rule of mixtures. This was due to the presence of voids. The void content was calculated from the difference between the measured and calculated relative densities.

When wood fiber was added in addition to the calcium carbonate, the density, strength and stiffness were lowered as we have seen before. The void content was again calculated as has just been described. Therefore, at the 30% filler loading level, the density obtained can be changed from 1.11 to 0.87 to 0.70 with a corresponding lowering of strength and stiffness by changing the filler from

Table 10.2 Strength and stiffness data of polypropylene with various calcium carbonate levels and combinations of calcium carbonate and wood fiber made by the continuous orientation process

Material	Wood fiber (wt%)	Calcium carbonate (wt%)	Relative density	Void content (vol.%)	Maximum flexural strength (psi)*	Flexural modulus (psi)*
Polypropylene	0	0	0.92	0	18 798	560 479
10CC/PP	0	10	0.97	8	19 779	675 470
20CC/PP	0	20	1.05	11	18 651	611 398
30CC/PP	0	30	1.11	17	16 958	543 972
10CC/10WF/PP	10	10	0.88	19	15 105	419 964
20CC/10WF/PP	10	20	0.87	29	14 307	446 264
10CC/20WF/PP	20	10	0.70	37	9 447	233 098

*1 psi = 6.89×10^3 Pa.

Oriented wood–polymer composites and related materials 221

10.10 Corner view of oriented 30% calcium carbonate/polypropylene.

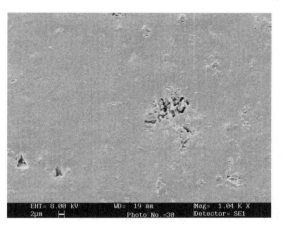

10.11 Enlargement of part of the end of Fig. 10.10.

30% calcium carbonate to 20% calcium carbonate/10% wood fiber to 10% calcium carbonate/20% wood fiber. All these compositions have 70% polypropylene. The SEM pictures show what is happening. Figure 10.11, which is an enlargement of the top of the corner view shown in Fig. 10.10, shows that the 30% calcium carbonate is largely bonded to the polypropylene. The small quantities of voids are due to cohesive failure or the separating of agglomerates not adhesive failure. In the case of wood fiber, it was the lack of adhesion that caused the void structure.

Figure 10.13 is an enlargement of the top of the corner view shown in Fig. 10.12 of the mixture of fillers used in this oriented composite. In this instance, the composite (by weight) is 20% calcium carbonate, 10% wood fiber in the oriented polypropylene. The calcium carbonate is largely bonded and the wood fiber is not. The void content is 29% (by volume).

222 Wood–polymer composites

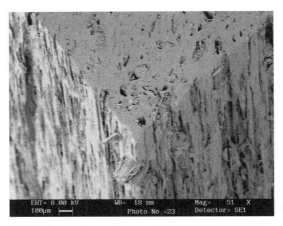

10.12 Corner view of oriented 20% CC/10% WF/PP.

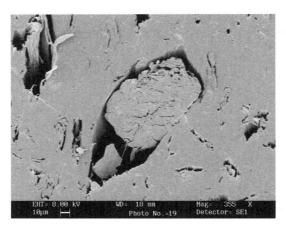

10.13 Enlargement of part of the end of Fig. 10.12.

In the physical property testing done on these composites, it was noted that on water soak tests, there was water uptake but that there was no change in physical properties and no swelling or dimensional change. It was also noted that on drying all the water left the composite. It was felt that this characteristic of these composites could be used to advantage; that chemicals could be added to the composite that would react with the filler. To test this hypothesis, the first chemical to be tried was water and hydration.

Reactive fillers

The two reactive fillers of interest in the construction market are Portland cement and gypsum. These were the first two reactive fillers to be considered. After initial work is completed on these two reactive fillers, others will be considered.

Portland cement is made from limestone, clay and sand as the primary ingredients in a rotating furnace called a kiln where temperatures reach 1500 °C (2732 °F). The intense heat causes chemical reactions that convert the partially molten raw materials into pellets called clinker. After adding some gypsum and other key materials, the mixture is ground to an extremely fine gray powder (75 μm) called 'Portland cement'. There are different types of Portland cement that are manufactured to meet various physical and chemical requirements. The American Society for Testing and Materials (ASTM) Specification C-150 provides for eight types of Portland cement. Type 1 Portland cement is a normal, general-purpose cement suitable for all uses and is the type that was used. The four major compounds in Portland cement have compositions approximating to tricalcium silicate C_3S, dicalcium silicate C_2S, tricalcium aluminate C_3A, and tetracalcium aluminoferrite C_4AF. Small variations in the lime content cause large alterations in the C_3S and C_2S contents of cements. The presence of an excess of uncombined or *free lime* must be avoided in cement clinker, since it undergoes an increase in volume during hydration, so weakening the hardened paste.

Portland cement has a relative density of 3.0 in contrast to polypropylene which has a relative density of 0.9. With this large difference, there is quite a difference between composites based on weight percent in contrast to volume percent. Weight percent is what is easily measured (and weighed) whereas the 'rule of mixtures' deals with volume percent. To illustrate how large this difference is, 40 wt% Portland cement/polypropylene is only 16.7 vol.% PC/PP. Fifty wt% PC/PP is 23.1 vol.%; and 60 wt% PC/PP is 31 vol.%. To get to 50 vol.%, 77 wt% PC/PP is needed. Die drawing experiments were done as well as free drawn PC/PP composites. Table 10.3 shows the results from free drawing 40, 50 and 60 wt% Portland cement in polypropylene. Having obtained porosity in the composite, as shown by the void content column, it was decided to attempt to hydrate the Portland cement in the oriented polypropylene. Hydration turned out to be more difficult than expected. Just soaking the composite sample in water had no effect. Elevated temperature and pressure had to be used to force water uptake. Successful hydration was obtained by placing the oriented PC/PP composite in a pressure cooker for 10 min, and then the samples were immersed in water for 3 days and then allowed to dry under normal room conditions. The

Table 10.3 Free drawn PC/PP composites

Cement (wt%)	Cement (vol.%)	Draw ratio	Relative density	Void content (%)
40	16.7	13.0	0.74	40.8
50	23.1	11.5	0.75	45.8
60	31.0	11.1	0.66	57.5

Table 10.4 Flexural strength and density of PC/OPP composites

Portland cement (wt%)	Unhydrated max. flexural strength (psi)*	Unhydrated relative density	Hydrated max. flexural strength (psi)*	Hydrated relative density
37.5	7200	0.90	8100	0.92
52.5	5600	0.85	6300	0.90
67.5	3500	0.82	4200	0.91

*1 psi = 6.89×10^3 Pa.

residual water uptake was calculated from the mass gain. Not surprisingly, the 60% PC/OPP sample absorbed more water than the 50% and that one more than the 40%.[28] Samples of the unhydrated and hydrated PC/OPP composites were then subjected to flexural testing and the testing results are shown in Table 10.4. Table 10.4 shows the increase in the maximum flexural strength and the relative density on hydration of the three die drawn PC/OPP composites. It is felt that this is a strong indication that a chemical reaction has been effected.

Agricultural fibers

Instead of using wood fiber as the filler, agricultural fibers are being investigated as a suitable filler. Already, rice husks have been successfully used as a filler in WPCs used for decking. Carney Timber Products in Barrie, Ontario is one company that uses rice husks in its manufacturing operations. There is a lot of waste material available, such as wheat straw, corn stover, and soy residue. If this waste agricultural fiber can be economically incorporated into a thermoplastic composite and produces a composite with adequate physical properties, it should be an additional source of income to farmers. A study led by the University of Guelph, and financed by the Ontario government, is currently underway to evaluate this hypothesis.

Applications

Besides the construction applications currently being manufactured (decking, railing, docks, fencing, and roofing shingles), there are new opportunities worth investigating. Siding is a major market application. James Hardie has introduced its fiber-cement siding in the 1990s and now commands 14% of the siding market. The oriented Portland cement/polypropylene composites that have been mentioned above could become a competitive product in this market. Landscape architecture is a broad market that includes decking and fencing but also adds several new market opportunities, such as gazebos, trellises, arbors, benches, and tables. Structural applications are also new market possibilities. The first oriented WPC bridge has been designed, fabricated and installed in the arboretum at the University of Guelph.

10.5 Future trends

With the discovery of a new class of low-density, high-strength materials made by die-drawing WPCs, a large number of new processes and products are possible.

10.5.1 Processes

Only one of the metal forming processes has been investigated in detail to date. There are several others worth considering. Sheet products made by rolling could possibly compete with oriented strand board (OSB). Thermoforming or compression moulding are processes used in the packaging and automotive markets that warrant investigation. Injection moulding of oriented WPCs, while challenging, could open large new market opportunities, especially in the automotive market. Hydroforming, a process used in the automotive industry to form steel parts, especially chassis frames, has been investigated at McMaster University with oriented polypropylene, where positive results in the research work has led to there being 'proof of concept'.

10.5.2 Applications

Whereas construction is the primary market being pursued by most industrial manufacturers currently, there are major opportunities in the two other large plastic markets, packaging and automotive. Plans are currently underway to explore these opportunities.

10.6 References

1. A. Ciferri, I.M. Ward (eds.), *Ultra-High Modulus Polymers*, Applied Science, Barking, Essex, England, (1979).
2. G. Capaccio, T.A. Crompton and I.M. Ward, 'The drawing behaviour of linear polyethylene. I. Rate of drawing as a function of polymer molecular weight and initial thermal treatment', *J. Polymer Sci.*, **14**, 1641–1658 (1976).
3. G. Capaccio, T.A. Crompton and I.M. Ward, 'The drawing behaviour of linear polyethylene. II. Effect of draw temperature and molecular weight on draw ratio and modulus', *J. Polymer Sci. Phys. Ed.*, **18**, 301–309 (1980).
4. A. Richardson, B. Parsons and I.M. Ward, 'Production and properties of high stiffness polymer rod, sheet, and thick monofilament oriented by large-scale die drawing', *Plastics and Rubber Processing and Applications*, **6**, 347–361 (1986).
5. G. Capaccio and I.M. Ward, 'Preparation of UHMWPE; effect of MW and MW distribution on drawing behaviour and mechanical properties', *Polymer*, **15**, 233–238 (1973).

6. G. Craggs, A.K. Powell and I.M. Ward, UK Patent Appl. GB 2 207 436 A (1989).
7. I.M. Ward and A. Selwood, United States Patent 4,801,419 (1989).
8. I.M. Ward, 'Control of strength and creep in high modulus polyethylenes', *33rd IUPAC International Symposium on Macromolecules*, Montreal (1990).
9. R.A. Panlasigui, S. DiPede, K.R. Tate and R.T. Woodhams, 'Solid phase processing of oriented plastics', Technical report for Alcan International Inc. (1988).
10. P.E. Burke, G.C. Weatherly and R.T. Woodhams, 'Uniaxial roll-drawing of isotactic polypropylene sheet', *Polymer Engineering and Science*, **27**, 518–523 (1987).
11. K.R. Tate, A.R. Perrin and R.T. Woodhams, 'Oriented polypropylene sheet by rolling-drawing', *Plastics Engineering*, **43**, 29–31 (1987).
12. R.T. Woodhams, 'Super-strong plastics', *Canadian Research*, 29–30 (1988).
13. G. Couval, L.-H. Wang and R.S. Porter, 'Properties of oriented polyethylene bars of large cross section', *Journal of Applied Polymer Science*, **52**, 1211–1215 (1994).
14. J. Francour and L. Morris, United States Patent 5,169,589 (1992).
15. G. Courval, D.R.M Thomas and S.G. Allen, United States Patent 5,204,045 (1993).
16. G. Courval, United States Patent 5,169,587 (1992).
17. I.M. Ward, P.D. Coates and M.M. Dumoulin (eds), *Solid Phase Processing of Polymers*, Hanser Gardner Publications (2000).
18. S. Dorval, M. Bolduc, D. Delfosse, W. Newson, E.M. Maine and F.W. Maine, 'The testing of oriented polyethylene for ballistic protection applications: Part A – High strain rate testing portion', *Proceedings of ICCM-10*, Whistler, BC, Canada, August 1995.
19. S. Dorval, M. Bolduc, W. Newson, E.M. Maine and F.W. Maine, 'The testing of oriented polyethylene for ballistic protection applications: Part B – The non-high strain rate testing portion', *45th Canadian Chemical Engineering Conference*, Quebec City, October 1995.
20. W.R. Newson and F.W. Maine, 'Oriented polyethylene and polypropylene sheet', *Composites '96 and Oriented Polymers Symposium*, Boucherville, October 1996.
21. R. Gould, 'Continuous roll orienting yields long-term performance values', *Modern Plastics*, 158–162, May 1988.
22. C. Zaza, M.S. Fleiszer, F.W. Maine and C. Mechefske, 'Beating injury with a different drumstick: a pilot study', *Medical Problems of Performing Artists*, 39–44 (2000).
23. W.R. Newson and F.W. Maine, 'High value products and applications for woodfibre–polymer composites', *Progress in Woodfibre–Plastic Composites–2000*, Toronto (2000).
24. W.R. Newson and F.W. Maine, 'Second generation wodfibre–polymer composites', *Progress in Woodfibre–Plastic Composites–2002*, Toronto (2002).
25. F.W. Maine and W.R. Newson, Canadian Patent 2,397,676.
26. F.W. Maine and W.R. Newson, European Patent EP 1,242,220.
27. F.W. Maine and W.R. Newson, United States Patent 6,939,496.
28. W.R. Newson and F.W. Maine, 'Advances in new WPC technologies', *Progress in Woodfibre–Plastic Composites–2004*, Toronto (2004).

11
Wood–polymer composite foams

G GUO, University of Southern California, USA,
G M RIZVI, University of Ontario Institute of Technology, Canada and
C B PARK, University of Toronto, Canada

11.1 Introduction

This chapter describes the unique advantages and applications of wood-polymer composite (WPC) foams, delineates the critical issues and processing strategies associated with foaming of WPC, and then introduces the state-of-the-art technologies used to produce the WPC foams.

11.1.1 Significance of wood–polymer composite foams

WPCs made from wood fibers and polymers have been commercialized and are enjoying rapid growth in wood-replacement applications. In addition to their usage as building products, WPCs are also used in many automotive, infrastructure and other consumer/industrial applications (Smith & Wolcott, 2006). In general, WPCs have greater durability and lower maintenance cost, improved dimensional stability, better resistance to moisture and biological degradation, lower cost (than composites with other fibers), and better recyclability. However, the scope of their use in many applications has been somewhat limited because of their low impact strength and high density as compared with natural wood. Generation of a foam-like structure in WPCs can decrease their density, improve their specific mechanical properties (such as impact strength, toughness, and tensile strength), and improve their nailing and screwing properties, all at a reduced material cost. Furthermore, the foaming of WPCs results in better surface definition, and sharper contours and corners than when they are not foamed (Schut, 2001). During production, the foamed composites run at lower temperatures and faster speeds than their un-foamed counterparts owing to the plasticizing effects of gas; consequently, the production cost is also reduced (Schut, 2001). One obvious disadvantage associated with WPC foams is that their stiffness and flexural strength are decreased owing to reduced density. This can be effectively compensated for either by producing a sandwich structure for WPCs with a foamed composite core inside and an unfoamed layer outside (e.g.

by co-extrusion), or by choosing a better ratio of polymer to wood fiber, and/or by adding a coupling agent to improve the wood fiber–polymer interfacial bonding.

WPC foams can be used in non-structural building products (decking, fencing, railings, window and door profiles, shingles, floating docks, etc.), automotive products (interior panels, rear shelves, tire covers, etc.), and numerous other industrial and consumer products. Since the mid-1990s, a continuing interest in WPC foams has been driven by a desire to decrease material costs, lower weight, improve properties, and meet strict environmental regulations on the use of chemicals in building materials (e.g. phasing out of chromated copper arsenate (CCA) treated lumber for residential use), as well as by advances in processing technology. Foamed WPCs are poised to penetrate the WPC market now worth more than a billion dollars. Many patents filed on foaming technologies for WPCs can be found: Park *et al.* (2005) filed a patent regarding foaming of WPCs; Andersen Corporation filed patents regarding manufacture of foamed WPC products (Finley, 2000, 2001); CertainTeed Corporation has one patent (Stucky & Elinski, 2002); Crane Plastics Company LLC has one patent regarding foam composite wood replacement material (Zehner, 2003).

11.1.2 Materials used in wood–polymer composite foams

All WPCs mentioned in other chapters, either thermoplastic based or thermoset based, can be foamed with chemical blowing agents (CBAs) or physical blowing agents (PBAs). They can also be foamed with heat expandable microspheres or with stretching-induced foaming mechanisms. These technologies will be discussed further in Sections 11.5–11.8. The wood used in WPCs is either in particulate form, very short fibers (wood flour), or long fibers with an aspect ratio higher than 10:1. The common species used include pine, maple, and oak. WPCs typically contain 30–70 wt% wood. However, a high wood content usually leads to processing difficulties mainly owing to increased effective viscosity and moisture/volatile emissions in foaming of WPC.

Many additives, such as coupling agents, UV stabilizers, pigments, lubricants, fungicides, nucleating agents, and flame retardants, are typically used to either improve processing or enhance the overall performance of WPC foams.

11.1.3 Processing methods for wood–polymer composite foams

The current production technologies for WPCs are all based on melt processing methods. They mainly include extrusion, injection molding, and compression molding. All these processes can be modified easily for producing WPC foams, but extrusion is the most commonly used processing method. Processing of WPC often includes two steps, i.e. compounding and product forming. In compounding, WF is mixed with a molten polymer to achieve thorough dispersion so that it is uniformly distributed in the polymer matrix. Compounding is often conducted using a batch mixer, a K-mixer, a twin-screw compounder or other mixing devices. The compounded material can be formed into granules or pellets for future processing, or can be immediately processed into a final product through a shaping die (e.g. sheet or profile dies in extrusion), a mold (injection molding), or a compression mold. Foaming typically occurs in the step of product forming. These two steps can also be combined into one process, and WPC foams can be produced in one continuous process, typically with a twin-screw compounder. Special processing equipment or method is often required to handle the blowing agents and make WPC foams. Examples would be: a proper processing temperature profile which allows CBAs to decompose fully while preventing WF degradation, or a gas injection system for PBA-based WPC foaming, or a pulling system for producing stretched WPC foams, etc.

11.2 Structure and characterization of wood–polymer composite foams

WPC foams, which contain a large number of cells, are often characterized by scanning electron microscopy to determine their cellular structures. The same characterizing terminology is often used for WPC foams as is typically used for polymeric foams, i.e., the foam density, the volume expansion ratio, the void fraction, the cell density, and the average cell size.

The foam density is one of the structural parameters that directly represent the density reduction of the un-foamed material. The foam density (ρ_f) can be calculated as:

$$\rho_f = \frac{M}{V} \qquad 11.1$$

where M and V are the mass and the volume of WPC foam sample, respectively. It can also be determined by water volume displacement of a known mass of WPC foam (ASTM C693).

The volume expansion ratio is also often used to describe the volume change of WPC foams, and it has an inverse relationship with the foam density as shown in Equation 11.2:

$$\phi = \frac{\rho_f}{\rho} \qquad 11.2$$

where ϕ is the volume expansion ratio, ρ is the density of un-foamed WPC sample, and ρ_f is the density of WPC foam sample. Another term frequently used to describe the amount of void in the WPC foams is the void fraction (V_f), defined as:

$$V_f = 1 \frac{\rho_f}{\rho} \qquad 11.3$$

The mechanical properties are generally proportional to the foam density (Landrock, 1995). It is important to determine the appropriate range of WPC foam density that is suitable for desired applications, and to control the production method to achieve it. Apart from the foam density and the volume expansion ratio, the cell geometry, such as open versus closed cell and uniformity of cell size and distribution, greatly affect the properties of WPC foams. WPC foams with a higher WF content typically display a greater open-cell structure.

The cell density, N_0, is defined as the number of cells per cubic centimeter relative to the un-foamed material. It is calculated by Equation (11.4) (Matuana et al., 1997):

$$N_0 = \left(\frac{NM^2}{a}\right)^{3/2} \left[\frac{1}{1 - V_f}\right] \qquad 11.4$$

where N is the number of bubbles in the micrograph, and a and M are the area and the magnification factor of the micrograph, respectively. From Equations 11.3 and 11.4, the average cell size d can be determined as (Matuana et al., 1997):

$$d = \left(\frac{6V_f}{\pi N_0 (1 - V_f)}\right)^{1/3} \qquad 11.5$$

Compared with the conventional foam products, microcellular plastics with a cell size less than 10 μm and a cell density higher than 10^9 cells/cm^3 demonstrate better properties. For example, the induced microcells that are smaller than the pre-existing flaws can serve as crack arrestors by offering a blunt edge to propagating crack tips, thereby greatly enhancing toughness (Suh, 1996). When properly produced, microcellular foams have an impact strength that is five times higher than that of their un-foamed counterparts (Shimbo et al., 1995; Park et al., 1998a). The fatigue life of microcellular foams is also found to be 14 times that of solid parts (Seeler & Kumar, 1993). Since microcellular plastics have demonstrated such superior properties, the research interest has shifted to producing foams with a smaller cell size and a higher cell density, such as nano-cellular foams with the cell size on the order of nanometers.

If the cell morphology of the foamed WPCs consists of a large number of uniformly distributed small cells, then the specific mechanical properties are significantly improved (Matuana et al., 1997, 1998, 2004). Matuana et al. (2004) studied microcellular foamed PVC–wood fiber composites and found that tensile and impact properties of microcellular foamed PVC–wood fiber composites were most sensitive to changes in the cell morphology and the surface modification of fibers.

11.3 Critical issues in production of wood–polymer composite foams

11.3.1 Dispersion of wood fiber for uniform morphology

To achieve consistent morphology and thus uniform property of WPC foams, it is necessary to disperse and distribute WFs uniformly in the polymer matrix. However, the difference in polarity between the hydrophobic polymer matrix and the hydrophilic WF makes it difficult to achieve this. WFs tend to adhere to one another via inter-fiber hydrogen bonding due to their polar structure. Therefore, it is typically necessary that the fibers be treated chemically and/or that external processing aids be used in order to facilitate the dispersion process. Coupling agents, or compatibilizing agents, are usually used to improve the interfacial adhesion by either forming chemical bridges between fibers and polymers or by modifying the fiber surface to make it compatible with the polymer matrix. They are crucial for both the processing and ultimately the performance of WPC foams.

11.3.2 Degradation and volatile emissions released from wood fiber

Wood has four basic constituents: cellulose, hemicelluloses, lignin, and extractives (Haygreen & Bowyer, 1996). Additionally, wood, like all living organisms, also contains water. Wood is also susceptible to degradation which can result from various factors such as plant (bacterial, fungi), animal (insects), climatic, mechanical, chemical, or thermal factors. It is manifested as changes in appearance, structure, and/or chemical composition of wood, and ranges from simple discoloration to rendering the wood completely useless for structural applications.

The production of foamed WPCs, normally involves high processing temperatures, which can easily lead to wood degradation and volatile emissions. Therefore, controlling the thermal degradation of WFs is one of the major factors in WPC production. During processing, as the temperature increases, the WF typically, experiences the following changes: evaporation of moisture (up to 100 °C), evaporation of volatile substance (95–150 °C or higher), superficial carbonization and slow exit of flammable gases (150C–200 °C), and faster exit of flammable gases (over 200 °C) (Tsoumis, 1991). Therefore, to prevent degradation of WF during WPC processing, it is necessary to lower the processing temperature, generally to below 200 °C.

During high-temperature forming processes, WF also releases moisture and other volatiles, which lead to deterioration of the cell structure of WPC foams by causing cell coalescence and cell collapse (Park et al., 1996; Rizvi et al., 2003). Rizvi et al. (2003) indicated that the large quantity of moisture in wood fiber may lead to the deterioration of the cell structure of foamed WPCs in terms of cell size, non-uniformity, and poor surface quality. Since these conditions can cause poor mechanical properties of the foamed composites, the removal or reduction of moisture/volatiles from WF becomes a critical issue for improvement of these properties. Thermogravimetric analysis (TGA) studies of WF indicated that at WPC processing temperatures, most of the volatiles were released from the extractives. Dried WF, undried WF, and extractive-free WF (with extractives removed using acetone extraction) were used and compared in the study. The results confirmed that removal of the adsorbed moisture from WF results in a better cell morphology. However, it seemed that some gaseous emissions released from WF are soluble in plastic and thereby favorably contribute to the development of the cell morphology, if proper processing conditions are used.

In order to yield a stable fine-celled morphology, the volatile emissions from the WF during the final stages of foam processing should be reduced to a bare minimum. This can be achieved by initially adopting any of the standard drying techniques, such as online devolatilization (Rizvi *et al.*, 2003; Zhang *et al.*, 2004), oven drying, hot air convective drying, drying in a K-mixer (Haygreen & Bowyer, 1996), and the like. Then, during the subsequent processing stages, the temperature should be maintained below the drying temperature (Rizvi *et al.*, 2003; Guo *et al.*, 2004a).

11.3.3 Increased melt viscosity

When the solid WF particles are introduced into the molten polymer, the viscosity of the mixture is increased. This effect becomes increasingly significant as the WF content increases. Guo *et al.* (2004b) examined the non-Newtonian behaviors of WPCs with different pinewood WF contents (0%, 10%, 30%, and 50%) with a rotational rheometer. The viscosity increased significantly as the WF content was raised (Guo *et al.*, 2004b). Li & Wolcott (2005) studied the shear and extensional flow properties of the melts of high-density polyethylene (HDPE)–maple composites with a capillary rheometry. It was found that both shear and extensional viscosities increase with WF content (Figs 11.1 and 11.2) but the filler content dependence is not as significant as for suspensions of inorganic fillers at similar filler loadings. The coupling agent, maleated polyethylene, played a role as an internal lubrication in both shear and extensional flow (Li & Wolcott, 2005). For extensional flow in convergent geometries, it provides lubrication between the wood particles and matrix polymer by enhancing the orientation of filler particles. During shear flow, the coupling agent also behaves as an internal lubricant as it reduces the yielding tendencies (Li & Wolcott, 2005).

The increased melt viscosity due to the presence of WF not only causes the increase of processing pressure, but also makes the dispersion of WF in the polymer matrix more difficult. In turn, this necessitates that special adjustments be made to the equipment and die design. The processing pressure can be reduced by one of the following three options: increasing the processing tem-

Wood–polymer composite foams   233

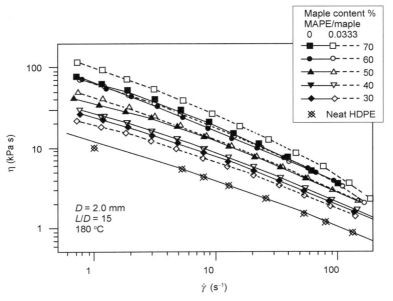

11.1 Shear viscosity of HDPE filled with different weight percentage of 40 mesh maple with and without maleated polyethylene (T.Q. Li *et al.*, Rheology of wood plastics melt. Part 1. Capillary rheometry of HDPE filled with maple, *Polymer Engineering and Science*, **45** (4), 549, 2005).

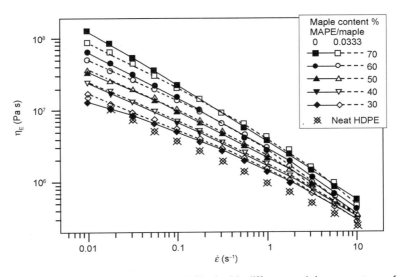

11.2 Extensional viscosity of HDPE filled with different weight percentage of 40 mesh maple with and without maleated polyethylene. (T.Q. Li *et al.*, Rheology of wood plastics melt. Part 1. Capillary rheometry of HDPE filled with maple, *Polymer Engineering and Science*, **45** (4), 549, 2005).

perature, using a higher melt index (MI) material, or employing a lubricating additive. However, increasing the processing temperature has its limitations. At higher temperatures, a larger amount of volatile emissions is released from the WF, which subsequently deteriorates the cell structure of WPC foams. Moreover, processing temperature should be kept to a minimum to prevent the temperature-induced WF degradation. WPC manufacturers tend to keep the WF content as high as possible to obtain maximum reduction in material costs. Thus, using a higher MI resin or using a lubricant may be a viable strategy for increasing WF loading while keeping the processing temperature low.

11.3.4 Critical processing temperatures and control of residence time

In order to minimize the deteriorating effects of volatile emissions from WF on the WPC foam morphology, the processing temperature should be kept to a bare minimum. However, lowering the temperature causes an increase in the 'apparent' viscosity of the molten mixture of plastic and WF, and thereby poses greater processing challenges. The determination of an optimum processing temperature – one that does not compromise satisfactory processing conditions – is therefore crucial to ensuring the formation of acceptable cellular structures (Guo *et al.*, 2004a).

The amount of volatiles released from WFs is influenced by both temperature and the exposure time at this temperature. Figure 11.3 (Guo *et al.*, 2004a) shows TGA thermograms of WF at different isothermal temperatures. As the TGA gives the time-dependent rate of weight loss at each condition, it can be used to predict the amount of volatile emissions that may be released from the WF during the WPC foam processing. Guo *et al.* (2004a) proposed a method of estimating the emissions from WF during an extrusion process by using TGA data. It also can be extended to other WPC processes.

Though influenced by temperature and time, the emissions from WF are primarily governed by the highest processing temperature. Studies were conducted (Guo *et al.*, 2004a) to determine the critical processing temperature in a tandem extrusion system, as shown in Fig. 11.3, above which the excessive volatile emissions would prevent the formation of a uniformly distributed fine-celled structure. The drying method employed was on-line devolatilization, which occurred through a vent. They did not use any blowing agent, and the resultant foam structure was only generated by the WF emissions that were given off during processing.

The experimental results show that the highest processing temperature of plastic–WF composites should be minimized, preferably below 170 °C, to avoid the adverse effects of the volatiles generated from the WF during processing. The cell morphology above this temperature was visibly irregular, and at lower temperatures, the foaming effect of volatiles from WF was insignificant.

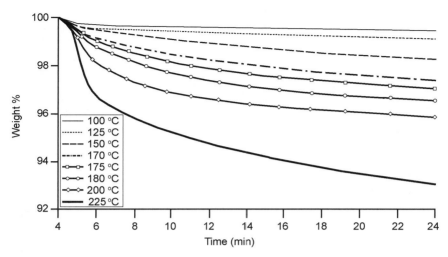

11.3 TGA thermograms of WF at isothermal temperatures (G. Guo, G.M. Rizvi, C.B. Park, and W.S. Lin, *J. Appl. Polym. Sci.*, **91**, 621, 2004).

Therefore, it would be desirable to process the foamed WF composite materials at as low a temperature as possible and with as short a residence time as possible, to avoid the adverse effects of volatile emissions.

11.4 Fundamental mechanisms in blowing agent-based foaming of wood–polymer composites

This section introduces the fundamental mechanisms in blowing agent-based foaming of WPCs. The ideal phase changes in foaming of WPC are presented, to facilitate understanding of WPC foaming process and its requirements. This also provides some insight into WPC foam processing system design, conventional system modification, and control of processing conditions, to be able to produce WPC foams with a fine-celled structure and the desired density reduction in a cost-effective way.

11.4.1 Phase changes in foaming of wood–polymer composites

Foaming of WPCs with either CBAs or PBAs can be implemented in various processes such as extrusion, injection molding, or compression molding. The fundamental mechanisms for various processes are common. Since extrusion foaming of WPCs is most widely used in industry, the phase changes in extrusion processing are explained. For illustration, Fig. 11.4 schematically depicts the morphological changes that should ideally occur in a gas–WF/plastic system during a continuous extrusion process for producing foamed WPC. Initially, the WF and the plastic (or other additives such as coupling agents, nucleating

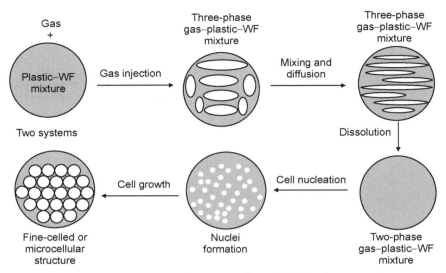

11.4 Ideal morphological change of gas–WF plastic system in extrusion foaming.

agents, UV stabilizers, colorants, etc.) are mixed and compounded. This mixture is fed into an extrusion system and brought to a molten state. A soluble amount of the blowing gas is then introduced into the WF/plastic melt stream in the extruder, forming a three-phase plastic/gas/WF mixture. The gas bubbles are broken into smaller bubbles and stretched through shear mixing. This increases the surface area through which gas can diffuse into the polymer matrix and also shortens the diffusion distance and consequently expedites the gas dissolution in the plastic. Eventually, at elevated temperature and pressure (i.e. above the saturation pressure), the gas will diffuse and dissolve into the plastic matrix to create a two-phase plastic–gas/WF mixture with a uniform gas concentration. The temperature of the mixture is then decreased, preferably without lowering its pressure, to increase its melt strength. Finally, it is subjected to a rapid pressure drop in the nucleating die, the solubility of gas in the plastic decreases rapidly, resulting in a thermodynamic instability, which becomes the driving force for creating a large number of nuclei homogeneously distributed in the plastic matrix. The dissolved gas diffuses into these nuclei, causing them to grow under the reduced pressure and temperature. The temperature of the melt is optimized, so that the melt strength of the polymer is sufficient to prevent the rupture of cell walls as the cells grow and the walls become thin, while at the same time it is not so stiff as to prevent the bubble from growing. The nucleated bubbles continue to grow under low pressure, and finally a fine-celled foam structure is formed in plastic–WF composites.

To sum up these morphological changes, there are three essential processing mechanisms involved: a mechanism for completely dissolving a soluble amount of a blowing gas into a polymer, under a high processing pressure; a mechanism for inducing a thermodynamic instability in the homogeneous polymer/gas solution formed earlier; and a mechanism for controlling the growth of bubbles, while preventing them from coalescing and collapsing.

11.4.2 Important mechanisms in foaming of wood–polymer composites

This section will describe the three important mechanisms in foaming of WPCs with blowing agents.

Polymer/gas solution formation

The formation of a uniform single-phase plastic/gas solution is essential in continuous fine-celled WPC foam processing. It is the dissolution of a large amount of gas in polymer that provides an opportunity for suddenly changing the solubility, which is the driving force for microcellular nucleation. The gas does not dissolve in wood, which retains its solid phase and is not plasticated during processing (Matuana *et al.*, 1997). This limits the amount of gas that can be dissolved in the mixture and utilized for homogeneous nucleation. Moreover, the interfaces of the solid WF particles and the plastic matrix provide a dominant force for heterogeneous nucleation (Throne, 1996).

A couple of factors that influence the dissolution and thereby the nucleation should be considered. Introducing a proper amount of blowing agent in WPC melt is critically important, since undissolved gas can generate undesirably large voids in the final WPC foams. This generally requires that the amount of injected gas should be below the solubility limit of gas in the polymer. However, this may not be true, as the addition of WF in the polymer may affect the solubility. It was reported (Guo *et al.*, 2007a) that the addition of WF increased the apparent solubility of gas in WPCs. This could be hypothesized as follows. The addition of WF in polymer matrix introduces some void at either the interfacial area between WF and polymer matrix, or within the WF itself (i.e. hollow voids in the cellular wood structure which do not completely collapse during processing). The void volume can accommodate more gas molecules and, thereby, 'apparently' increase the solubility.

For hydrophobic polymers such as PP and PE, the interfacial regions between the WF and the polymers are not wetted. The 'inter-phases' may provide channels for fast gas movement (Matuana *et al.*, 1996) and, consequently, the apparent/effective diffusion is enhanced. This holds true even when a coupling agent is employed to provide linkages between the hygroscopic WF and the hydrophobic polymer chains. The linkages only act locally and do not cause wetting of the entire fiber by the polymer and, thus, the channels for gas flow are not blocked. However, for polymers that wet the WF completely, the WF strands hinder the diffusion process and thereby increase the diffusion distance and diffusion time for complete dissolution of gas into polymer.

Cell nucleation

The driving force for microcellular nucleation is the thermodynamic instability induced by a sudden drop in the gas solubility. In extrusion, this thermodynamic instability of gas/polymer solution can be achieved through a rapid-pressure-drop-rate die, which initiates nucleation of a large number of microcells (Park & Suh, 1996). During foaming of WPCs, a large number of solid WF particles are present in the polymer matrix, which makes it very unlikely that homogeneous nucleation would dominate over heterogeneous nucleation. Heterogeneous nucleation can occur, either due to an increase in the free energy of the system caused by reduced surface tension at the interface of the molten polymer and the solid WF, or due to the entrapped gas in the micro-voids at the interfaces (Throne, 1996).

Another factor affecting cell nucleation is that WF particles continuously release volatiles at the elevated processing temperature. These volatiles contain H_2O, CO_2, and other constituents (Mohanty & Misra, 1995; Pan & Richards, 1990; Scheirs *et al.*, 2001; Orfao *et al.*, 1999). Although CO_2 is soluble in polymer, H_2O has very low solubility and no solubility data are available for the other volatiles. The presence of undissolved H_2O and possibly other volatiles is also bound to reduce the homogeneous nucleation potential of the processing system.

If the system pressure is high enough, i.e. higher than the solubility pressure of all the volatiles, and if the processing time is long enough, then all the volatiles may dissolve in the polymer matrix. However, cell nucleation will still occur at the boundary of WF or the extractive residues. It is still possible to generate a large number of nuclei by ensuring extensive dispersion and uniform distribution of the WF particles or extractive residues.

Cell growth control

After the cells have been nucleated at or near the extrusion die exit, they will continue to grow as long as gas is still available in the polymer matrix to diffuse into the nucleated cells, and the cell walls/membranes are pliant enough to sustain expansion. The cell growth stops when all the gas dissolved in the plastic matrix is depleted or the polymer matrix is crystallized and becomes too stiff to allow further growth. The cell growth mechanism is governed by a number of kinetic and system parameters such as the viscosity, the diffusion coefficient, the gas concentration, and the number of nucleated bubbles.

There are two critical issues involved in cell growth: *cell coalescence* and *cell collapse*. When a large number of nuclei are generated in a continuous process, they begin to grow very quickly owing to the fast diffusion of gas from the plastic matrix into the cells at high temperatures. In addition, the plastic matrix is soft at the elevated temperature, thus offers lower resistance against the cell growth. When the expanding cells come in contact with each other, adjacent cells tend to coalesce because the total free energy will be reduced through coalescence of cells (Klempner & Frisch, 1991; Adamson, 1990). Cell rupture is promoted when the stretched thin cell membrane separating two cells is not strong enough to sustain the tension developed during cell growth. It should be noted that the shear field generated during the shaping process also tends to stretch the nucleated bubbles, so accelerating cell coalescence further. When cells are coalesced, the initial cell population density is deteriorated. The other critical issue, cell collapse, is caused by the gas escaping into the atmosphere, owing to high diffusivity when the plastic melt is at a high temperature. In other words, most of the dissolved gas tends to diffuse out into the atmosphere rather than diffusing internally into the nucleated cells. Park *et al.* (1996, 1998b) developed a process that can prevent cell coalescence and gas escape in the cell growth stage. Cell coalescence was suppressed by cooling the plastic/gas solution homogeneously to increase the melt strength, and gas escape was controlled by cooling the surface of the extrudate to form a solid skin layer.

In summary, the factors to be considered for achieving uniformly controlled cell growth are: (1) use of appropriate amount of blowing gas (below the solubility limit), (2) minimize the diffusion of gas out to atmosphere, to prevent cell collapse and improve the blowing agent efficiency, and (3) suppress cell coalescence due to rupturing of cell walls. All these conditions are achieved through proper control of material and processing parameters (Park *et al.*, 1998b).

11.5 Foaming of wood–polymer composites with chemical blowing agents

11.5.1 Chemical blowing agents: endothermic and exothermic

Chemical blowing agents (CBAs) are thermally unstable chemicals, which, on being heated, decompose and release blowing gases at a certain decomposition temperature. Compared with PBAs, CBAs are typically more expensive, but they do not require the special processing equipment (e.g. gas injection system and storage equipment) normally needed for using PBAs. The conventional plastic processing equipment can be used with CBAs. One advantage of CBAs is that they can be used either under high pressure or at atmospheric pressure. For example, foaming in rotational molding is typically conducted at atmospheric pressure. In addition, the residues from the decomposition of CBAs can serve as a nucleating agent for improving cell nucleation. The nucleating effect is obvious in foaming of WPCs when the WF content is low (Rodrigue *et al.*, 2006).

CBAs can be divided into two major categories: endothermic and exothermic. Endothermic CBAs absorb heat and usually generate carbon dioxide (CO_2), whereas the exothermic CBAs release heat and produce mainly nitrogen (N_2) (sometimes in combination with other gases) during their chemical decomposition reactions. Compared with CO_2, N_2 is a more efficient blowing gas because of its slower rate of diffusion through polymers. Therefore, it is easier to control cell growth during the foaming process when N_2 is used as the blowing gas. It is worth noting that once an exothermic CBA begins to decompose, it continues to do so spontaneously until the CBA is entirely exhausted. This results in a faster decomposition over a narrower temperature range. The heat generated by exothermic CBAs may soften the polymer matrix, which favors the occurrence of cell coalescence during the foaming process, resulting in a poor cellular structure with large bubbles (Klempner & Frisch, 1991). In contrast, endothermic CBAs require additional heat to support their continuing decomposition. This helps to cool the polymer, increase the viscosity of the melt, stabilize the cellular structure, and reduce cell coalescence (Klempner & Frisch, 1991). Consequently, they have a longer decomposition time and a broader temperature range.

CBAs can be used either in their pure form or in the form of master-batch (i.e., with or without polymer carriers). It has been reported that the polymer carriers in a CBA master-batch can improve the compatibility between the CBA and the polymer matrix, which may improve cell morphology (Reedy, 2000). Exothermic CBAs have been shown to produce a smaller cell size in wood composites than endothermic CBAs do in the extrusion foaming of wood/PVC composites (Mengeloglu & Matuana, 2004). However, it has also been reported by Li & Matuana (2003) that CBA types (endothermic vs exothermic) and forms (pure or master-batch) do not affect the void fractions achieved in both neat high-density polyethylene (HDPE) and WF/HDPE composite foams. This conclusion was arrived at by studying three endothermic CBAs (BIH40, sodium bicarbonate (SB), and FP) and three exothermic CBAs (Celogen-OT, Celogen-AZNP, and EX210). Figure 11.5 shows the effect of CBA types on the void fraction of foamed neat HDPE and WF/HDPE composites with and without coupling agent. It also indicates that the use of coupling agent increased the void fraction of WPC foams. The average cell size achieved in the WF/HDPE composite foams was insensitive to the CBA contents irrespective of CBA type (Li & Matuana, 2003). Rizvi et al. (2007) employed six different CBAs (Table 11.1) in foaming of WPCs in extrusion: the results show that regardless of the CBA type, nearly all the CBAs display similar density reduction behaviors.

The following factors should be kept in mind while choosing a CBA for foaming of WPCs:

- The decomposition temperature of a given CBA should be higher than the polymer melting temperature, but lower than the degradation temperature of

Table 11.1 Properties of the chemical blowing agents used in extrusion foaming of WPCs (G.M. Rizvi, Ph.D. Thesis, University of Toronto, 2003)

Supplier	CBA	From data sheets							Measured at MPML*	
		Active ingredient (%)	Gas/CO$_2$ evolved (ml/g)	Main gas/es	Other gas/es	Onset of decomp. (°C)	Operating T (°C)	Total gas yield (ml/g)	Onset of decomp. (°C)	Peak of decomp. (°C)
Endex International	Endex ABC 2750	50	50–55	CO$_2$	H$_2$O	148	up to 315		133	155
Clariant	Hydrocerol BIH-40E	40	50–60	CO$_2$	H$_2$O, CO	140	180–210		157	165
Clariant	Hydrocerol CT 1401	70	120–130	CO$_2$, N$_2$	H$_2$O, CO, NH$_3$	140–150	190–210		149	167
Reedy International	Safoam PE-20	20	15–25	CO$_2$	H$_2$O	158	183	80	157	168
Reedy International	Safoam FPE-20	20	15–25	CO$_2$	H$_2$O	158	183	67	156	165
Reedy International	Safoam RIC-10FP	10	5–10	CO$_2$	H$_2$O	140	165		151	161

* Microcellular Plastics Manufacturing Laboratory

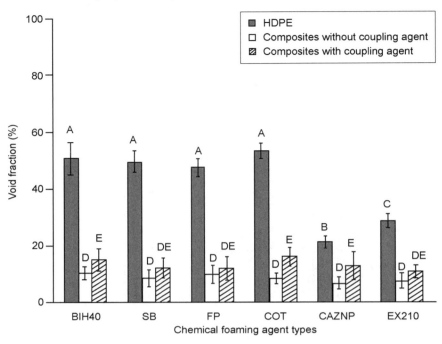

11.5 The effect of CBA types on the void fraction of foamed neat HDPE and WF/HDPE composites with and without coupling agent. The letters in the graph represent the ANOVA results. The same letter denotes that the difference between these two treatments is not statistically significant ($p > 0.1$). Otherwise, the difference is statistically significant ($p < 0.05$) (Q. Li and L.M. Matuana, *J. Appl. Polym. Sci.*, **88**(14), 3139, 2003).

WF, to ensure that the CBA will fully decompose to produce cellular structure without causing WF degradation.
- The processing temperature should also be kept as low as possible to suppress the generation of volatiles from WF.
- Therefore, one important criterion is to choose a CBA with a decomposition temperature just above the melting temperature.
- In addition to these, the factors normally considered for foaming of polymers in general, such as the rate of gas yield, gaseous composition, ease of dispersion, and cost should also be taken into account when choosing a CBA.

11.5.2 Foaming of wood–polymer composites with chemical blowing agents in extrusion

Most of the research work on extrusion foaming of WPCs has been conducted with CBAs. Mengeloglu & Matuana (2004) studied foaming of rigid PVC/WF composites with CBAs in extrusion. They examined the effects of CBA types (endothermic versus exothermic) and concentrations as well as the influence of all-acrylic processing aid on the density and cell morphology of neat rigid PVC foams and rigid PVC/WF composite foams. They found that the density reduction of foamed WPCs was not influenced by the CBA content, regardless of CBA type. They also found that exothermic CBAs produced smaller average cell sizes compared with the endothermic counterparts.

Li & Matuana (2003) studied extrusion foaming of HDPE and HDPE/WF composites by using different CBAs. The effects of the CBA contents, the CBA type, and the coupling agents on the density reduction and cell morphology were examined with a one-way analysis of variance and thermal analysis. Their studies demonstrated that the density reduction was not affected by the CBA types for both the neat HDPE and HDPE/WF composite foams. They also indicated that the use of coupling agent in extrusion foaming of HDPE/WF composites was favorable for achieving a high void fraction. Zhang et al. (2005) studied PP/WF composites foams extruded with CBAs. The effects of the coupling agent (maleated PP), WF content (20%, 30%, and 40% by weight), and blowing agent content on the cell structure were reported. It was found that the addition of a coupling agent increased the blowing agent efficiency and helped to reduce the density of the composites.

Rizvi et al. (2007) developed strategies for reducing volatile emissions during WPC extrusion with CBAs. Drying or devolatilizing at higher temperatures, increasing residence time during drying/devolatilizing stage, and limiting the highest barrel temperature during the final processing stage helped to reduce volatile emissions. To apply these strategies, they proposed a two-stage process. In the first stage PWC pellets containing a coupling agent were produced using a twin-screw compounder which was kept at 175 °C. The volatiles generated during the compounding process were purged to atmosphere during pelletizing. These pellets were then mixed with CBA master-batches and extruded in a single-screw extruder to obtain the foamed WPCs. A very fine-celled structure, with an average cell size smaller than 100 μm and a desired density range of 0.6–0.8 g/cm^3, was successfully obtained for all the CBAs used.

11.5.3 Foaming of wood–polymer composites with chemical blowing agents in injection molding

Injection molding is one of the most widely used plastic processing techniques for mass-producing complex parts which typically cannot be produced by extrusion. Molten plastic is injected at high pressure into a mold, which shapes the product. Products made by injection molding can range from very small components to large body panels of cars. WPC foams can be processed in injection molding to make lightweight complex parts.

Bledzki & Faruk (2005, 2006) produced PP-based wood composites in injection molding with different CBAs, and found that the CBA and WF content strongly affected the microcellular structure of WPC foams, and thereby their mechanical properties such as Charpy impact strength and impact resistance. It is also worth noting that foaming of WPCs reduced the odor concentration and improved surface roughness. They studied five different types of WFs (hard wood fiber, finer hard wood fiber, soft wood fiber, finer soft wood fiber, and long wood fiber) (Bledzki & Faruk, 2006) and found that the finer WFs produced better microcellular structures and the higher WF contents resulted in smaller cells at a constant CBA content. WPC foams with finer WFs also led to better physico-mechanical properties because of their finer cellular structure. Bledzki & Faruk (2004) have also compared foaming of PP-based WPCs with CBAs in both injection molding and extrusion, and found that injection molding yields better results than extrusion, and that exothermic foaming agents perform better than endothermic ones in terms of cell morphology and density reduction. Using injection molding they successfully produced PP/WF composite foams, with a density reduction of 24%, having a cell size in the range of 10–50 μm (Bledzki & Faruk, 2004).

Gosselin *et al.* (2006a, 2006b) also studied WPC foams produced in injection molding. In their studies, 0–40 wt% birch WFs were blended with a post-consumer recycled HDPE/PP matrix (85:15 ratio), and then foamed with CBAs in injection molding. Maleic anhydride–polypropylene copolymer (MAPP) was used as a coupling agent at 0–10 wt% of WF content. Their results show that both cell size and skin thickness of WPC foams increased significantly with the WF content, but were relatively unaffected by mold temperature and CBA concentration. A cell density of about 10^6 cells/cm^3 was achieved and it varied inversely with cell size. Increasing MAPP content (in the range of 0–10 wt%) reduced skin thickness and skin thickness variability, and improved WF dispersion in the melt. It was found that both the specific complex moduli in flexion and torsion increased significantly with the WF content, but were reduced slightly by foaming. The addition of MAPP (up to 1–5%) improved both the properties. The specific Young's moduli obtained in tensile tests behaved similarly, but were optimized at the WF content of 20–30%. Elongation at break was decreased with foaming and the addition of MAPP.

11.6 Foaming of wood–polymer composites with physical blowing agents

11.6.1 Physical blowing agents

PBAs are volatile liquids that undergo a change of state during processing to form a cellular structure in plastics or composites. PBAs can also be in the form of compressed gases which are injected during processing. Compared with CBAs, PBAs have the following advantages: they are more economical, yield better cell morphology, generate larger volume expansion ratio in general, and have no processing temperature limits. However, their use requires modifications of conventional equipment and is technologically more challenging. Conventional PBAs such as halocarbons (chlorofluorocarbons (CFCs) and hydrochlorofluorocarbons (HCFCs)) and hydrocarbons are subject to strict environmental regulations today. An alternative group of inert-gas blowing agents such as carbon dioxide (CO_2), nitrogen (N_2) and argon (Ar) are gaining acceptance, since they are more 'environmentally friendly' and do not deplete the ozone layer (Dey *et al.*, 1995).

11.6.2 Foaming of wood–polymer composites with physical blowing agents in a batch process

Some of the first fine-celled/microcellular WPCs were produced using a batch processing technique (Matuana *et al.*, 1997, 1998, 2004). The process involves saturation of compression-molded WPC samples with a physical blowing agent (CO_2 or N_2) in a high-pressure chamber, in which a large number of voids are nucleated by inducing a thermodynamic instability, which is affected by a pressure drop and is followed by a temperature increase to enable cell growth. The cell nucleation is governed mainly by the saturation pressure which is suddenly reduced to the atmospheric pressure, whereas the cell growth is dictated by the heating temperature and time. Thus, cell nucleation and the foam density are controlled independent of each other, which means that the foaming temperature can be set as low as possible for easy control of the cell growth stage.

A long time is needed to saturate the WPC with the PBA owing to the slow rate of gas diffusion into the polymer at room temperature. Therefore, batch processing is in general not commercially viable. But as a research method, it is extremely useful for obtaining critical processing parameters, such as foaming time, foaming pressure and temperature, and gas content. Using batch processing, microcellular foaming of WF/PVC composites has been achieved, as have ten-fold expansions in the fabrication of WPCs (Matuana *et al.*, 1997, 1998, 2004). It has also been reported that the solubility and diffusivity of gas are affected by the treatment of WF with coupling agents (Matuana *et al.*, 1997, 1998, 2004).

11.6.3 Foaming of wood–polymer composites with physical blowing agents in extrusion

Commercialization of WPCs foamed with PBAs calls for development of efficient processing systems. An innovative tandem extrusion system was developed for fine-celled foaming of WPCs using a PBA (Rizvi *et al.*, 2003; Zhang *et al.*, 2004; Park *et al.*, 2005). This system (Fig. 11.6) is capable of continuous on-line moisture removal and PBA injection. It consists of two extruders: the first twin-

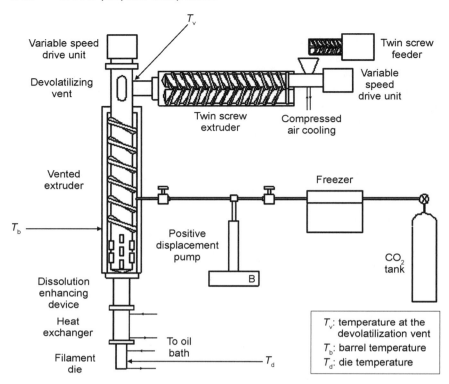

11.6 Tandem extrusion system for WPC foam processing (G. Guo, G.M. Rizvi, C.B. Park, and W.S. Lin, *J. Appl. Polym. Sci.*, **91**, 621, 2004).

screw extruder is used to compound the WF and the plastic, and the second extruder is used to foam the WPC. There is a vent at the interconnection of the two extruders that is used for on-line removal of moisture and volatiles.

Compared with CBA-based processing, PBA-based processing has no decomposition temperature requirements, reduces cost, and generally produces better cell morphology. However, proper capital investment (i.e. a gas injection system) is required. The gas injection system typically consists of a syringe pump, pressure transducers, thermocouples, freezers, high pressure valves, etc. The main features of PBA processing are that PBAs (such as N_2 or CO_2) are injected into the extruder and uniformly dispersed into the plastic matrix under high pressure using high shear and elevated extrusion temperatures. The WF/plastic–gas solution is then homogeneously cooled in a heat exchanger to increase the melt strength, which is necessary to prevent cell coalescence. Finally, the WF/plastic–gas solution passes through the die, where foaming occurs (Park *et al.*, 2002, 2005; Guo *et al.*, 2004a).

Zhang *et al.* (2004) investigated the foamability of HDPE/WF composites using a continuous extrusion process, and a fine-celled structure was obtained at a WF content (up to 40 wt%). WPCs with a fine-cell structure (i.e. cell sizes less than 100 μm) have been produced in extrusion foaming with N_2 (Guo et al., 2005). A further advantage of using PBA is that the processing window for the die temperature is larger than when CBA is used (Guo et al., 2005).

Solubility and diffusivity are important properties which affect foaming of wood composites; and a few studies have been conducted to investigate these. Doroudiani et al. (2002) studied the effects of pressure and WF content on the sorption and diffusion of CO_2 in WF/PS composites. They found that an increase in saturation pressure leads to an increase in the solubility and diffusion coefficients, whereas the addition of WFs decreases both of these properties. Guo et al. (2007a) studied the crystallization temperature, crystallinity, solubility and diffusivity for mPE, mPE/nano, mPE/WF composites, and mPE/WF/nanocomposites, as shown in Table 11.2. Figure 11.7 shows the CO_2 desorption curves of the samples. It indicates that the crystallinity of the mPE/WF/nanocomposites varied significantly with the WF content and the clay content. The solubility was well correlated to the crystallinity. The addition of clay did not significantly change the diffusivity of CO_2 in the composites.

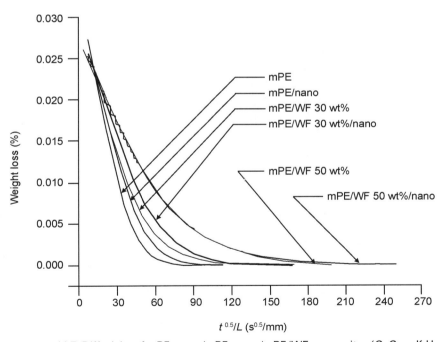

11.7 Diffusivity of mPE, nano/mPE, nano/mPE/WF composites (G. Guo, K.H. Wang, C.B. Park, Y.S. Kim, and G. Li, 'Effects of nano-particles on density reduction and cell morphology of extruded mPE/wood-fiber/nano composites', J. Appl. Polym. Sci., **104**, 1058–1063, 2007).

Table 11.2 Crystallization temperature, crystallinity, solubility and diffusivity of mPE, mPE/nano, and mPE/WF/nanocomposites (G. Guo, K.H. Wang, C.B. Park, Y.S. Kim, and G. Li, 'Effects of nano-particles on density reduction and cell morphology of extruded mPE/wood-fiber/nano composites', *Journal of Applied Polymer Science*, **104**, 1058–1063, 2007.

Sample	T_c (°C)	Crystallinity of composites (%)	Crystallinity of polymer (%)	Solubility (g-gas/g-composites)	Solubility (g-gas/g-polymer)	Diffusivity (cm²/s)
mPE	86.65	34.56	34.56	0.03092	*0.03092*	11.7 × 10⁻⁷
mPE/nano(5%)	91.33	32.18	33.87	0.03063	*0.03222*	8.0 × 10⁻⁷
mPE/30%WF	87.98	28.47	40.67	0.02891	*0.04130*	6.3 × 10⁻⁷
mPE/30%WF/nano(3.5%)	91.76	20.43	30.72	0.02847	*0.04217*	4.8 × 10⁻⁷
mPE/50%WF	88.58	15.12	30.24	0.02756	*0.05512*	2.6 × 10⁻⁷
mPE/50%WF/nano(2.5%)	89.27	10.95	23.05	0.02626	*0.05528*	2.8 × 10⁻⁷

Note: Italic data are calculated based on the measured data.

11.7 Foaming of wood–polymer composites with heat expandable microspheres

Thermoplastic microspheres are small spherical particles, consisting of a polymer shell encapsulating a gas. Upon heating, the outer themoplastic shell softens and the gas inside the shell expands, resulting in a dramatic increase in the volume of the microsphere. The particle size of the microspheres ranges from 10 to 32 μm in their unexpanded form. When fully expanded, the microspheres will expand approximately four times in diameter or up to 60 times in volume (Ahmed, 2004). There are two types of microspheres: unexpanded and expanded microspheres. Unexpanded microspheres are used as blowing agents in many areas such as printing inks, paper, textiles, polyurethanes, PVC. The expanded microspheres are used as lightweight fillers in various applications (e.g., fiber-reinforced composites, syntactic foams).

The expandable beads are used as blowing agents in many foaming applications. Ahmed (2004) investigated the use of the hollow thermoplastic microsphere (TMS) as a foaming agent for producing WPC foams in extrusion. He found that about 3% TMS could lead to a density reduction of 38% for PP-based WPCs with the WF contents of 20–30%. Flexural modulii increased with an increase of WF content, and it nearly doubled at the 50% WF loading compared with that of pure PP. It is also reported that the microspheres in the WPC also behave like reinforcing additive (Ahmed, 2004). Figure 11.8 shows SEM micrographs of PP/50%WF composites foamed with 5% TMS. A number of WPC processors in the United States have successfully used heat expandable microspheres as a foaming agent in their commercial products (Ahmed, 2004).

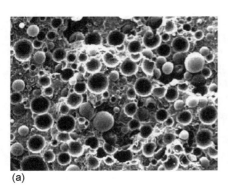
(a)

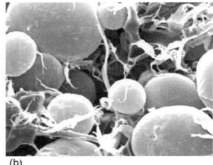
(b)

11.8 SEM micrographs of PP/50%WF composites foamed with 5% thermoplastic microspheres: (a) magnification ×29; (b) magnification ×100 (M. Ahmed, 'Thermoplastic microspheres as blowing agents for wood filled plastics', *Wood-Plastic Composites 2004*, Vienna, Austria, September 14–16, 2004).

11.8 Void formation in wood–polymer composites using stretching technology

Maine & Newson (2005) and Newson & Maine (2002) have recently developed a technology for orienting the polymer molecules and WFs by stretching, which also induces the generation of voids at the WF and polymer matrix interface. The advantages of oriented WPC are that they demonstrate: (1) almost the same texture as wood; (2) the same density range as natural wood; (3) excellent nailing and screwing abilities; and (4) better properties than un-oriented WPCs (Maine & Newson, 2005; Newson & Maine, 2002). Plastic deformation of polymeric materials results in orientation of molecular chains, which, in most cases, leads to an increase in the material's toughness and strength (Ciferri & Ward, 1979; Zachariades & Porter, 1983). A high degree of orientation of the polymer chain can be achieved by special processing techniques, such as ram extrusion (piston-cylinder), hydrostatic extrusion, and die drawing (solid-state extrusion) (Coats & Ward, 1980, 1981; Coats et al., 1980). Since the temperature impacts chain mobility, the processing temperature in these procedures is usually maintained above the glass transition temperature but below the melting temperature. In this temperature range, the chains have limited mobility, and once stretched, they cannot curl up again. However, low temperature extrusion is very difficult to implement.

The WF dispersion in the plastic matrix is facilitated by using coupling agents, which also improve the mechanical properties of WPCs by strengthening the interfacial bonding between the fibers and the matrix. The molecular orientation that occurs as a result of stretching also acts to improve the mechanical properties of WPCs. The stretching ability of WPCs is affected by their composition, the use of coupling agents, and the processing conditions. When the WF content increases above 40%, the stretching technology becomes difficult to implement, probably owing to the increased brittleness of WPCs when the WF content is high. Kim et al. (2004a, 2004b) investigated the stretched WF/PP composites with 30% WF in a batch process. A schematic of a WPC processing system is shown in Fig. 11.9. The three stages of WPC processing are melt-blending, extrusion and sizing, and stretching. The stretched WF/PP composites achieved approximately 30% density reduction, and the tensile strength and the elongation at break of the composites were increased five times.

11.9 Effects of additives on wood–polymer composite foams

Use of various additives in WPCs is often required for enhancing performance, improving aesthetics, reducing cost, improving processability and productivity (Sherman, 2004). Lubricant packages are used to improve fiber dispersion,

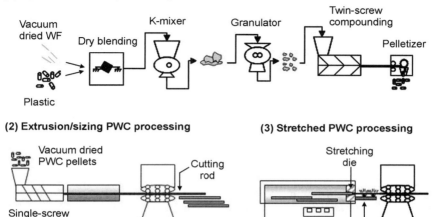

11.9 Schematic stretching of WPCs (Y.S. Kim, G. Guo, K.H. Wang, C.B. Park, and F.W. Maine, *SPE ANTEC Technical Papers,* Paper #809, May 16–19, 2004).

reduce melt viscosity, and improve throughput, and thus reduce manufacturing cost. Wood contains naturally soluble materials. They come to the surface and cause discoloration. Colorants are widely used in WPCs to prevent staining and providing color stability. The use of colorants and lubricants is well established. As durability and strength requirements become more critical, other additives such as antimicrobials, ultraviolet light (UV) stabilizers, and coupling agents are also used. Antimicrobials are used to maintain appearance and physical property. They are effective against all forms of fungi, various molds, and other microorganisms. Plastic in WPCs generally ages rapidly under the effects of light, oxygen, and heat. UV radiation can break down the chemical bonds in a plastic, a process called photodegradation. This leads to loss of strength and discoloration. The use of UV stabilizers can improve the performance of WPCs and extend their lifetime. Coupling agents, or compatibilizers, are used to improve the interfacial bonding between WF and the plastic, as well as the dispersion of WF in the plastic matrix.

Additive formulations for WPCs are getting more sophisticated to meet all these requirements. However, the interactive effects of additives have not been well studied, especially for foaming of WPCs. The following sections introduce some effects of additives on WPC foams.

11.9.1 Talc

Talc, as an inorganic filler, is often added to WPCs to enhance the modulus of elasticity and modulus of rupture, increase heat distortion temperature, reduce weight gain and swell in water, reduce warpage, etc. (Clark & Noel, 2005). In WPC foam processing, talc, as a nucleating agent, is also added to enhance the bubble-nucleating capability. However, when the WF content increases, the effects of talc on cell morphology of WPC foams are not significant (Guo et al., 2006). In this case, it functions more as a reinforcing filler rather than as a nucleating agent.

11.9.2 Nano-particles

The use of nano-sized layered silicate particles (e.g. clay) to reinforce polymers has drawn a great deal of attention in recent years. Adding a small amount of clay can dramatically improve a wide variety of properties of the polymer/clay nanocomposites, including the tensile modulus and strength, flexural modulus and strength, thermal stability, flame retardancy, and barrier properties while maintaining impact strength and allowing for recycling (Giannelis, 1996; Wang et al., 2002). Guo et al. (2007a) introduced nano-sized clay into WPC, and demonstrated that the addition of nano-clay facilitated the foam expansion and improved the cell morphology. Flame-retardant abilities of WPC foams were also improved by adding nano-clay (Guo et al., 2007a, 2007b). It is indicated that achieving a higher degree of exfoliation of nano-clay is key to enhancing the flame retarding properties of nanocomposites when a very small amount of clay (i.e., less than 1%) is used. Gilman et al. (2000) also reported that montmorillonite (MMT) clay must be nano-dispersed to enhance the flame-retarding properties of nano-composites.

11.10 Summary and future trends

Foaming of WPCs offers unique advantages such as weight reduction, material cost savings, mechanical property enhancement (i.e. impact strength, toughness), etc. The physical and mechanical properties of WPC foams are closely related to their cellular structure. Incorporating a fine-celled structure into WPCs typically is more desirable. However, a number of factors influence WPC foam processing when WF is used. Increasing the WF content increases the apparent viscosity, which necessitates either the use of a higher melt flow index (MFI) resin or the use of a lubricating additive. Use of coupling agents facilitates the dispersion and bonding of hydrophilic WF with hydrophobic resins. Suppression of the volatile emissions from the WF can be achieved by drying the WF, but this places restrictions on the highest allowable processing temperatures, in conjunction with minimizing the residence times at these temperatures. A uniform fine-celled WPC structure can be obtained if all these factors are controlled. Active research is being carried out to develop new technologies for producing WPC foams with higher performance.

11.11 References

Adamson, A W (1990), *Physical Chemistry of Surface*, John Wiley International, New York.
Ahmed, M (2004), 'Themoplastic Microspheres as Blowing Agents for Wood Filled Plastics', *Wood-Plastic Composites 2004*, Vienna, Austria, September 14–16, 2004.
ASTM (2003), ASTM C693-93 *Standard Test Methods for Density of Glass by Buoyancy*, ASTM International.
Bledzki, A K & Faruk, O (2004), 'Extrusion and Injection Moulded Microcellular Wood Fibre Reinforced Polypropylene Composites', *Cellular Polymers*, 23(4), 211–227.
Bledzki, A K & Faruk, O (2005), 'Microcellular Wood Fiber Reinforced PP Composites: Cell Morphology, Surface Roughness, Impact, and Odor Properties', *Journal of Cellular Plastics*, 11(41), 539–550.
Bledzki, A K & Faruk, O (2006), 'Microcellular Injection Molded Wood Fiber-PP Composites: Part I – Effect of Chemical Foaming Agent Content on Cell Morphology and Physico-mechanical Properties', *Journal of Cellular Plastics*, 1(42), 63–76.
Ciferri, A & Ward, I M (1979), *Ultra-High Modulus Polymers*, Applied Science Publishers, London.
Clark, R J & Noel, O (2005), 'Recent Advances in Talc-Reinforced Wood-Plastic Composites', *8th International Conference on Woodfibre-Plastic Composites*, Madison, Wisconsin, May 23–25, 2005.
Coats, P D & Ward, I M (1980), 'Neck Profiles in Drawn Linear Polyethylene', *J. of Material Sci.*, 15(11), 2897–2914.
Coats, P D & Ward, I M (1981), 'Die Drawing: Solid Phase Drawing of Polymers Through a Converging Die', *Polym. Eng. Sci.*, 21(10), 612–618.
Coats, P D, Gibson, A G & Ward, I M (1980), 'An Analysis of the Mechanics of Solid Phase Extrusion of Polymers', *J. of Material Sci.*, 15(2), 359–375.
Dey, S K, Jacob, C & Xanthos, M (1995), 'Inert-Gas Extrusion of Rigid PVC Foam' *SPE ANTEC Tech. Papers*, 4138–4143.
Doroudiani, S, Chaffey, C E & Kortschot, M T (2002), 'Sorption and Diffusion of Carbon Dioxide in Wood-fiber/polystyrene Composites', *Journal of Polymer Science Part B: Polymer Physics*, 40(8), 723–735.
Finley, M D (2000), US Patent No. 6054207.
Finley, M D (2001), US Patent No. 6342172.
Giannelis, E P (1996), 'Polymer Layered Silicate Nanocomposites', *Adv. Mater.*, 8(1), 29–35.
Gilman, J W, Jackson, C L, Morgan, A B, Harris, R H, Manias, J E, Giannelis, E P, Wuthenow, M, Hiltion, D & Philips, S H (2000), 'Flammability Properties of Polymer-layered-silicate Nanocomposites. Polypropylene and Polystyrene Nanocomposites', *Chem. Mat.*, **12**, 1866–1873.
Gosselin, R, Rodrigue, D & Riedl, B (2006a), 'Injection Molding of Postconsumer Wood-Plastic Composites I: Morphology', *Journal of Thermoplastic Composite Materials*, 19(6), 639–657.
Gosselin, R, Rodrigue, D & Riedl, B (2006b), 'Injection Molding of Postconsumer Wood-Plastic Composites II: Mechanical Properties', *Journal of Thermoplastic Composite Materials*, 19(6), 659–669.
Guo, G, Rizvi, G M, Park, C B & Lin, W S (2004a), 'Critical Processing Temperature in Manufacture of Fine-Celled Plastic/Wood-fiber Composite Foams', *J.of Appl.Polym.Sci.*, 91, 621–629.

Guo, G, Wang, K H, Rizvi, G M, Kim, Y S & Park, C B (2004b), 'Effects of Wood-Fiber Content on the Density and Morphology of mPE/Wood-fiber Composites', *PPS-20*, Paper #105, Akron, Ohio, June 20–24, 2004.

Guo, G, Lee, Y H, Rizvi, G M & Park, C B (2005), 'PBA-Based Extrusion Foaming of HDPE/Wood-Fiber Composites', SPE, *ANTEC, Technical Papers*, Paper #102038, Boston, MA, May 1–5, 2005.

Guo, G, Rizvi, G M, Lee, Y H & Park, C B (2006), 'Morphology in WPC during Extrusion Foaming with N_2', Society of Plastics Engineers, *ANTEC, Technical Papers*, Paper # 103795, Charlotte, NC, May 7–11, 2006.

Guo, G, Wang, K H, Park, C B, Kim, Y S & Li, G (2007a), 'Effects of Nano-Particles on Density Reduction and Cell Morphology of Extruded mPE/Wood-Fiber/Nano Composites', *Journal of Applied Polymer Science*, 104, 1058–1063.

Guo, G, Park, C B, Lee, Y H, Kim, Y S & Sain, M (2007b), 'Flame Retarding Effects of Nano-Clay on Wood-fiber Composites', *Polymer Engineering and Science*, 47(3), 330–336.

Haygreen, J G & Bowyer, J L (1996), *Forest Products and Wood Science: An Introduction* Iowa State University Press, Ames.

Kim, Y S, Guo, G, Wang, K H, Park, C B & Maine, F W (2004a), 'Processing/Structure/Property Relationships for Artificial Wood Made from Stretched PP/Wood-fiber Composites', *SPE ANTEC Technical Papers*, Paper #809, May 16–19, 2004.

Kim, Y S, Guo, G & Park, C B (2004b), 'Effects of the Coupling Agent on the Structure and Property of Stretched PP/Wood-fiber Composites-Based Artificial Wood', *Foams 2004*, Wilmington, DE, October 5–6, 2004.

Klempner, D & Frisch, K C (1991), *Handbook of Polymeric Foams and Foam Technology*, Hanser Publishers, New York.

Landrock, A H (1995), *Handbook of Plastic Foams: Types, Properties, Manufacture, and Application*, Noyes Publications, Bracknell.

Li, Q & Matuana, L M (2003), 'Foam Extrusion of High Density Polyethylene/wood-flour Composites using Chemical Foaming Agents', *J. Appl. Polym. Sci.*, 88(14), 3139–3150.

Li, T Q & Wolcott, M P (2005), 'Rheology of Wood Plastics Melt. Part 1. Capillary Rheometry of HDPE Filled with Maple', *Polym. Eng. Sci.*, 45 (4), 549–559.

Maine, F W & Newson, W R (2005), 'Method and Apparatus for Forming Composite Material and Composite Therefrom', US Patent 6,939,496 B2.

Matuana, L M, Park, C B & Balatinecz, J J (1996), 'Characterization of Microcellular Foamed PVC/Cellulosic-Fiber Composites', *Journal of Cellular Plastics*, 32(5), 449–469.

Matuana, L M, Park, C B & Balatinecz, J J (1997), 'Processing and Cell Morphology Relationships for Microcellular Foamed PVC/Wood-Fiber Composites', *Polym. Eng. Sci.*, **37(7)**, 1137–1147.

Matuana, L M, Park, C B & Balatinecz, J J (1998), 'Cell Morphology and Property Relationships of Microcellular Foamed PVC/wood-fiber Composites', *Polym. Eng. Sci.*, 38(11), 1862–1872.

Matuana, L M, Park, C B & Balatinecz, J J (2004), 'The Effect of Low Levels of Plasticizer on the Rheological and Mechanical Properties of Polyvinyl Chloride/newsprint-fiber Composites', *Journal of Vinyl and Additive Technology*, 3(4), 265–273.

Mengeloglu F & Matuana, L M (2004), 'Foaming of Rigid PVC/woodflour Composites Through a Continuous Extrusion Process', *J. Vinyl. Add. Technol.*, 7(3), 142–148

Mohanty, A K & Misra, M (1995), 'Studies on Jute Composites – A Literature Review', *Polym.-Plastic Thechnol. Eng.*, 34(5), 729–792.

Newson, W R & Maine, F W (2002), 'Second Generation Woodfibre-Polymer Composites', *Progress in Woodfibre-Plastic Composites*, May 23–24, Toronto, 2002.

Orfao, J J M, Antunes, F J A & Figueiredo, J L (1999), 'Pyrolysis Kinetics of Lignocellulosic Materials – Three Independent Reactions Model', *Fuel*, 78, 349–358.

Pan, W P & Richards, G N (1990), 'Volatile Products of Oxidative Pyrolysis of Wood: Influence of Metal Ions', *J. Anal. Appl. Pyrol.*, 17, 261–273.

Park, C B & Suh, N P (1996), 'Filamentary Extrusion of Microcellular Polymers using a Rapid Decompressive Element', *Polym. Eng. Sci.*, 36(1), 34–48.

Park, C B, Behravesh, A H & Venter, R D (1996), 'Chapter 8 – A Strategy for Suppression of Cell Coalescence in the Extrusion of Microcellular HIPS Foams', in Khemani, K, *Polymeric Foams: Science and Technology*, American Chemical Society, Washington, 115–129.

Park, C B, Doroudiani, S & Kortschot, M T (1998a), 'Processing and Characterization of Microcellular Foamed High-Density Polyethylene/Isotactic Polypropylene Blends', *Polym. Eng. Sci.*, 38(7), 1205–1215.

Park, C B, Behravesh, A H & Venter, R D (1998b), 'Low-Density, Microcellular Foam Processing in Extrusion Using CO_2', *Polym. Eng. Sci.*, 38(11), 1812–1823.

Park, C B, Rizvi, G M & Zhang, H (2002), 'Plastic Wood Fiber Foam Structure and Method of Producing Same', *Canadian Patent*, Application No. CA2384968.

Park, C B, Rizvi, G M & Zhang, H (2005), 'Plastic Wood Fiber Foam Structure and Method of Producing Same', *U.S. Patent* 6,936,200 B2.

Reedy, M E (2000), 'How Chemical Foaming Agents Improve Performance and Productivity', *Plast. Eng.*, 56(5), 47–50.

Rizvi, G M, Park, C B, Lin, W S, Guo, G. & Pop-Iliev, R. (2003), 'Expansion Mechanisms of Plastic/Wood-Fiber Composite Foams with Moisture, Dissolved Gaseous Volatiles, and Un-dissolved Gas Bubbles', *Polym. Eng. Sci.*, 43(7), 1347–1360.

Rizvi, G M, Park, C B & Guo, G (2007), 'Strategies for Processing Wood Plastic Composites with Chemical Blowing Agents', *Journal of Cellular Plastics*, 44(2), 125–137.

Rodrigue, D, Souici, S & Twite-Kabamba, E (2006), 'Effect of Wood Powder on Polymer Foam Nucleation', *Journal of Vinyl and Additive Technology*, 12(1), 19–24.

Scheirs, J, Camino G & Tumiatti, W (2001), 'Overview of Water Evolution During the Thermal Degradation of Cellulose', *European Polymer Journal*, 37(5), 933–942.

Schut, J H (2001), 'Foaming Expands Possibilities for Wood-Fiber Composites', *Plastics Technology*, July 2001.

Seeler, K A & Kumar, V (1993), 'Tension-Tension Fatigue of Microcellular Polycarbonate: Initial Results', *Journal of Reinforced Plastics and Composites*, 12(3), 359–376.

Sherman, L M (2004), 'Wood-Filled Plastics – They Need the Right Additives for Strength, Good Looks & Long Life', *Plastics Technology*, July 2004.

Shimbo, M, Balwin, D F & Suh, N P (1995), 'The Viscosity Behavior of Microcellular Plastics with Varying Cell Size', *Polym. Eng. Sci.*, 35(17), 1387–1393.

Smith, P M & Wolcott, M P (2006), 'Opportunities for Wood/Natural Fiber-Plastic Composites in Residential and Industrial Applications', *Forest Products Journal*, 56(3), 4.

Stucky, D J & Elinski, R (2002), US Patent No. 6344268.

Suh, N P (1996), 'Microcellular Plastics', in Stevenson J F (Ed.), *Innovation in Polymer Processing*, Munich, Hanser, 93–149.

Throne, J (1996), *Thermoplastic Foams*, Sherwood Publishers, Hinckley, Ohio.

Tsoumis, G (1991), *Science and Technology of Wood: Structure, Properties, Utilization*, Van Nostrand Reinhold, New York, 198–199.

Wang, K H, Chung, I J, Jang, M C, Keum, J K & Song, H H (2002), 'Deformation Behavior of Polyethylene/Silicate Nanocomposites As Studied by Real-Time Wide-Angle X-ray Scattering', *Macromolecules*, 35(14), 5529–5535.

Zachariades, E & Porter, R S (1983), *The Strength and Stiffness of Polymers*, Marcel Dekker, New York.

Zehner, B E (2003), US Patent No. 6590004.

Zhang, H., Rizvi, G M & Park, C B (2004), 'Development of an Extrusion System for Producing Fine-Celled HDPE/Wood-fiber Composite Foams Using CO_2', *Adv Polym Techn.*, 23(4), 263–276.

Zhang, S, Rodrigue, D & Riedl, B (2005), 'Preparation and Morphology of Polypropylene/Wood Flour Composite Foams via Extrusion', *Polymer Composites*, 26(6), 731–738.

12
Performance measurement and construction applications of wood–polymer composites

R J TICHY, Washington State University, USA

12.1 Introduction

There has been a rapid increase since the mid-1990s in the number of producers of wood and natural fiber polymer composites in North America and the trend is expanding in Europe, with recent interest from Asia, Australia, South America, and South Africa. The North American experience began with the simple combination of wood particles and with a variety of thermoplastics resulting in the terminology wood–plastic composites (WPCs). Since those early product introductions the level of sophistication has expanded to include a wide range of wood, natural fiber, and non-wood fibers as reinforcements combined with, for the most part, one of three polymers – polyethylene (PE), polypropylene (PP), or polyvinyl chloride (PVC).

Because of WPC popularity in the consumer markets (primarily as outdoor lumber substitutes) building code officials and even the lumber industry are beginning to take notice. Although the entire WPC industry represents barely a percentage point of the total lumber consumption in North America (Fig. 12.1) (Bright and Smith, 2002; Smith and Bright, 2002), its attributes attracted the attention of two of the world's largest building products retailers (Home Depot, Lowes). Provided the manufacturers of WPCs embrace the code requirements for these products, there appears to be no reason this trend will cease.

In the United States there has been significant change in the building code landscape over the past few years. The International Code Council (ICC) has been formed. The intent was to replace the three codes that governed US building construction with a single code for the entire nation. Unfortunately, the consolidation process is slow. The country now has the International Building

12.1 Total US wood consumption vs WPC materials.

Code (IBC, 2006), but many states and local jurisdictions continue to cling to their more familiar predecessors, resulting in not one code, but four. To further complicate the situation, a second national code has been promulgated by the National Fire Protection Association (2003), adding a fifth choice to the building code list. The good news is that the ICC no longer publishes the three former codes and the NFPA code does not appear to be providing any noticeable competition to the ICC. In time, there will be a single building code for the country – a welcome relief to the current situation.

For those nations where building codes are written by the government, it is important to point out that the US process is very different. In the United States, building codes are written by non-profit corporations such as the ICC. The contents of the code are open to public scrutiny and discussion. The process in some ways resembles that of a consensus standards writing organization where all stakeholders can contribute to code development. This process is both a blessing and a curse. It is a blessing because anyone can contribute and new technology can be incorporated in a reasonably timely fashion. As such, the code is in a constant state of renewal. This is also the curse. The constant state of renewal is rife with controversy and often time-consuming discussions resulting in protracted times to completion.

In support of the building code are numerous standards, many of which are developed within consensus-based standards writing organizations such as the American Society for Testing and Materials. Design standards, performance-based specifications, product standards, and many more are also in a constant state of development. The relationship between standards and the codes is one where significant guidance is provided by the standard writing groups, from which the building code organizations draw the salient parts. A good example of this is the ICC Evaluation Service, Inc. (2005) publication *Acceptance Criteria for Deck Board Span Ratings and Guardrail Systems*.

In Canada, similarities exist between the US and Canadian systems, except that the product acceptance guidelines in Canada are created primarily by experts within or associated with the central government. Even with the numerous commonalities between these two North American neighbors, there are enough code-related differences to make it difficult to move products across the borders.

Globally, what is needed is a unified approach to assess fitness-for-use. This is a performance-based approach where a product, intended for a specific application (e.g. deck boards or guardrails), is assessed according to building code requirements for that particular application. The advantage of this system is that it makes the manufacturing process irrelevant and product material content relatively unimportant. Being directly linked to building code requirements, a performance-based system simplifies the acceptance process and minimizes the games competitors can play with the acceptance system. Since the objective of building codes worldwide is to provide safe and healthy structures, a performance-based approach to product acceptance makes sense internationally.

Unfortunately, what is occurring in many countries is a perpetuation of the traditional process of establishing basic material properties and then extrapolating from this information performance measures for actual products. While this process has a historical precedent, the engineering community is moving toward performance-based design and building codes. For WPCs and its successors to be considered true engineered materials, it is recommended that performance-based measures be included in all product assessments.

12.2 Performance measures and building codes

Successful structural products require an in-depth evaluation of product performance. For most structural applications both mechanical properties and long-term resistance to environmental assault are important. To be truly successful in the marketplace, each product must be able to maintain adequate performance levels throughout its intended lifetime. The only real question is, how to measure adequate performance. While the following discussion will focus on performance attributes important to code-listed structural products, appearance issues can be equally important from a market acceptance standpoint. For example, color change due to ultraviolet exposure is typically not a code concern, but it is certainly important to the customer.

In general, performance guidelines for building construction materials are well understood and the methods for evaluation exist. However, when a new class of material arrives, a complete set of adequate test methods often do not exist and, therefore, must be developed. The responsibility for developing the appropriate measures of performance usually begins with the product developer. However, this is where a close relationship between the product development experts and the building code organizations becomes critical. Performance measures must be jointly developed that *assess performance relative to building code requirements*. Performance measures based on material property norms can be misleading and time-consuming. While material property assessments may be useful process control measures employed by the product manufacturer, they are of little value in judging fitness for use in building code regulated applications. In the United States, performance-based assessments are gaining favor. Development of *ASTM D 7032 Standard Specification for Establishing Performance Ratings for Wood–Plastic Composite Deck Boards and Guardrail Systems* (ASTM International, 2008) was such a case.

ASTM D 7032 is a performance specification that provides a list of test procedures to assess conformance to building code requirements. This ASTM specification ultimately became the basis for the ICC ES acceptance criteria for deck boards and guardrail systems – AC174 (ICC Evaluation Service Inc., 2005). Although the title of ASTM D 7032 might suggest a limited scope (i.e. wood–plastic composites), the performance measures are specific to products used as deck boards and guardrail systems – and not specific to a particular material type. With only a few exceptions, material composition is irrelevant.

However, a simple set of performance measures is not the answer to all code-related issues. For true structural materials, performance measures coupled with allowable engineering design stresses are required. To establish allowable design stresses, product-specific testing is conducted at statistically rigorous levels to generate baseline performance along with appropriate adjustment factors. Relying on code prescribed levels of safety and a quality assurance program that involves ongoing independent third-party inspection, allowable design stresses can be assigned that provide architects and engineers confidence in their designs. Currently under development is an ASTM standard specification for the establishment of engineering design values for WPCs. This standard is based on reliability-based design concepts (similar to load and resistance factor design) that will maintain performance levels at least equivalent to those of existing wood-based structural products.

It is important to note that more than consensus-based standards are needed for use of a product in code-regulated building construction. The standard test methods and specifications must be synthesized into code recognized acceptance criteria (used by the building code organizations to assess compliance with code requirements) or the test methods and specifications for these products must be adopted directly into the building code as an alternate material.

For many construction materials, acceptance criteria set the basis for code recognition, resulting in a code report called an evaluation services report (ESR). For deck boards and guardrail systems AC174 is the appropriate Acceptance Criteria (ICC Evaluation Service, Inc., 2005). If product evaluations according to these criteria are met, the product is eligible to receive an ESR number, identifying that specific product as being acceptable for use in that specific application. The influence of an ESR is *significant from both a building code and a market perspective*. Without an ESR, any code official in the United States can ban the product from use. The ESR also provides some level of protection from litigation. In addition to the legal requirements of a code-listed product, there are also significant market advantages to be able to state your product is code listed.

12.3 Wood–polymer composite properties

By varying raw material content, manufacturing process, and/or product geometry, product possibilities are nearly endless. It therefore becomes important to understand the property relationships for the basic product components. Several years ago, a Simplex model was developed (Adcock and Wolcott, 1999) to predict the effect of the various raw material components for a wide range of composite formulations. The model demonstrated the ability to predict basic

Performance measurement and construction applications 261

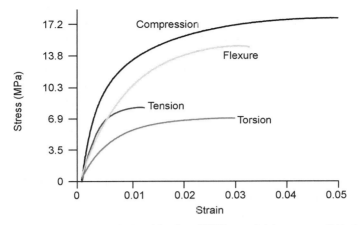

12.2 Basic property relationships for a WPC material (courtesy of Washington State University).

constitutive relationships to within a few percent. Models such as these will play an important role in product development activities in the future.

Figure 12.2 shows the relationship between several primary mechanical properties of a wood–HDPE composite formulation. Note the difference between the tensile and compressive properties. This radically deviates from tradition linear elastic materials such as steel. Utilizing these constitutive relations, the flexural performance of complicated cross-sections can be accurately predicted. This suggests that pure material properties (usually from coupon-sized specimens) can be used for product design – a valuable product development tool.

In addition to the basic structural properties, other property relationships of importance include the effect of natural exposure conditions such as temperature, moisture, ultraviolet radiation, freeze–thaw, and biological deterioration. Depending upon material composition and the end-use application, some or all of these effects may be important. These effects can also be ones without meaningful procedures to quantify product property changes. It is here that meaningful performance measures need development.

Combining natural fibers or wood with polymers is not new. Industrialized nations have decades of experience with composites using thermoset and thermoplastic polymer systems. However, their use as lumber substitutes in outdoor structural applications is drawing much attention. One of the most useful attributes of many WPCs is their ability to resist moisture intrusion. The hygroscopic nature of wood and its resultant effects are well documented. And thermoplastic's general lack of interest in water is equally understood. The combination of two materials, with no particular attraction for one another, would initially seem counterproductive. However, the salient mechanism from this combination is the rather elementary result of interrupted passages for moisture to travel through the composite. The impact of this attribute goes well beyond the visible effects of dimensional change and mechanical property reduction. If one were to wait long enough (typically many months), moisture absorption of the wood component can achieve levels similar to that of solid wood; however, the practical implication is that in the natural environment, sustained levels of excessive moisture are not common. Consequently, timing is everything. Because the transpiration rate is slow, the deleterious effects of moisture, in a majority of this planet's climatic regions, are minimal. While the industry is still relatively young, there exists no conclusive evidence that wholesale failure of these products is imminent. So, the question remains, what are meaningful measures of long-term moisture performance for this class of material? And, what collateral effect(s) might be associated with moisture absorption?

An appropriate performance measure is one that stresses the product to conditions likely to occur in service. This is where the discussion begins. However, for some, there appears to be an insatiable desire to attack a product at extreme levels with extreme measures, regardless of whether or not the attack simulates anything that could be expected in service. There is endless debate on what level of attack is appropriate. Because patience is generally lacking when it comes to long-term assessment techniques, attention automatically turns toward 'accelerated' methods. Accelerated test methods are often used as 'screening' tools. If it is presumed that an accelerated test does in fact evaluate a meaningful performance parameter, the most that can be expected from a screening tool is that a truly poor product can be identified. Accelerated tests provide little useful information about the actual long-term performance of the product, but they can be useful to help identify the 'bad actors' in a population.

Until the scientific relationships between the test results of accelerated tests and actual exposure tests are identified and meaningful mechanisms of failure are determined, accelerated test methods can only be used as screening tools. The good news is that in-service exposure histories in excess of a decade are beginning to emerge for several commercial products. And from this experience, it may be possible to develop appropriate performance measures. Unfortunately, until these relationships are established, the development of appropriate performance measures will remain a process of educated guessing.

For example, traditional wood composites such as particleboard or MDF used in interior applications have established numerous performance measures related to moisture intrusion. These test methods typically prescribe a time period for exposure. Compared with the traditional wood composites, WPCs measured with these same procedures perform quite well. However, many WPCs are used in exterior applications. As such, water-soak and rain simulation tests are considered. The point is that performance measures must relate to some practical in-service concern.

Pursuant to appropriate performance measures, it is incumbent upon those skilled in this area to explore the most probable product degradation

Performance measurement and construction applications 263

mechanisms. As new scientific information becomes available, performance measures can be added or deleted from the current list. Based on the current science, performance tests (beyond temperature and moisture effects on mechanical properties) have been developed that include creep–rupture (duration of load), ultraviolet light exposure, freeze–thaw effects, and biological deterioration. Several of these mechanisms are briefly discussed below.

Creep–rupture was deemed particularly critical for structural components where long-term loads could be applied. Building code organizations are very wary of new construction materials, particularly when life safety issues are involved. Their only recourse is to demand product performance equivalent to that of other products in the same application. Based on this notion, a standard was developed, *ASTM D 6815 Standard Specification for Evaluation of Duration of Load and Creep Effects of Wood and Wood-based Products* (ASTM International, 2002), to assess product performance relative to solid wood in structural applications. The basic premise was that if the new product behaved enough like wood, it could utilize the duration of load factors for wood. An experiment at Washington State University was performed on two distinctly different wood fiber–thermoplastic composite formulations (Brandt and Fridley, 2002). These preliminary data showed that in this case the PVC formulation passed the three performance criteria of ASTM D 6815 while the three HDPE formulations did not. Notice that both the PVC formulation and HDPE formulations had larger duration of load effects than for solid wood (Fig. 12.3) according to the *National Design Standard for Wood Construction* (American Forest and Paper Association, 2006). However, this test showed that the PVC

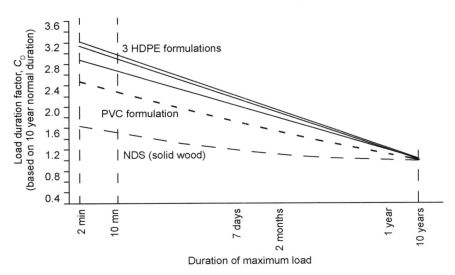

12.3 Creep–rupture performance of WPC products (HDPE, high-density polyethylene; PVC, polyvinyl chloride; NDS, National Design Standard).

formulation was not significantly different from solid wood, while the HDPE formulations were. Although this is not an exhaustive study, it indicates that the standard was effective at differentiating product performance. Clearly, additional creep–rupture study is warranted, particularly as these products move into more demanding structural applications.

In North America, freeze–thaw was perceived to be an important consideration for exterior applications. For deck boards it was decided that water submersion followed by freezing would be a severe treatment. The actual cycle is described in ASTM D 7032. After the cycle was determined, the debate quickly degenerated to how many freeze–thaw cycles were appropriate. A US code organization felt 50 cycles might be acceptable, while the product manufacturers argued for significantly fewer. Considering the fact that one cycle requires 3 days to conduct, one cycle per week could be accomplished. In an attempt to answer this question a study was initiated to establish a meaningful number of cycles. Figure 12.4 shows the effect of freeze–thaw on flexural strength. Three commercially available WPCs were subjected to 1, 5, 10 and 50 cycles. From this plot it was shown that nearly 90% of the strength loss occurred within the first five cycles. Ultimately the manufacturers and the code organizations agreed on three cycles as a reasonable number as a screening test.

There is no shortage of potential tests to consider. However, nearly all of these performance measures can only estimate the potential for a problem. For example, no one can definitively relate laboratory UV exposure to actual product longevity. Even natural UV exposure experiments are at best an estimate. A great example of this uncertainty occurred following the eruption of Mt. Pinatubo. A US car manufacturer was evaluating a new coating system at natural exposure facilities around the country. Nearly 2 years of data suggested excellent performance. Based on these data, thousands of cars were manufactured, only to find premature coating failures, leading the car manufacturer to a

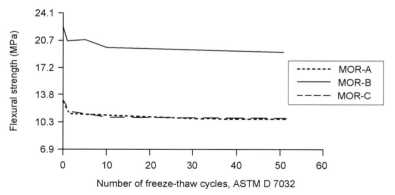

12.4 Freeze–thaw effect on the flexural strength of WPCs (courtesy of Washington State University).

financially unpleasant result. It was later concluded that the ash circling the globe from the eruption disrupted the actual exposure providing misleading performance results. The point is, accelerated exposures (and even natural exposures) can only hope to identify the 'bad actors' in a population. *Accelerated tests provide no useful information about product longevity.* Therefore, time spent developing accelerated torture tests has limited value. This information is only useful as a screening tool.

12.4 Building construction applications

As builders, consumers, and code officials gain experience and confidence with WPCs, their influence in building construction will grow. Previously discussed was the influence this industry is beginning to have on lumber consumption in the United States. Just a few years ago WPCs did not appear on the lumber industry piechart (Fig. 12.1). If the trend continues, and there is no technical reason why it cannot, the 'pie' will continue to change, as WPCs expand, and possibly create, new markets. The following discussion will focus on several building construction applications that are currently, or soon to become, part of the WPC industry.

Marine applications are an obvious place to start. In a recent survey (Bright and Smith, 2002; Smith and Bright, 2002), product specifiers and project engineers were asked to rank the key attributes driving material selection for waterfront facility construction. The most striking result shown in Fig. 12.5 was the *lack* of importance initial product cost had on the buying decision. While it

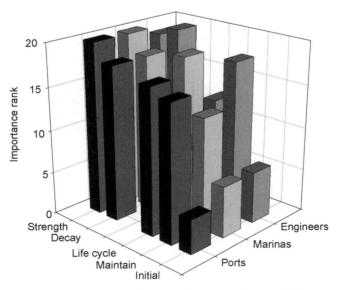

12.5 Key attributes for product specifiers – waterfront facilities.

was no surprise that strength, durability, and maintenance costs appeared to be universally important; the next most revealing result was the general lack of trust in life-cycle cost information. This could be due to the dearth of meaningful substantiating information for life-cycle claims made by material suppliers.

12.4.1 Product opportunities

In addition to the decking product applications (Fig. 12.6), which have been so successful in North America, numerous other opportunities await. Fencing systems are beginning to appear in the marketplace (Fig. 12.7). WPC use in windows and doors is slowly growing in several markets around the world, and at least two manufacturers of roofing materials are selling products in the United States. One siding/cladding manufacturer is in business; and roofline/exterior trim products are of particular interest to several current manufacturers worldwide. At some point interior products for applications such as molding and millwork, furniture parts, and industrial flooring will likely arrive.

With the exception of decking products (deck boards and guards), most of the current and potential market opportunities demand minimal structural perform-

12.6 Decking application in North America (courtesy of Composatron).

Performance measurement and construction applications 267

12.7 Fence system (courtesy of Composatron).

ance requirements. However, engineers and architects are already asking for products and systems that can be used in more demanding structural applications. Pursuant to this request several research organizations have developed WPCs with remarkable properties.

Even something as simple as a sill plate in residential construction (Fig. 12.8) is an excellent opportunity for WPCs. For this application the WPC has compression strength perpendicular to the long axis (extrusion direction) and nail-holding capacity that exceeds that of lumber. The composite restricts water transpiration; and termites are not interested. And, as an added advantage, the natural fiber composite has a higher degree of damping and vibration control than traditional products – all characteristics of an excellent sill plate. Washington State University is currently developing an innovative sill plate

268 Wood–polymer composites

12.8 Sill plate application in residential construction (courtesy of TM&I, Inc.).

product that provides superior attachment to the wall elements, thereby substantially improving the dynamic performance of wall systems.

The important question is whether or not WPCs can compete with current product offerings in a like application. Research has shown that many of these composites possess structural properties equivalent to those of solid wood products. For example, Table 12.1 shows estimated WPC design values compared with common structural grades of lumber (Bender *et al.*, 2006). Note that the values shown in Table 12.1 include appropriate safety factor reductions to make them comparable with the structural lumber values.

Several conclusions can be drawn from the comparisons in Table 12.1. First, within the WPC products it is not surprising to see that the HDPE products exhibit the lowest allowable design values with the PVC products providing the highest design values. With the exception of modulus of elasticity, many PVC products possessed design values exceeding those of No. 2 grade Douglas fir-larch and southern pine. Several PP products also compared favorably to the structural lumber. In fact, the PVC and PP products generally showed superior shear strength parallel, compression strength perpendicular, and dowel-bearing strength. Manipulating product geometry and the use of reinforcement (beyond the wood fiber) can improve these mechanical properties even further. The remaining unknown is the cost/benefit ratio between competing products.

Recent cooperation between WSU and the Strandex Corporation resulted in a 100 × 300 mm (4 × 12-inch) PVC deck board profile. This large deck board was commercially manufactured and used in a pedestrian/recreation vehicle bridge spanning 27.5 m (90 feet) (Fig. 12.9). Deck boards 7.4 m (24 feet) long were installed over multiple 1.8 m (6 foot) spans and designed to resist 455 kg (1000 lbf) concentrated loads. This same deck board was also installed at the US Naval Academy in Annapolis Maryland to serve as a wave screen. Both applications demanded extreme weather resistance and significant load-carrying capacity.

In New York State, McLaren Engineering Group designed and built a reinforced plastic lumber demonstration bridge capable of carrying commercial traffic across a small creek (Fig. 12.10). As the mechanical properties of these

Table 12.1 Design value property comparison between WPCs and visually graded structural lumber (WWPA rules)

Product	Bending MPa (psi) F_b	Tension parallel[a] MPa (psi) F_t	Shear parallel[a] MPa (psi) F_v	Compression perpendicular[b] MPa (psi) $F_{c\perp}$	Compression parallel[a] MPa (psi) $F_{c\|}$	Modulus of elasticity MPa (psi) E	Dowel bearing strength MPa (psi) $F_{e\|}$
Wood-plastic composite							
HDPE	1.4–3.4 (200–490)	0.8–2.0 (110–290)	1.0 (150)	1.6–3.6 (230–520)	1.6–3.6 (230–520)	1790–5170 (260 000–750 000)	35.7 (5180)
PP	3.0–8.1 (430–1180)	2.7 (390)	2.9 (425)	7.4 (1070)	2.8–7.4 (410–1070)	3520–6000 (510 000–870 000)	84.8 (12 300)
PVC	6.1–9.2 (885–1340)	4.3 (620)	5.2 (750)	9.8 (1420)	9.7–10.4 (1400–1510)	4830–7590 (700 000–1 100 000)	72.4–128.3 (10 500–18 600)
Structural lumber (No. 2 Joist and plank)							
Douglas fir–larch	7.4 (1080)	4.8 (690)	1.2 (180)	4.3 (625)	9.8 (1418)	11 000 (1 600 000)	38.6 (5600)
Hem–fir	7.0 (1020)	4.3 (630)	1.0 (150)	2.8 (405)	9.4 (1365)	8960 (1 300 000)	33.1 (4800)
Southern pine	8.3 (1200)	4.5 (650)	1.2 (175)	3.9 (565)	10.7 (1550)	11 000 (1 600 000)	42.4 (6150)
Spruce–pine–fir	7.2 (1050)	3.7 (540)	0.9 (135)	2.9 (425)	8.3 (1208)	9650 (1 400 000)	32.4 (4700)

[a] Parallel to the extrusion direction for WPC, parallel to the grain for lumber.
[b] Perpendicular to the extrusion direction for WPC, perpendicular to the grain for lumber.

270 Wood–polymer composites

(a)

(b)

12.9 Bridge deck surface using 100 × 300 mm² (4 × 12 inch) WPC deck boards: (a) Rattlesnake Creek bridge, Montana, USA; (b) deck board connection system (courtesy of Washington State University).

12.10 Reinforced plastic lumber bridge.

composites improve, it is fully expected that even more sophisticated structural applications will follow.

12.5 Conclusions

As WPCs move into new applications, disciplined and focused research effort will be required to develop commercially successful products. Concurrent with new product development, meaningful performance measures must also be defined. Depending upon customer expectations or building code requirements, performance assessments will necessarily change. It is, therefore, incumbent upon those charged with developing product performance measures to look carefully at appropriate end-use requirements for a given application. Use of existing product assessment measures can be a starting point. However, depending upon the end-use requirements, additional performance assessments may be required.

For structural applications, the product needs to be measured against the relevant building code provisions. For non-structural applications the product needs to be measured against customer expectations. In either case *product performance measures* should be the target, and not an arbitrary or traditional material property measure. Material property assessments are useful for product development activities, but carry little significance in the marketplace. Product performance is key to success, not material performance.

The important point is to develop performance measures that relate to actual in-service conditions. However, measuring in-service conditions is not trivial. This typically involves some assessment of long-term performance, which invariably leads to accelerated test methods. Caution must be exercised when using accelerated test methods. Accelerated test methods provide *very little information* beyond a simple screening assessment. Estimates of product longevity are tenuous at best and quite possibly irresponsible. Thinking beyond the status quo, and the courage to accept the risk of making engineering judgments, will be required. Once product performance is adequately measured, do not underestimate the market importance of a code-listed product. Products that have passed code-related assessments carry significant marketplace credibility.

Research has shown that WPCs possess structural performance comparable to that of some lumber products. It has also been shown that these composites can be engineered to withstand significant environmental and structural loads. Building on these developments, the ability to design formulations and manipulate product geometry offer exciting opportunities for the future.

It is clear that the industry is exploring more sophisticated applications for these unique composites. There is little doubt that this emerging class of materials has numerous current and potential opportunities in both structural and non-structural applications. WPCs, as a lumber substitute in several outdoor applications, have already captured a noticeable market share; and there appears to be no compelling reason why this trend should cease. As more demanding applications arise, more sophisticated and meaningful performance measures must be developed.

12.6 References

Adcock, T. and M.P Wolcott. 1999. *The Influence of Wood Plastic Composite Formulation: Studies on Flexural and Moisture Related Properties.* Project End Report, Naval Waterfront Facilities. Office of Naval Research

American Forest and Paper Association. 2006. *National Design Specification for Wood Construction.* AF&PA, Washington, DC.

ASTM International. 2002. *ASTM D 6815 Standard Specification for Evaluation of Duration of Load and Creep Effects of Wood and Wood-based Products.* American Society for Testing and Materials.

ASTM International. 2008. *ASTM D 7032 Standard Specification for Establishing Performance Ratings for Wood–Plastic Composite Deck Boards and Guardrail Systems.* American Society for Testing and Materials.

Bender, D.A., M.P. Wolcott and J.D. Dolan. 2006. Structural design and applications with wood-plastic composites. *Wood Design Focus – A Journal of Contemporary Wood Engineering*, **16**(3): 13–15.

Brandt C. and K. Fridley. 2002. Effect of load rate on flexural properties of wood–plastic composites. *Wood and Fiber Science*, 35(1): 135–147.

Bright, K.D. and P.M. Smith. 2002. Perceptions of new and established waterfront materials by US marine decision makers. *Wood and Fiber Science*, **34**(2): 186–204.

IBC 2006. *International Building Code 2006.* International Code Council. Whittier, California.

ICC Evaluation Service, Inc. 2005. *Acceptance Criteria for Deck Board Span Ratings and Guardrail Systems (AC174).* International Code Council, Whittier, California.

National Fire Protection Association. 2003. *NFPA 5000: Building Construction and Safety Code*, 2003 edition. NFPA, Quincy, Massachusetts.

Smith, P.M. and K.D. Bright. 2002. Perceptions of new and established waterfront materials: US port authorities and engineering consulting firms. *Wood and Fiber Science*, 34(1): 28–41.

13
Life-cycle assessment (LCA) of wood–polymer composites: a case study

T THAMAE and C BAILLIE, Queens University, Canada

13.1 Introduction: comparing wood–polymer and glass-fiber reinforced polypropylene car door panels

Life-cycle assessment (LCA) is a tool used to evaluate environmental impacts associated with a particular product throughout its entire life cycle (from extraction of raw materials to the end of life; from cradle to grave) (Murphy, 2003; SETAC, 1993). LCA can be used to compare two or more products to find out which of them is more preferable from an environmental point of view. For a fair comparison, products should meet the same service requirements. Also, since different products have different impacts at different stages in their lifetimes, comparing them at only one stage can give misleading results. Their whole life cycles should be considered.

The automotive industry makes a significant contribution to environmental pollution, especially in emitting greenhouse gases. But researchers in the field of natural fiber composites (NFCs) suggest that these products can help reduce the environmental impact of this industry. Several advantages of these composites over conventional materials include low cost, light weight, increased biodegradability and abundance of natural fibers (Torres and Cubillas, 2005). The worldwide production of lignocellulosic materials, the primary sources of natural fibers, is estimated to be 2×10^{11} tonnes annually (Hinrichsen et al., 2000) and the automotive industry is increasingly applying these composites in parts manufacturing. In North America, the transportation sector shares 10% of the use of wood fiber composites and 16% of the use of other NFCs (Suddell and Evans, 2005). In Europe, the automotive industry had a share of 55% of the 65000 tonnes of wood fibers in the markets in 2003 (Markarian, 2005).

One of the main reasons for a current surge in the research and application of NFCs is that they are viewed as eco-friendly (Li and Wolcott, 2004). Nevertheless, environmental impacts of a product can be case specific. For instance, one study which analyzed impacts of hemp fiber plastic door panel against glass fiber plastic door panel for housing using LCA showed that both panels had almost similar impacts on the environment (Murphy, 2003). According to Baillie (2004), LCA reports from EU CRAFT projects showed that NFCs had more impacts on the environment than glass fiber plastic composites except for automotive applications where NFCs performed better due to fuel savings owing to their light weight.

There are other reports showing LCA studies in automotive applications (Corbiere-Nicollier *et al.*, 2001; Diener and Siehler, 1999; Schmidt and Beyer, 1998; Wotzel *et al.*, 1999). Review of these studies by Joshi *et al.* (2004) concluded that NFCs were more eco-friendly than glass fiber composites in automotive applications because (1) natural fiber production has lower impacts than glass fiber production; they depend mainly on solar energy to grow and they absorb CO_2 during growth, (2) natural fibers take more volume per unit weight of a composite, therefore NFCs require less polymer matrix (which has more impacts on the environment) than glass fiber composites, and (3) owing to their light weight, NFCs improve automotive fuel economy, thereby reducing emissions during use (Sivertsen *et al.*, 2003); it is possible to recover and use energy by incinerating natural fiber but not glass fiber.

If the above observations hold true, these findings have important implications for the automotive industry. They may encourage policies and voluntary adjustments that act in favor of NFCs as opposed to other conventional composites in this industry. But LCAs make a large number of assumptions. This factor causes high uncertainties in data and can make LCA studies unreliable (Ayres, 1995; Joshi *et al.*, 2004). Therefore there is a need to confirm or refute previous studies by carrying out further LCAs under different assumptions.

This study compares life-cycle impacts of wood fiber reinforced polypropylene (PP) and glass fiber reinforced PP car door panels. Environmental areas in which both panels have major impacts are identified. The two panels are compared on the basis of impacts on these environmental areas. Further, the study carries out sensitivity analyses to determine how environmental impacts of the two panels change with changes in initial assumptions which we call a reference scenario. The study concludes by looking at the recent European Union (EU) environmental legislation on the end-of-life vehicles (ELVs) and its possible impacts on the automotive life cycles.

13.2 The life-cycle assessment process

13.2.1 The data collection

One of the strengths of LCA is its attempts to consider impacts of 'all' inputs and outputs during the entire life cycle of a product. But this can also be a weakness. LCAs require much time and resources in order to collect data. As a result, some authors argue this might have contributed to a decline in LCA studies with time (Sivertsen *et al.*, 2003). The problem has been addressed by the creation of commercial public databases. Nevertheless, these databases cannot be specific enough to address every situation. The normal problem of lack of data for some unit process of a system is resolved by carefully selecting and using similar processes or by leaving out less significant processes (Corbiere-Nicollier et al., 2001).

Furthermore, public databases can have several deficiencies. They may have data that are aggregated into categories that are too broad. Ayres (1995) cites a database that has a category 'hydrocarbons'. The author argues that hydrocarbons may be as different as their environmental impacts. Failure to distinguish among compounds can lead to questionable results. For example, it is argued that Cr^{3+} has equivalent impact on human toxicity of 6.7 while that of Cr^{6+} is 47 000 (Krozer and Vis, 1998). Yet some databases do not distinguish between compounds of these ions. Granted, the attempt to satisfy this means more details, more analyses, more time, and, perhaps, less cost-effective databases.

13.2.2 Modeling life-cycle assessment

LCAs normally begin with goal and scope definition. This is the stage that defines the goals of an LCA and delineates the boundaries of the study. It determines the outcome of the LCA. Without such boundaries, there is no end to the amount of data any single study can include.

An LCA goal must be stated precisely to leave no room for ambiguity (UNEP, 1996). For most studies, goals may fall within the following areas: comparisons between two products, product and process development, decisions on buying, structuring and building up information, eco-labeling, environmental product declarations and decisions on regulations (UNEP, 1996). One may ask specific questions to formulate goals. Will the study be for internal or external applications? Who will form the audience for the results of this study? To what level of complexity shall the LCA be confined (Murphy, 2003)?

The scope includes the functional unit, functions of the system, system boundaries, allocation procedures, types of impact and methodology for impact assessment and interpretation that follows, data requirements, assumptions, limitations, and so on (UNEP, 1996). This is the most subjective part of LCA. Consider a functional unit. It is defined as a measure of performance of a product (SETAC, 1993). It is a standard into which all inputs, outputs, and processes of a system are related. In practice, it measures the amounts of a product or service needed to perform a particular function. However, one product can have different functions. For instance, a technical function may differ from a social one but both are functions (Krozer and Vis, 1998).

Then system boundaries determine the extent of assessment. LCA practitioners decide which inputs, outputs, and processes to include or exclude (Murphy, 2003). Where will boundaries be drawn and how? Goedkoop and Oele

(2004, p. 5) illustrate, 'In an LCA of milk cartoons, trucks are used ... to produce trucks, steel is needed, to produce steel coal is needed, to produce coal trucks are needed one cannot trace all inputs and outputs ...'. Undoubtedly, the question of where to draw a line in this endless chain is a problematic one.

The list of emissions and use of resources per functional unit is called an inventory. The inventory is normally followed by impact assessment. This part determines impacts of products or services on predetermined impact categories affecting areas of ecosystem quality, human health, and resource depletion (Daniel and Rosen, 2002). Different substances have different impacts on the environment. So it is important to know their relative contributions to a specific environmental problem. For instance, the global warming potential of nitrous oxide (N_2O) is 310 times that of carbon dioxide (CO_2). So the impact assessment goes beyond quantities of these substances to their significance.

There is a challenge. Whenever practitioners try to connect the inventory results to the impacts they have on the environment, these results begin to lose spatial, temporal, dose–response relationships (change in effect on an organism resulting from differences in levels of exposure), and threshold dimensions (Daniel and Rosen, 2002). Most impact assessment methods do not reflect these complex realities and they may not do so in the near future (UNEP, 1996). Also, the majority of impact assessment methods use the criterion that more mass of a substance means more impact (Daniel and Rosen, 2002). However, beyond certain thresholds in the environment, whether an impact is 1 or 10 units does not change the damage made. Despite the limitations, LCAs provide insights into life-cycle impacts of products that no other tool can provide. The interpretation stage facilitates estimation of uncertainties.

13.3 Goal and scope definition

13.3.1 Goals

The main goal of this LCA is to compare the environmental impacts of replacing a glass fiber reinforced PP car door panel with wood fiber reinforced PP car door panel. The secondary goals are: to identify the impact of glass fiber replacement on major greenhouse gases and major sources of air pollution, to identify major environmental impact categories to which these two panels have significant impacts, to identify processes and substances that contribute more to the main impact categories for both panels, and to determine changes in LCA results due to changes in basic assumptions or reference scenario.

13.3.2 Functional unit and system boundaries

The functional unit will be a car door panel of volume 992 cm^3 for a service life of 200 000 km. This lifetime duration is recommended by the European Council

Life-cycle assessment (LCA) of wood–polymer composites 277

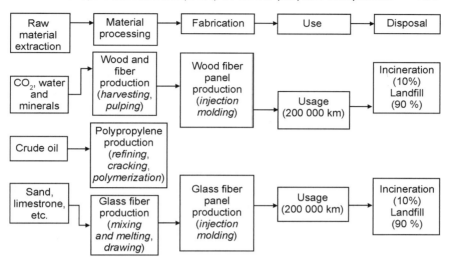

13.1 The system boundaries.

of Automotive Research (EUCAR) for petrol and diesel cars of weight greater than 1500 kg assumed in this study (Ridge, 1998).

One of the door panels under comparison is made up of glass fibers and PP (GFP). The other one is made up of wood fibers (mechanical pulp) and PP (WFP). In both panels, fiber component contributes 40% and PP component contributes 60% by weight.

WFP and GFP must perform the same function to be compared. This means they should have the same volume and mechanical properties. In practice, glass fiber composites will have higher mechanical properties than wood fiber composites. But the panels do not experience high stresses in use (stresses during use are normally lower than the strengths of wood fiber panels (Suddell and Evans, 2005). Therefore differences in their strength properties have little impact on their functions.

The impacts will be considered in the following three stages: assembly (raw material extraction, raw material processing, and panel fabrication), use, and end of life (Fig. 13.1).

13.3.3 Sources of data

Background data were used to estimate quantities of inputs and outputs for different unit processes within a system. The BUWAL 250 (Renilde, 2004) database was selected because it covers more data related to the products under study. This database is based on European scenarios. The choice of data is limited to BUWAL 250 to avoid uncertainty caused by using data from different sources which are based on different assumptions.

13.3.4 Assembly

Polypropylene

PP is produced by cracking of natural gas or light oils. The cracking results in the monomer propylene, which is then polymerized into PP. There are several impacts associated with production of PP. There is much energy used in the extraction process (drilling for oil, or mining of coal). The process of refining is also energy intensive. Hydrocarbons are combined with steam and heated under temperatures of beyond 900 °C.

Some of the main impacts also include release of nitrogen oxides (impacts on eutrophication) and hydrocarbons (impacts on photochemical oxidation) (Honngu and Phillips, 1997). Other emissions include CO_2 and sulfur oxides which have impacts on global warming and acidification respectively. In BUWAL 250 database, producing 1 tonne of PP releases 1800 kg of CO_2, 11 kg of SO_x, and 10 kg of NO_x.

Glass fiber

Glass fiber is made up of sand, chalk, soda, and other components. It is mainly based on silica (SiO_2) but includes impurities. Na_2CO_3 is included to lower melting temperatures by about 1000 °C. $CaCO_3$ or CaO is used to make silica less water soluble. The production process is highly energy intensive. These raw materials are normally melted in a platinum–rhodium alloy bushing and passed through orifices with minute diameters to make fiber rovings (Chakravorty *et al.*, 1981). In this study, data used for production of glass sheets rather than fibers are used as a similar process owing to lack of data in available databases. Corbiere-Nicollier *et al.* (2001) used the impacts of glass wool production for glass fiber production to address the same problem of lack of data.

Wood fiber (mechanical pulp)

Mechanical pulp can be used in the production of wood plastic composites. The process of obtaining pulp begins with tree felling (using tree harvesters), and piling and hauling the trees to the roadside for further processing (Smook, 2002). Debarking and delimbing can be done before transportation to the mill to reduce weight and costs. The wood can then be transported to a sawmill where the logs are turned into small chips. The chips are used in mechanical pulping to produce fiber. All these processes require machinery: sawing machines, mechanical debarkers, sorting machines, chipping machines, and transportation. Therefore they are heavily dependent on energy.

Other environmental impacts include production of lignin, which adds to biological oxygen demand, volatile organic compounds which pollute air, water and soil bodies, and reduction in forest resources (Das and Patnaik, 2000). In cases where wood used for mechanical pulping is a waste of another process, use of wood fiber could have fewer environmental impacts. Such a scenario is not modeled in this study.

Transport of raw materials

The transport distances are already included in the BUWAL data. The rest of the distances such as moving raw materials to make panels are not included in this study. However, transport of raw materials can contribute very little impacts in life cycles of many products (Corbiere-Nicollier et al., 2001; Schmidt and Beyer, 1998).

Panel production

It is assumed that the panels are injection moulded. The fabrication of the panel is assumed to use 2.88 MJ/kg during injection molding (Corbiere-Nicollier et al., 2003). Other processes are ignored.

13.3.5 Use phase

To model this phase a specific car was selected and some of its features are shown below. Equation 13.1 by Keoleian et al. (1998) was used to determine the contribution of door panel weight to the fuel consumption of a car. This equation assumes that fuel consumption has a linear relationship with weight. No other weight savings are modeled.

$$F = M_t \times L \times \left(\frac{FE}{M_v}\right) \times \left(\frac{\Delta f}{\Delta M}\right) \qquad 13.1$$

where F is the fuel consumed over the entire life of a panel (litres), M_t is the mass of the door panel (kg), L is life of a door panel which is assumed to be the same as life of the car (200 000 km), FE is fuel economy (11.9 l/100 km for city and 7.8 l/100 km for highway), M_v is mass of the vehicle (1504 kg). $\Delta f/\Delta M$ is a fuel consumption correlation with mass. It is assumed that fuel reduction of 4.38% (Δf) follows every 10% reduction in weight (ΔM) (Keoleian et al., 1998). Combined fuel economy, FE_{comb}, which is 9.62 l/100 km, is calculated using the equation 13.2 by Sullivan and Hu (1995).

$$FE_{comb} = \frac{1}{\left(\frac{0.55}{FE_{city}}\right) + \left(\frac{0.45}{FE_{hwy}}\right)} \qquad 13.2$$

where FE_{city} is city fuel economy and FE_{hwy} is highway fuel economy.

13.3.6 End of life

The data for recycling (see Section 13.6.1), landfill, and incineration of wood fiber is taken from recycling, landfill, and incineration of newsprint. Newsprint is made of mainly mechanical fiber similar to that used in composites (Lundquist et al., 2003). In the present practice, very little plastic waste is either recycled or incinerated (Bruce et al., 1999). Most of the waste goes to landfills and a small fraction may be incinerated. It is assumed that there is 10% incineration and 90% landfill for both panels for this reference scenario.

13.3.7 Allocation

For incineration, it is assumed that some energy is recovered during the burning of wood fiber and PP. No energy is recovered during glass incineration. The energy generated from incineration replaces the energy that does not have to be produced within the system (expansion of system boundaries).

13.3.8 Impact assessment methods

This study uses Simapro software to make the analyses. It also uses the impact assessment method selection criteria (Table 13.1) as provided by Simapro method selector (http://www.pre.nl) to choose the best among available impact assessment methods. Table 13.2 shows that the Ecoindicator99 egalitarian version (EI99E/E) is the most suitable method that meets the criteria. It scores the highest number of points (18 points). The description of this method and

Table 13.1 Criteria for choosing impact assessment method

Choice area	Expected method characteristics
Single scores	Able to switch between single scores and separate impact category indicator
Weighting set	Uses panel (group of experts) to determine weighting factors
Time perspective	A very long time perspective should be applied in the modeling (future generations are very important)
Geographic coverage and acceptance	Valid in Europe (almost all foreground data are European based)
Simplicity versus scientific quality	Uses advanced scientific models to calculate characterization values
Completeness	As complete as possible, including land-use, small particulates, radioactive substances, solid waste, etc.

Source: http://www.pre.nl

Table 13.2 Results of Simapro method selector

Method	Points	Single scores	Weighting set	Time perspective	Geographic coverage and acceptance	Simplicity versus scientific quality	Completeness
EI99E/E	18	3	3	3	3	3	3
EI99 H/A	17	3	3	2	3	3	3
EI99 I/I	16	3	3	1	3	3	3
EPS	14	3	0	3	2	3	3
EDIP	14	3	3	3	1	3	1
Impact 2002+	13	2	0	2	3	3	3
CML2000	10	0	0	3	3	3	1
EI95	9	3	2	1	2	1	0
GWP	6	1	0	1	3	1	0
CML92	5	0	0	2	2	1	0
UBP	5	2	0	0	1	1	1

Source: http://www.pre.nl

other possible methods can be found on the relevant impact assessment methods manual (Goedkoop and Oele, 2004).

In short, EI99E/E method calculates impacts in terms of damage to the environment; a damage-oriented approach. It assesses damage to human health, ecosystem quality and resources. Other methods, EPS (environmental priority strategy in design), CML method and Ecoindicator95 (EI95) method are used to check the results. These methods are considered to be reasonably documented and most commonly used (Ridge, 1998).

13.4 Inventory

This section focuses on identifying major contributing substances to the use of energy and emissions to air in the life cycle of the two panels. The stage gives attention to the quantities of substances rather than their environmental impacts.

13.4.1 Use of energy

Figure 13.2 sums use of energy for different stages in the life cycles of the two panels. To produce 1 kg of wood fiber requires more energy (15.55 MJ) than to produce 1 kg of glass fiber (12.08 MJ). The energy used in logging, processing of wood chips and the process of mechanical pulping can be very high (Das and Patnaik, 2000). However, a little more energy is used in assembly of 1 kg of GFP than for the same amount of WFP. GFP has a higher volume of PP which needs more energy to produce. It also has a higher weight which uses more energy during fabrication. The same thing applies to energy for the entire life cycle (assembly, use, and end of life). Owing to higher weight, GFP uses more energy especially during use.

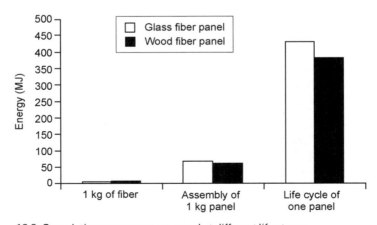

13.2 Cumulative energy use per panel at different life stages.

13.4.2 Major greenhouse gases

Although CO_2 has the least global warming potential, it clearly surpasses others as the major greenhouse gas produced during the life cycles of the two panels (Table 13.3). As shown in Fig. 13.3, replacing GFP with WFP leads to highest benefits in reducing N_2O (which has global warming potential of 296) emissions by 57%. There is a reduction in all greenhouse gases due to this replacement.

13.4.3 Major air pollutants

Substances shown in Table 13.4 can be identified as major air pollutants (Harrison, 2001). For all panels, these pollutants reduce as a result of replacement of GFP by WFP. This reduction ranges from 8.5% for cadmium, 100.4% for lead, to 595.5% for ammonia (NH_3). The data analysis shows that it is the production of glass fiber that results in large amounts of NH_3.

Table 13.3 Amounts of life-cycle emissions of major greenhouse gases (Harrison, 2001; Ramaswamy et al., 2001)

Substance	Global warming potential			Atmospheric lifetime (years)	Life cycle of WFP	Life cycle of GFP
	Time horizon (years)					
	20	100	500			
CO_2	1	1	1	50–200	24.9 kg	28.3 kg
CH_4	62	23	7	12–17	9.62 g	11.3 g
N_2O	275	296	156	120	1.82 g	2.85 g

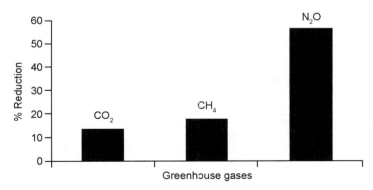

13.3 Reduction in greenhouse gases due to replacement of glass fiber by wood fiber.

Table 13.4 Amounts of life-cycle emissions of major air pollutants

Substance	Amounts per life cycle			Description	Areas of negative environmental impacts
	WFP	GFP	% saving due to replacement		
Ammonia (NH_3)	3.11 mg	21.6 mg	594.53		Vegetation, acidification, eutrophication
Non-methane volatile organic compounds (NMVOCs)	176 g	200 g	13.64	Compounds with high vapor pressure that makes them escape easily to the atmosphere	Human health, form photochemical oxidants, plants and human health
Nitrogen oxides (NO_x)	9.62 g	11.5 g	19.54		Human health, acidification, eutrophication
Sulfur oxides (SO_x)	47 g	52.8 g	12.34		Human health, corrosion, crops and forests
Carbon monoxide (CO)	509 g	574 g	12.77		Human health, modify climates
Particulates	5.37 g	6.55 g	21.97	Liquid or solid particles suspended in air with diameter from 10 nm to 100 μm	Affect health, crop quality, visibility
Heavy metals				Atomic weights between 63.546 and 200.590 and specific gravities greater than 4.0	Human health, vegetation, food chain
Cadmium (Cd)	725 μg	787 μg	8.55		
Mercury (Hg)	102 μg	114 μg	11.76		
Lead (Pb)	24.4 mg	48.9 mg	100.41		
Zinc (Zn)	6.99 mg	7.91 mg	13.16		

13.5 Impact assessment

This section uses the EI99E/E impact assessment method to identify main impact categories and impacts of different substances or processes in these categories. Characterization is a step that quantifies impacts of substances on the environment in each impact category. Since characterization values are given in different units, this part presents normalized values to compare relative impacts of different impact categories. Normalization is done by calculating the contribution of each impact category to the overall environmental problem. Different methods use different normalization criteria. Results using EI99E/E showed that categories described on Table 13.5 receive comparatively more significant impacts as will be shown.

13.5.1 Production of fiber

Production of wood fiber has less environmental impact in most of the impact categories except carcinogens (Fig. 13.4). Wood fiber has more effects in the carcinogens category because data shows that wood fiber production emits more than four times nickel to the air, and three times arsenic (carcinogens) to water than glass fiber production. Glass fiber production has twice the impacts of wood fiber production in the fossil fuels category.

It was shown in Section 13.4.1 that production of wood fiber consumes more energy than production of glass fiber. Therefore production of wood fiber could be expected to have more impact on fossil fuel category than production of glass fiber. But from the BUWAL library, use of energy in Section 13.4.1 includes energy from non-fossil sources such as hydro and nuclear power which are not

Table 13.5 Main impact categories for the life cycles of the two panels (Goedkoop and Oele, 2004)

Category	Definitions	Unit
Fossil fuels	Surplus energy per extracted MJ fossil fuel as a result of lower quality resources	MJ surplus
Respiratory inorganics	Respiratory effects resulting from winter smog caused by emissions of dust, sulphur and nitrogen oxides and other gases to air	DALY/kg
Climate change	Damage resulting from an increase in disease and death caused by climate change	DALY/kg
Ecotoxicity	Damage to ecosystem quality, as a result of emission of ecotoxic substances to air, water and soil	PAF $\times$ m^2 $\times$ yr/kg

DALY = disability adjusted life years.
PAF = potentially affected fraction.

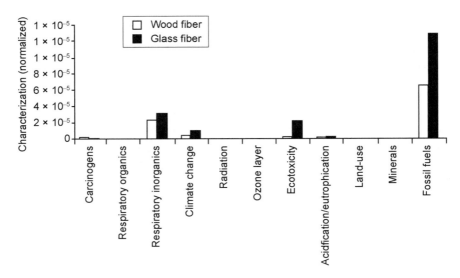

13.4 Comparison between production impacts of 1 kg of glass fiber and 1 kg of wood fiber.

part of the fossil fuel category (see Table 13.5). Thus wood fiber performs better in terms of impacts on fossils fuels only, not over all energy sources.

13.5.2 Assembly of the panels

Assembly of the two panels involves fiber production (Fig. 13.4), PP production, and panel fabrication. In this stage, the two panels have little differences in their environmental impacts although GFP still has more impacts than WFP in most categories (Fig. 13.5). The highest impacts are in the fossil fuel category. While production of wood fiber had almost half of the impacts of glass fiber production on this category (Fig. 13.4), production of WFP has only 15% less impact compared with GFP production.

13.5.3 End of life

At the end of life, the climate change category sees the highest impacts for both panels (Fig. 13.6). This is due to emission of CO_2 and NH_4 in both incineration and landfill in both panels. In contrast, fossil fuel category gets the highest environmental benefits. (Note: just as positive characterization values indicate a degree of negative impact of a particular process on the environment, negative values indicate a degree of positive impacts of a process on the environment; see Section 13.6.1.) This is partly due to incineration of both panels which releases heat that can be used to replace heat produced by burning fossil fuels within the system. There are more benefits in incineration of WFP than in incineration of

Life-cycle assessment (LCA) of wood–polymer composites 287

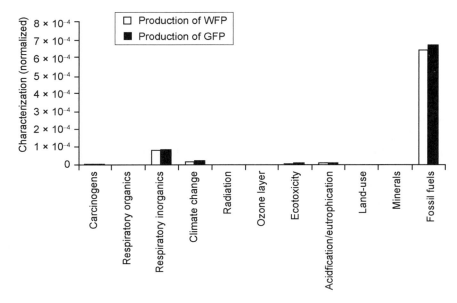

13.5 Comparison between assembly impacts of WFP and GFP.

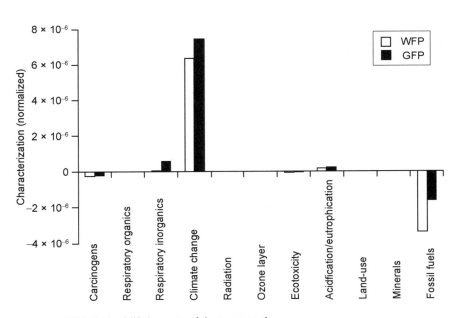

13.6 End-of-life impacts of the two panels.

GFP in this category. This is because in burning GFP, only PP releases energy whereas in burning WFP, both PP and wood fiber release energy.

13.5.4 Life-cycle comparisons of the panels

This is the stage where we compare the impacts of the whole life cycles of the two panels (assembly, use and end of life together). The fossil fuel category sees the highest impacts (Fig. 13.7). In this category, WFP has 12% less impact than GFP. Except for the fossil fuel category, there are few differences in both panels for other categories.

Figure 13.8 illustrates degree of impact of different life stages on the life cycle of WFP. The use stage, which contributes 86% of the life-cycle impacts for WFP and 85% for GFP, uses petrol, a fossil fuel, as the car moves around. The assembly stage contributes 13.9% for WFP and 14.6% for GFP on the life-cycle impacts. Production of PP contributes 90% of assembly impacts of WFP and 86% of assembly impacts of GFP. At the end of life, end of life contributes 0.09% for WFP and 0.15% for GFP on their life-cycle impacts. Thus the end-of-life impacts are insignificant compared with the other two life stages for both panels. The impacts of natural fibers in reducing the use of PP per unit volume of composite (due to their low density) and fuel consumption of the car during use is what makes them more attractive than glass fibers in this application (see Section 13.1).

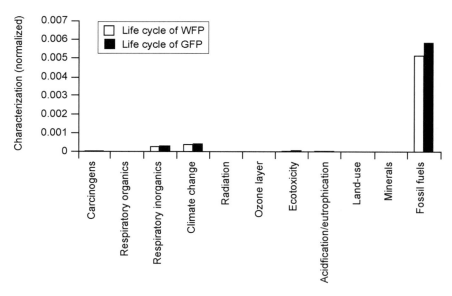

13.7 Comparison between life cycle (assembly, use, and disposal) impact of the two panels.

Life-cycle assessment (LCA) of wood–polymer composites

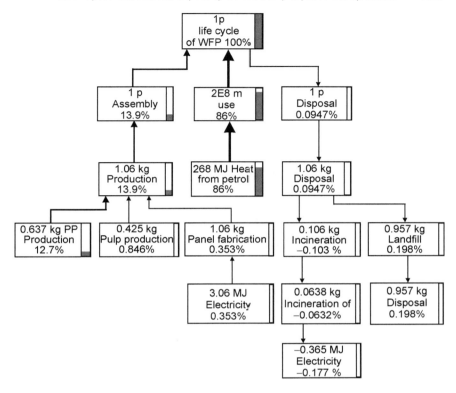

13.8 Impacts of different life stages in the life cycle of WFP.

13.5.5 Substances' contribution to impact categories in life cycles of the panels

Fossil fuels

Crude oil has the highest contribution in this category, followed by natural gas (Fig. 13.9). Coal has the lowest impact. This is true for both panels and all three fuel sources contribute more impact in the GFP life cycle than in the WFP life cycle.

Climate change

In this category, CO_2 has the most impacts (Fig. 13.10). This is despite the fact that it has the least global warming potential. CO_2 is emitted in greater amounts than any of the major greenhouse gases (see Table 13.3). Methane is the next highest contributing factor and N_2O is the least.

290　Wood–polymer composites

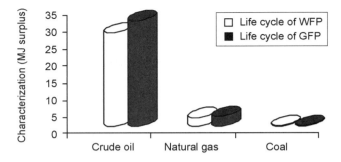

13.9 Contribution of fossil fuels to fossil fuel category.

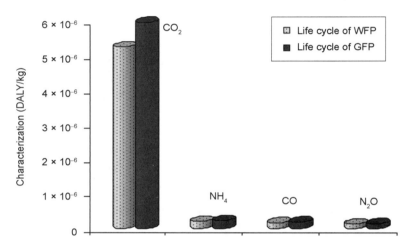

13.10 Impacts of airborne emissions on climate change category.

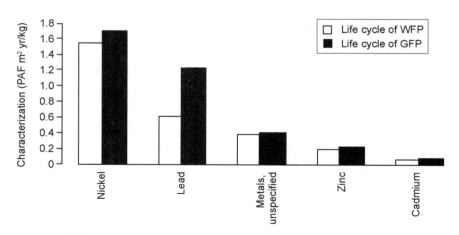

13.11 Impacts of airborne metal emissions on ecotoxicity category.

Life-cycle assessment (LCA) of wood–polymer composites 291

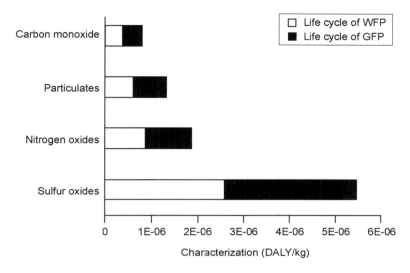

13.12 Impacts of airborne emissions on respiratory inorganics category.

Ecotoxicity

For metal emissions to air, the most dominant element is the emission of nickel, while cadmium has the least impacts for both panels (Fig. 13.11). The impact of GFP still dominates in all elements, especially lead.

Respiratory inorganics

Sulfur oxides emissions dominate this category followed by nitrogen oxides, particulates, and carbon monoxide respectively (Fig. 13.12). GFP slightly dominates impacts for all emissions.

13.6 Interpretation

13.6.1 Sensitivity analysis

Basic assumptions were made in Section 13.3. These assumptions were referred to as reference scenario. The purpose of sensitivity analysis is to check how LCA results change in response to changes in some of these assumptions. Figures 13.13 and 13.14 summarize environmental benefits or burdens that result from changes in the reference scenario. Negative percent values show the degree of environmental benefits obtained by deviating from this reference scenario. Likewise, positive percent values indicate degree of environmental burdens incurred due to these changes. Extreme cases (ultimate possible deviation from reference scenario) are considered in some cases.

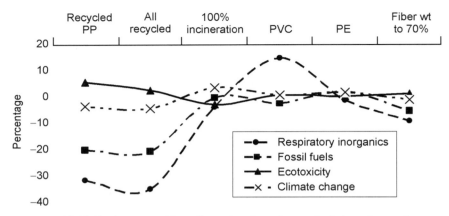

13.13 Environmental benefits or burdens due to deviations from reference scenario in life cycle of WFP.

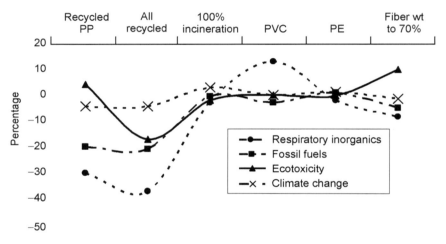

13.14 Environmental benefits or burdens due to deviations from reference scenario in life cycle of GFP.

Incineration and use of recycled materials

Wood fiber/plastic composites can range from the use of 100% virgin plastic and fiber to the use of 100% recycled plastic and fiber (Godavarti, 2005). Three categories – respiratory inorganics, fossil fuels and climate change – benefit as a result of using 100% recycled PP for both panels (Figs 13.13 and 13.14). Most benefits are realized in the category of respiratory inorganics. However, use of recycled PP increases impacts on ecotoxicity a little.

'All recycled' materials involves using 100% recycled PP and 100% recycled wood fiber or glass fiber as reinforcement. This scenario results in slight improvement compared with when only 100% recycled PP is used in all categories for WFP. There is a significant improvement in the category of ecotoxicity for GFP. In this panel, the improvement compared with when only recycled PP is used is 21% (Fig. 13.14). Recycling reduces impacts because some unit processes such as raw material extraction and processing (e.g. polymerization) are cut.

For 100% incineration (incineration of the whole panel), there are few benefits for either panel in three categories. The fourth category, climate change, experiences more burdens due to this incineration. This is because of the greenhouse gas emissions, especially CO_2, that result from burning of these products.

Replacing the matrix (PP)/increasing fiber weight fraction

Polyvinyl chloride (PVC) and polyethylene (PE) are other plastics normally used in wood fiber composites (Godavarti, 2005). In using PVC, the only slight benefit is in the category of fossil fuels for both panels. Burdens increase by about 14% for both panels as a result of this replacement in the category of respiratory inorganics. The other categories experience little or no change. So in cases where fossil fuels are a category of more importance, replacement by PVC will be slightly beneficial. Any decisions on replacement in this case would depend on which category gets higher priority between the respiratory inorganics and fossil fuels categories. For both panels, substitution by PE makes very little difference in all categories.

Most wood–plastic composites in car applications can use up to 70% by weight of fiber (Suddell and Evans, 2005). Increasing fiber weight fraction to 70% benefits all the categories except ecotoxicity. This shows slight increase in burdens (0.7%) for WFP and higher increase in burdens (11%) for GFP. The category most benefited is respiratory inorganics which has impact decreased by 9.8% for WFP and 8.75% for GFP.

Fiber weight fraction impact on distributions of main life stages

Typical changes in life-cycle impact distribution of three main life stages (assembly, use and end of life) due to changes in glass fiber weight fraction are shown in Fig. 13.15. Similar changes occur in WFP due to changes in wood fiber weight fraction. These changes in impact distribution are a result of reduced impacts that happens when PP is substituted by fiber. PP production is identified as the main polluter in the assembly stage (see Fig. 13.8). Therefore, reduction in PP weight in the composite leads to less and less impact contribution of the assembly stage.

Influence of different impact assessment methods

The three methods in Figs 13.16, 13.17, and 13.18 show that for all categories, life-cycle impact of GFP is greater than life-cycle impact of WFP.

294 Wood–polymer composites

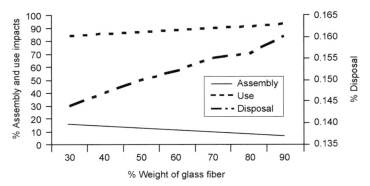

13.15 Impacts of weight variation on the contribution of three life stages to life cycle impacts of GFP.

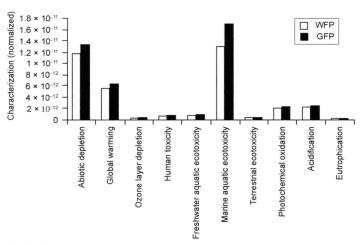

13.16 CML baseline method.

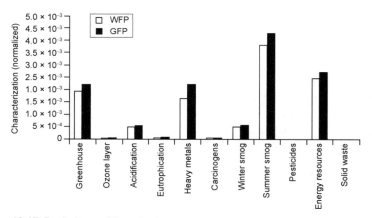

13.17 Ecoindicator95 method.

Life-cycle assessment (LCA) of wood–polymer composites 295

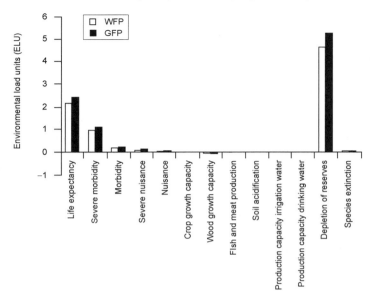

13.18 EPS 2000 method.

13.7 The possible effect of the European Union legislation on the end-of-life vehicles

The chapter concludes by looking at the recent EU environmental legislation on ELVs and its possible impacts on automotive life cycles. By the year 2000, the fifth EU Environmental Programme issued an environmental directive for the ELVs. The purpose of this mandatory directive was to better manage ELV waste and improve the environmental impacts (Marsh, 2005). To meet the directive targets, 95% and 85% by weight of vehicles should be recoverable/reusable and recyclable respectively by the year 2015 (Markarian, 2007; Smink, 2007). Interestingly, the current ELV recycling rate is already high. It is estimated that 70–80% of the ELVs is recycled in the EU jurisdiction (Gerrard and Kandlikar, 2007). While most of the recycled material is ferrous metal, the remaining 20–25% consists of mainly plastics and their composites which are hard to recycle. So the EU Directive is more likely to have a direct impact on the use of these materials.

Plastics and their composites have continued to replace metals in automotive applications due to their desirable properties such as light weight and flexibility in processing (Deanin and Srinivasan, 1998). However, despite breakthroughs in recycling technologies for plastics and plastic composites, these materials are yet to be fully recycled. There are several obstacles to overcome. For ELV plastics and composites, the difficult task is getting all stakeholders involved to work together to ensure a smooth and stable flow of materials (Ferrão and Amaral, 2006; Marsh, 2005). Also, recycling may be difficult because of plastic physical problems, such as aging and contamination, and complications due to the presence of undocumented fibers and plastics (Deanin and Srinivasan, 1998).

Nevertheless, the industry is already making progress to meet the targets set by the directive. Gerrard and Kandlikar (2007) observed that the players in the automotive industries have begun taking concrete steps to recycle and disassemble the vehicles. They noticed increased innovations in methods of recycling and separation of residues. Automotive companies such as Ford in the United Kingdom already have large recycling facilities for ELVs and there is a progress towards recycling of fiber reinforced plastics (Marsh, 2005). However, Gerrard and Kandlikar (2007) also noticed a slow progress towards designing for reuse and remanufacturing in this industry. This is possibly due to high cost of labor and a need for significant organizational changes necessary to achieve these.

On the other hand, some authors argue that the directive puts so much emphasis on meeting recycling targets that automotive manufacturers may miss other, greener, alternatives. This is especially important in cases where using these alternatives would be cheaper and perhaps environmentally better (Handley *et al.*, 2002). For instance, it is reported that DaimlerChrysler is already making 140 auto parts containing natural fibers (Gerrard and Kandlikar, 2007). Given the pressure to recycle, companies could opt for use of more metals that are easy to recycle but lead to high fuel consumption. Therefore the benefits of using natural fiber composites and bioplastics in automotives could be forfeited since these materials are presently hard to recycle (Gerrard and Kandlikar, 2007; Markarian, 2007). This could reverse the progress made in their use.

13.8 Conclusions

A life-cycle assessment carried out has shown that replacement of glass fiber by wood fiber in PP composites used as car door panels reduces environmental impacts of these panels. Air pollution, especially emission of greenhouse gases, is an important environmental topic due to the suspected impacts of these gases on global warming. Replacing GFP with WFP can reduce major greenhouse gases significantly.

The life cycles of the two panels mainly have impacts on fossil fuels, ecotoxicity, climate change, and respiratory inorganics categories according to the EI99E/E method. They have most impacts on fossil fuels. This is because the use stage, which dominates the impacts of life cycles by 86% and 85% for WFP and GFP respectively, depends heavily on fossil fuels (petrol). Production of both glass fiber and mechanical pulp, and PP, are also highly energy intensive and use much fuels. PP production dominates the impacts in the assembly stage for both panels.

Replacing PP with PVC slightly benefits the category of fossil fuels which is the major impact category although it increases impacts on respiratory inorganics significantly. Any replacement in this case would depend on which category gets higher priority between the two. Replacement by PE results in almost no changes in environmental impacts. However, use of recycled PP and recycled fiber results in generally high improvements on the main categories. Some benefits are realized by increasing fiber weight fraction from 40% to 70% for both panels in the main categories. But this increase brings more burdens in GFP in the category of ecotoxicity.

13.9 Acknowledgements

The authors would like to acknowledge support of the following people and organizations in this study: Dr Mohini Sain and Dr Suhara Panthapulakkal, University of Toronto, Department of Forestry, for providing details related to the production of the door panels, Dr Heather MacLean and Sabrina Spatari, University of Toronto Department of Civil Engineering, for helping with their experience on LCA, Queens University, Ontario Ministry of Agriculture and Food (OMAF), and Auto 21 for financial assistance.

13.10 References

Ayres R U (1995), 'Life cycle analysis: a critique,' *Resources, Conservation and Recycling*, **14** (3–4), 199–223.

Baillie C (2004), 'Why green composites?' in Baillie C, *Green Composites: Polymer Composites and the Environment*. Cambridge, Woodhead Publishing Limited.

Bruce M J, Behie L A and Berruti F (1999), 'Recycling of waste plastics by ultrapyrolysis using an internally circulating fluidized bed reactor,' *Journal of Analytical and Applied Pyrolysis*, 51 (1–2), 157–166.

Chakravorty D, Bhatnagar R and Sharma B (1981), 'Ceramic bushing for glass fibre production,' *Materials & Design*, **2** (3), 118–122.

Corbiere-Nicollier T, Laban B G, Lundquist L, Leterrier Y, Manson J A E and Jolliet O (2001), 'Lifecycle assessment of biofibers replacing glass fibers as reinforcement in plastics,' *Resource Conservation Recycling*, 33 (4), 267–287.

Daniel J and Rosen M (2002), 'Exergetic environmental assessment of life cycle emissions for various automobiles and fuels,' *Exergy, An International Journal*, **2** (4), 283–294.

Das C P and Patnaik L N (2000), 'Removal of lignin by industrial solid wastes practice,' *Periodical of Hazards, Toxic, and Radioactive Waste Management*, **4** (4), 156–161

Deanin R D and Srinivasan B (1998), 'Recycling of automotive plastics,' The 1998 56th Annual Technical Conference, ANTEC, Part 3 (of 3), Atlanta, GA, USA, 26–30 Apr., 3497–3499.

Diener J and Siehler U (1999), Okologischer Vergleich von NMT- und GMT-Bauteilen, *Angew Makromol Chem*, **272**, No. 4744.

Ferrão P and Amaral J (2006), 'Assessing the economics of auto recycling activities in relation to European Union Directive on end of life vehicles,' *Technological Forecasting & Social Change*, 73 (3), 277–289.

Gerrard J and Kandlikar M (2007), 'Is European end-of-life vehicle legislation living up to expectations? Assessing the impact of the ELV Directive on "green" innovation and vehicle recovery,' *Journal of Cleaner Production*, **15** (1), 17–27.

Godavarti S (2005), 'Thermoplastic wood fiber composites,' in Mohanty, A K, Misra M, Drzal L T, *Natural Fibers, Biopolymers, and Biocomposites*. Oxford, Taylor and Francis Group, CRC Press, Boca Raton, FL.
Goedkoop M and Oele M (2004), *Introduction to LCA with Simapro*. Amersfoort, Product Ecology Consultants.
Goedkoop M, Oele M and Effting S (2004), *Simapro 6 Database Manual Methods Library*. Amersfoort, Product Ecology Consultants.
Handley C, Brandon N P and van der Vorst R (2002), 'Impact of the European Union vehicle waste directive on end-of-life options for polymer electrolyte fuel cells,' *Journal of Power Sources*, **106** (1–2), 344–352.
Harrison R M (2001), *Pollution: Causes, Effects and Control*. Cambridge, Royal Society of Chemistry.
Hinrichsen G, Mohanty A and Misra M (2000), 'Biofibres, biodegradable polymers and biocomposites: an overview,' *Macromolecular Materials and Engineering*, **276–277** (1), 1–24.
Honngu T and Phillips G O (1997), *New Fibers*. Cambridge, Woodhead Publishing.
Joshi S V, Drzal L T, Mohanty A K and Arora S (2004), 'Are natural fiber composites environmentally superior to glass fiber reinforced composites?' *Composites Part A: Applied Science and Manufacturing*, **35** (3), 371–376.
Keoleian G A, Spatari S, Beal R, Stephens R D, Williams R L (1998), 'Application of life cycle inventory analysis to fuel tank system design,' *International Journal of Life Cycle Assessment*, **3** (1) 18–28.
Krozer J and Vis J C (1998), 'How to get LCA in the right direction,' *Journal of Cleaner Production*, **6** (1), 53–61.
Li T Q and Wolcott M P (2004), 'Rheology of HDPE–wood composites. I. Steady state shear and extensional flow,' *Composites Part A: Applied Science and Manufacturing*, **35** (3), 303–311.
Lundquist L, Marque B, Hagstrand P O, Leterrier Y and Månson J A E (2003), 'Novel pulp fibre reinforced thermoplastic composites,' *Composites Science and Technology*, **63** (1), 137–152.
Markarian J (2005), 'Wood–plastic composites: current trends in materials and processing,' *Plastics, Additives and Compounding*, **7** (5), 20–26.
Markarian J (2007), 'Strengthening compounds through fibre reinforcement,' *Reinforced Plastics*, **51** (2), 36–39.
Marsh G (2005), 'Recycling collaborative combats legislation threat,' *Reinforced Plastics*, **49** (8), 24–28.
Murphy R (2003), 'Life cycle assessment', in Baillie, C, *Green Composites: Polymer Composites and the Environment*. Cambridge: Woodhead Publishing Limited.
Ramaswamy V, Boucher O, Haigh J, Hauglustaine D, Haywood J, Myhre G, Nakajima T, Shi G Y and Solomon S (2001), *Climate Change 2001: The Scientific Basis*. Paris, UNEP.
Renilde S (2004), *Simapro 6 Database Manual: The BUWAL 250 Library*. Amersfoort, Product Ecology Consultants.
Ridge L (1998), *EUCAR – Automotive LCA guidelines – Phase 2*. Society of Automotive Engineers, SAE Technical Paper Series, no. 982185.
Schmidt, W and Beyer H (1998), *Life Cycle Study on a Natural-fiber-reinforced component*. Society of Automotive Engineers, SAE Technical Paper Series, no. 982195.
SETAC (1993), *Guidelines for Life-cycle Assessment: A 'Code of Practice'*, from the SETAC Workshop held at Sesimbra, Portugal, 31 March–3 April. Pensacola, FL:

Society of Environmental Toxicology and Chemistry: SETAC Foundation for Environmental Education.

Sivertsen L K, Haagensen Ö J and Albright D (2003), *A Review of the Life Cycle Environmental Performance of Automotive Magnesium*. Society of Automotive Engineers, SAE Technical Paper Series, no. 2003-01-0641.

Smink C K (2007), 'Vehicle recycling regulations: lessons from Denmark,' *Journal of Cleaner Production*, **15** (11–12), 1135–1146.

Smook G A (2002), *Handbook for Pulp & Paper Technologists*. Vancouver, Angus Wilde Publications.

Suddell B C and Evans W J (2005), 'Natural fiber composites in automotive applications,' in Mohanty, A K, Misra M, Drzal L T, *Natural Fibers, Biopolymers, and Biocomposites*. Oxford, Taylor and Francis Group, CRC Press, Boca Raton, FL.

Sullivan J L and Hu J (1995), *Life Cycle Energy Analysis for Automobiles*, Warrendale, Pennsylvania, USA, Society of Automotive Engineers, Inc.

Torres F G and Cubillas M L (2005), 'Study of the interfacial properties of natural fibre reinforced polyethylene,' *Polymer Testing*, **24** (6), 694–698.

UNEP (United Nations Environment Programme) (1996), *Life Cycle Assessment: What It Is and How To Do It*. Paris, UNEP.

Wotzel K, Wirth R and Flake R (1999), 'Life cycle studies on hemp fiber reinforced components and ABS for automotive parts,' *Angew Makromol Chem*, **272** (4673), 121–127.

14
Market and future trends for wood–polymer composites in Europe: the example of Germany

M CARUS and C GAHLE, nova-Institut, Germany and
H KORTE, Innovationsberatung Holz & Fasern, Germany

14.1 Introduction

According to the German definition, wood–plastic composites (WPCs) are thermoplastically processible composites that consist of varying contents of wood, plastics and additives, and are processed by thermoplastic shape-forming techniques such as extrusion, injection moulding, rotomoulding or pressing. The wood content, usually wood chips, flour or special wood fibres, can amount to more than 70%; plastics such as polyethylene (PE), polypropylene (PP) or, less frequently, polyvinyl chloride (PVC), and to a minor degree biopolymers or others serve as a matrix (Vogt et al., 2006). Additives common in plastics processing, for example bonding agents, UV protective agents, biocides, lubricants and pigments, determine the properties or colour of the material compound.

In contrast to North America, where WPCs are usually low priced and profiles are easily made from recycled plastics and wood fibres, in Germany and the rest of Europe WPC is becoming a more advanced speciality material, normally made from new materials, and used in many specifications. The potential price advantage, though hard to achieve, plays a minor role for WPCs, with technical and marketing aspects being more important. Ecological advantages are an important sales argument for WPC products for the end customers: for example, tropical timber, typically used for veranda decking in Germany, can be replaced by WPCs. Treatment with chemical wood preservatives can be omitted for the most part and WPCs can benefit from an extended useful life. The 'wood fibre as CO_2 repository' and the improved CO_2 balance in the life cycle of the products are good arguments against today's solely mineral oil-based plastics.

With WPCs, production can be carried out with almost no waste – a novelty for the timber industry, which always has to deal with large amounts of cut-offs and chips. The ease of processing to complex three-dimensional shapes is another argument in favour of WPCs. The usual processes of the plastics industry are used, so do not prove advantageous here. Technical advantages, such as greater stiffness or good acoustic properties as well as lower cost, and more stable raw material prices are important to the WPC industry.

14.2 The development of the European market: the example of Germany

The difference between the market in Germany and the rest of Europe in comparison with North America becomes clear in the choice of the basic materials. In the United States plastics and wood particles are mainly used at a ratio of 50:50; Asian countries are banking on rice husks for future developments which are locally available in large amounts. In Europe composites with a wood content of around 70%, mainly bound with new PP materials, are dominant.

In Europe the manufacturer can choose from three categories of different wood fibres, all of them by-products of saw and planer mills and native coniferous woods:

- Wood chips (sawdust and planer chips) with varying geometries and fine particulate matter/dust contents, usually purified by means of dry milling processes and sieving.
- Wood flour, approximately equal in length and width, slenderness ratio thus approximately 1, typical diameter 0.3–0.4 mm (T. Walter, personal communication, April 2005).
- Wood fibres from the refiner, typical length 2.5–3.0 mm, diameter 0.03 mm (slenderness ratio up to 100); however, these are comparatively expensive and difficult to process – and are still under development.

A specific characteristic of the German market is the competing energetic use of wood chips for the production of pellets. This has been heavily promoted for years with public funds for achieving more independence from mineral oil. Thanks to these funds and a comprehensive marketing campaign, the demand increased in 2005/2006 much faster than the production capacities of wood pellets and the mobilisation of respective wood qualities. This resulted in a shortage and thus a price increase of saw by-products, as needed for the production of WPC. The mild winter of 2006/2007 and wind damage caused by the 'Kyrill' storm in January 2007 temporarily eased the situation.

As long as the needs of the WPC industry are low, it is expected that enough wood material will continue to be available for this sector: at present the German WPC market is estimated at a total sales volume of about 20 000 tonnes (authors' estimates) – however, the results of an up-to-date survey among the manufacturers are yet to come. Thus up to 14 000 tonnes sawdust would be required here depending on the mixing ratio – but the German timber industry processes a total of approximately 10 million tonnes of log wood, round wood or round timber and other qualities (a total of 19.5 million m^3 in 2005 according to

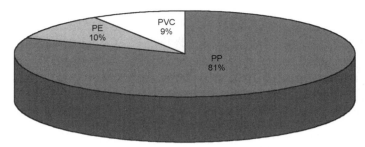

14.1 The use of polypropylene (PP) as matrix is dominant in Germany (2005) (source: Vogt *et al.*, 2006).

the VHI; Marutzky, 2007). Hence the raw material supply for WPC can be considered as long-term assured – although one has to face short-term price increases.

With regard to plastics, distinct differences from the American market can be found. In Germany and the rest of Europe, PP and PE are dominant (Fig. 14.1). In the past few months, PVC has attracted manufacturers' interest (see Section 14.5).

WPC granulates are priced according to their properties (Fig. 14.2). In general they have somewhat inferior material properties to natural fibre reinforced or even glass fibre reinforced plastics and therefore find themselves at the lower end of the price structure – comparable to non-reinforced PP. A price reduction, as for example in case of talcum fills, is not expected for the time being, owing to

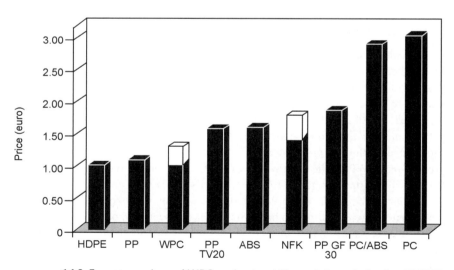

14.2 European prices of WPC and natural fibre reinforced plastics (NFRP) compared with conventional plastics (all prices in Euros per kg, potential economies of scale with WPC and NFRP are not yet included, as of 2005/06) (source: Ortmann *et al.*, 2005).

Market and future trends for wood–polymer composites in Europe 303

technological restraints and small capacities – but the customer gets a stiffer and more ecological material. In addition it is expected that the price will moderate for mineral oil-dependent plastics.

On the whole, the market for WPCs in Europe has developed very well in recent years: several manufacturers doubled their production within a year. It should not be forgotten that many smaller manufacturers have already disappeared from the market. It takes 'staying power' – and a lot of funds – to survive, particularly in the first years. The development from granulate to a market-ready product still takes a very long time. Technical aspects play a role here, but even more so do persuasive marketing and suitable distribution channels. This was clearly underestimated by many small companies.

Initially the first generation decking had not been fully developed and it did not meet the demands of consistent colouring and form stability, with the consequence of greying and distortion; leading to complaints, to which most of the manufacturers reacted by optimising the formula and technique. The same applies to injection moulding applications: many promising products were present on the market for a short time, because they could not prove themselves in practice. For example, a toy car – after initial euphoria about procedural advantages over elaborately milled and grinded parts – was taken off the market, as the corners of the racing car could not withstand the high mechanical load that emerged during play and the adhesion of the lacquering was insufficient (J. Hertenberger, personal communications, nic Spiel & Art GmbH, 2005 and 2006). But this did not discourage other companies, the materials were optimised and thus new manufacturers have been gradually starting trials or development and production. Particularly encouraging in this context is the market entry of furniture giant Ikea which has added two injection-moulded chairs to its global product range (see Fig. 14.6). Within the European countries, as measured by the number of manufacturers, active research institutions and mechanical engineering companies, Germany is the most important location (Eder, 2007).

Many factors count in favour of Germany: there are both a major wood industry with the respective global players and renowned research institutions (for example the Fraunhofer Wilhelm-Klauditz-Institut) – the same, of course, applies to the German plastics industry, speciality chemistry and mechanical engineering sector (for example Coperion Werner & Pfleiderer, Pallmann or Reifenhäuser) – and concomitantly the German-speaking area is the largest end customer market in Europe (Tables 14.1 and 14.2).

In addition, the activities are publicly sponsored in many ways – there are different programmes from pre-competition research through to market introduction. Finally the players in Germany are very innovative and dedicated, as reflected in the number of patent applications. So it does not come as a surprise that more interested parties keep trying to 'jump on the bandwagon'.

Table 14.1 Gross domestic product (GDP)/purchasing power parity (PPP) of selected European countries (as of 2005)

Country	GDP (PPP) (in million international $)*
Germany	2 498 471
France	1 811 561
Italy	1 694 706
United Kingdom	1 825 837

Source: *The World Factbook 2005*, Central Intelligence Agency (CIA), May 2005.
* International comparative value of the University of Pennsylvania depending on PPP.

Table 14.2 Estimates of the development of the markets in Germany and Europe over 10 years

	Europe (tonnes)	Germany (tonnes)
2000	3000^2–$50 000^3$	—
2002	$15 000^1$	—
2003	$20 000^4$ $25 000^1$ $30 000^2$	—
2004	—	5000^8
2005	$40 000^{11}$ $100 000^{10,14}$	$10 000^9$
2006	$50 000^{11}$	—
2007	—	$20 000^{13}$
2010	$270 000^{14}$	$116 000^{15}$

Sources: 1. Kaczmarek and Wortberg (2003); 2. AMI (2003); 3. Eder (2003); 4. Kirsch and Daniel (2004); 5. www.american-recycler.com (01/2004); 6. Kikuchi, T., personal communication (EIN) (2005); 7. Kikuchi, T. (EIN) 2002; 8. Vogt *et al.* (2006); 9. nova 2005: WPC-Studie (Prognose); 10. Hackwell and Pritchard (2005), Update 2006; 11. Nash (AMI), zitiert nach: Holz- und Kunststoffverarbeitung (HK) 1–2/06 (personal communication); 12. Kikuchi (2006); 13. nova/Korte 2006 (forecast) (personal communication); 14. Eder (2007); 15. Karus and Müssig (2007).

14.3 The most significant wood–polymer composite products in the European market

The decking market is a particular focus of discussion. This is justified from a single segment point of view, without considering the conversion parts from the automotive industry. By 2004, it was obvious that there were many other applications (Fig. 14.3). These have gained in importance meanwhile, but because the decking segment has further developed, the relative distribution should have broadly remained the same. The most important markets are presented below.

Market and future trends for wood–polymer composites in Europe 305

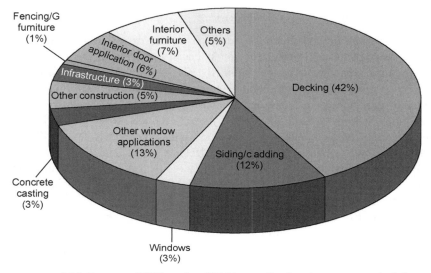

14.3 European WPC market 2004 by application (non-automotive) (source: Pritchard and Hackwell, 2006).

Table 14.3 presents the curent most important industries and their favoured techniques: it can be seen that the most important markets are the building industry, the automotive industry, the furniture industry and the consumer goods industry. As for the techniques, extrusion and injection moulding are dominant (Karus and Müssig, 2007).

Table 14.3 The most important industries and their favoured techniques

Industry/ technique	Building industry	Automotive industry	Furniture industry	Consumer goods
Extrusion	++	0	++	0
Injection moulding	0	++	+	++
Compression moulding and press flow-moulding	0	++	+	0
Current market size	++	++	+	+
Market trend	++	+	++	+
Remarks	End customer market, new products	Industry market price	End customer market, new products	End customer market, different haptics & optics

++ Very important and very large/increasing, respectively.
+ Important and large/increasing, respectively.
0 Less important and small, respectively.
Source: Carcello, F., personal communication, April 2007.

14.3.1 Building industry/deckings

Decking profiles are produced primarily with the extrusion technique. In Europe, these are usually panels of comb-shaped construction and differently structured surfaces. The structure is directly specified by the tool shape at the extruder and, depending on the manufacturer, in an additional step brushed or milled. Substructures, end strips and technical profiles, however, remain plain. Fixing materials for decking have already been produced from WPC with the injection moulding technique for a long time – a manufacturer is newly offering coffered decking that fits into the existing decking programme.

On the American market and also with the first generation of European decking there were occasionally considerable quality problems, so most of the manufacturers keep issuing unusual guarantees, for example 15 years against rotting. In order to win customer confidence, a European standard and a national, voluntary seal of approval are being worked on at the time of writing (2007). Because the product does not only combine the advantages of wood (mainly haptics and stiffness) with those of plastics (weathering resistance, processing), but also the disadvantages (moisture absorption, greying), the customer, and also the retailer and processor, have to be comprehensively trained and informed so that overreaching expectations and laying mistakes by the customer as well as processing mistakes by the manufacturer can be avoided.

A short time ago, deckings with an aluminium look emerged from close development cooperation between a customer and manufacturer – another innovative product that draws a lot of attention at international fairs. Other products for the building industry are garden fence systems, facade elements, elements for hydraulic engineering and profiles for concrete formwork. Noise protection walls, formwork panels, board materials and other products are still being developed. There is a total of about 25 decking manufacturers in Europe, most of them being located in Germany. It is a growing local market – with opportunities for foreign licensers and producers.

14.3.2 Automobiles

In the automotive sector there are three techniques for which wood fibre reinforced plastics are used. A classic case is long-fibre, press-moulded extensive parts (preliminary product: wood fibre non-wovens) such as interior door panels and parcel shelves, for which, however, thermosets are used for the most part. According to the German definition of WPC (see Section 14.1), these shall not be further looked at here.

For more than 20 years, an Italian manufacturer has been supplying thermoplastic WPC board materials for the German and Italian automotive industry (F. Carello, personal communication, April 2007). These boards are heated and post-formed, and theoretically can also be used for applications beyond automotive construction. In recent years, other board manufacturers of automotive applications in East Europe have entered the market.

Injection moulded parts made from WPCs are new to the automotive industry. These are used for glove boxes, fixing hooks and sound systems. Fan boxes and other applications are under preparation. The technical advantages are evident: for example, the better acoustical properties or the use of production residues, such as occurs with the above-mentioned press-moulded parts. In addition, there is the advantage of fewer sharp breaking edges compared with glass fibre reinforced parts; also the odour emissions frequently complained about by automotive manufacturers can be reduced to an acceptable level.

Details on this market are hard to find because of the monopoly-like supplier structure. According to estimates, in Europe about 10% of all thermoplastic WPC products were sold into this market in 2005 (Ortmann *et al.*, 2005).

14.3.3 Furniture manufacturing

German furniture companies are considered to be very successful and innovative in Europe – although Italy is still the benchmark in the field of high-class design. In this industry, many experiments are performed on details and new materials. Nevertheless up to now only a few products have established themselves on the market. The reasons for this are the high customer expectations by the (German) customer, but also the tough price competition and the initial bad materials that kept some manufacturers from further developments. The fact that, of all companies, it was Ikea to take the lead in 2006 in this field, was not expected at first, but at least it gave the whole WPC a 'moral' impetus. The WPC chair, consisting of six parts, is produced by means of injection moulding, can be assembled without any tools and consists of a mixing ratio of approximately 50:50 wood and plastics (see Section 14.2).

In addition there are examples of furniture made with extruded WPC parts, for example chests of drawers (Orgafile, South Africa, developed in Germany), seat shells for desk chairs (Werzalit, Germany), a shelving system (Maschinenbau Kitz, Germany) and lawn seats from different manufacturers. Injection moulding applications are still unusual, but also in this field there are applications for lawn chairs (e.g. produced by Steiner Gartenmöbel, Austria), grip plates and connecting materials that are produced by several manufacturers or at least offered as a sample.

Other production techniques are still in the development phase and market entry phase respectively. This particularly applies to board materials that could be used as surfaces for conventional wood materials in the future, or, thermally post-formed, to form a whole piece of furniture. The rotomoulding technique was also tested with cubic armchairs (PHK, Polymertechnik, Germany). It is especially suitable for one-piece parts that are designed as hollow parts.

14.3.4 Consumer goods

The market with the largest, still unexplored, potential is in the consumer goods sector. In particular the injection moulding technique is just starting to tap its potential. In this sector, the first manufacturers are entering the market. In contrast to extruded building materials that are mainly produced by the wood material industry, in the field of injection moulding it is the classic plastics processors who supplement petrochemical plastics with WPC, partly in combination with bioplastics. Advertising material, office supplies, flowerpots and even biodegradable urns are among the first products.

A Finnish development of a series of drinking vessels of varying size that are dishwasher and food safe and which can be used for hot drinks – and that with a wood content of at least 50% – is spectacular. Intended to be a substitute for traditional hand-carved products in Finland, the cups have difficulties on markets outside Scandinavia because of their special design and very high (because of the low quantities produced) price.

There are a total of 2000 manufacturers of consumer goods in Germany (Pritchard and Hackwell, 2006); the applications of durable plastics extend from electronic boxes to toys and convenience goods. In particular, the functionality, colouring and durability of the products – and of course the price! – must be comparable to present-day products in order to survive international competition. Owing to the large number of manufacturers and products, it should be possible on a structural level to experiment with WPC materials, i.e. to introduce single products for design or marketing reasons, thus establishing WPC in other consumer goods product lines in the coming years.

14.3.5 The appearance of wood–polymer composite products

The appearance of WPC products depends very much on the wood fraction's particle form and size. The range of particles or particle agglomerations spans from barely indistinguishable by eye from the matrix up to 8 mm, with an average size in the range of about 1–3 mm (Fig. 14.4). The kind of processing, e.g. direct extrusion or two-step processing with compounding as a first step determines the homogeneity of particle distribution from separating each particle from another or leaving bigger agglomerates or nests in the polymer matrix.

Most products have a smooth surface when they come from the tool. With decking it is different. Decking is mostly roughly corrugated at one side and fine corrugated on the other side to make it look like standard wooden decking. Some producers (e.g. Kosche and Novo-Tec, both German) offer their deckings without an after-treatment. Other producers (such as Werzalit, TechWood and Häussermann, all German) offer their deckings brushed to give the surface a wood-like appearance. In some cases the surfaces are embossed first with a

Market and future trends for wood–polymer composites in Europe 309

14.4 Size of wooden particles in different WPC products (source: Dr Hans Korte).

wood-like structure before brushing (eco Profil, Belgium). Deckings are mostly stained with grey or brown as dominant colours. In most cases only the polymer is stained and the wood particles are left in their original light beige brown colour. The wood particles set apart from the darker stained polymer matrix. In some special profiles for furniture applications the wood particles are stained too, giving the product a very homogeneous appearance.

14.4 Future trends – markets

'WPCs are a future trend of the wood material industry' (Marutzky, 2007). Although it still is a very small market, the industry considers WPC as one of the most important future technologies, in addition to the modern lightweight construction boards. What makes WPC so interesting? The material features a high technical substitution potential (Fig. 14.5): depending on the application, conventional materials can be replaced in almost any sector so that the wood material industry can open up markets that, up to now, were inaccessible: in the veranda decking sector, mainly tropical timbers and pressure-impregnated timbers (for ecological and weight reasons) are replaced, while, when it comes to replacing aluminium profiles, the non-metallic properties and energy-saving aspect are particular foci of production.

The most important trends, as partly already addressed in Section 14.3, are as follows:

- More suppliers, producers, traders; better distribution channels.
- Existing (extrusion) capacities becoming fully utilised.
- The end customer is fully informed and accepts the product on the market.

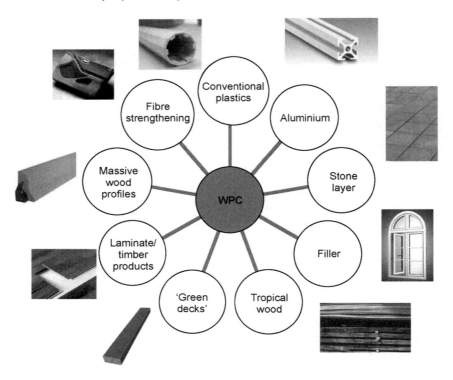

14.5 Substitution of conventional materials by WPC (source: Gahle, 2006).

- The quality becomes better and the price more attractive (economies of scale, amortised investments).
- Further extension of the range of products: in recent years, coffered decking produced by injection moulding, fixing materials, strips, underlayments, etc.
- The door stands open for further building applications. Other manufacturers will make use of this: facades, fence systems etc. However, the bureaucratic obstacles for construction parts such as in North America (e.g. bridges or simple residential buildings) are higher in Germany.
- The automotive industry offers potential, but only for big and established manufacturers with top qualities and long-term assured distribution channels.
- The furniture construction sector manages to enter global markets and offers big potential.
- Numerous developments in the consumer goods sector (today: office supplies, advertising materials, outdoor products, urns, flowerpots, but also packing materials and pallets; in the future other mass-produced articles, toys and disposable products).

14.5 Future trends: processing and materials

In Europe WPC development started with decking comparable to North America with the difference that Europe prefers hollow over solid profiles. Hollow profiles are preferred due to raw material cost savings, in contrast to North America where Trex Company started its WPC business with cheap recycled PE as a polymer matrix. In Europe WPC pioneers started with virgin PP, making solid profiles that were too expensive to sell. With upcoming new players in the past few years some solid profiles have come on the market and recycled polymer grades entered the market too. There is a trend from the early enthusiastic formulations with 70–80% of wood to lower wood contents, especially if the product is used for outdoor applications. If recycled polymer is in the focus, e.g. from CVP (clean value plastics), the wood content is less than 50%. Werzalit published its experience of 60% wood content being the reasonable upper limit for outdoor application if problems with water swelling and biodegradation are to be avoided (Schulte, 2006).

In 2006 Ikea introduced its 'light' rocking chair PS Ellan (Fig. 14.6a) made of six injection-moulded WPC parts. The material is classified to be wood and PP. The new attempt is to use the very small thermal shrinkage of WPC as an

(a)

(b)

14.6 (a) Six-part injection moulded rocking chair PS Ellan and (b) Ögla (eight parts) from IKEA made from wood/polypropylene (source: Inter IKEA Systems BV).

312 Wood–polymer composites

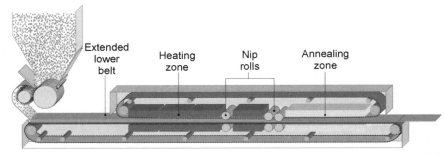

14.7 Continuous double belt press for WPC panel production from TPS (source: Technopartner Samtronic (TPS), Göppingen, Germany).

advantage. All parts of the chair are solid, with material thicknesses of up to 30 mm. To produce such a design with unfilled plastics would result in sunken surfaces. The selling price is €29.95 (US$39.99) for PS Ellan and even less for Ögla (€14.95/US$29.99).

TPS Technopartner Samtronic started from its continuous double belt pressing device for PVC floorings to develop a WPC panel pressing device (Fig. 14.7). The production width is 2400 mm and the production capacity is about 2700 kg/h with a panel thickness of 3–15 mm. The raw material in use is a pre-compounded pellet made by the Palltrusion process, which is known by plastic manufacturers as Plastagglomerator. The Palltrusion process is the most cost-effective compounding process to combine wood with plastic particles but the homogeneity of the pellets is supposed to be much less than from parallel double screw compounders.

Interestingly, the forming of panels from wood particles and PP is not an idea of the 21st century, but was invented in 1977 by Bison-Werke, Germany. The process, described in DE2743873A (Anon., 1977) and by Korte (2007), is based on standard wood panel production processes of its time. In contrast to TPS, Bison succeeded in scattering wood and polymer particles simultaneously without a pre-compounding step. After heating and pressing the panels were stored. In a second step the panels were post-formed under heat to produce three-dimensional shaped parts, e.g. automotive door skins, seat shells etc.

The use of shavings from wood planing as a wood source in WPC is difficult because of low bulk density of 60–70 kg/m and entanglement of shavings among each other. Entanglement poisons gravimetrical dosing systems and low bulk density impedes filling enough material into bulk mixing systems like hot cool mixing devices. Reimelt-Henschel MischSysteme overcomes this problem by compressing shavings to bricks or pellets before compounding these pellets with plastics in its combination of high intensity and cooler mixers. Owing to the higher shear energy at mixing, external heating of the high-intensity mixer's heating jacket can be omitted.

Another difficulty in preparing WPC is using wood with high moisture content. Most systems do not allow moisture contents of more than 8–10%. The

Market and future trends for wood–polymer composites in Europe 313

Finnish company Conenor Ltd developed a device called Conex® Wood Extruder and claims that wet materials (wood and/or washed plastics) with water contents up to 50% can be used, that coarse material will be milled down, fibre content may be up to 80% and the compound will be homogeneously mixed. Owing to special tool geometries, densities below 1.0 g/cm can be achieved.

Another attempt to achieve a very homogeneous mixture in WPC compounding is through using a planetary roller extruder which is delivered by Entex Rust & Mischke GmbH and used by Fawo®Wood, Germany. The use of such an extruder for compounding, e.g. from Bausano, Italy, is also recommended from WPC technology licensing company Strandex Europe, United Kingdom.

The development of WPC began in Europe with formulations using polyolefins or starch and colophony-based resins (Fasalex). Today in general PP is the dominant resin in Europe followed by PE. Latest developments show a big increase in PVC triggered by DeCeuninck, Belgium. AMI Agrolinz Melamine Industries GmbH, Austria, developed a thermoplastic resin based on typical thermoset resin melamine. Hype®wood is an advanced wood melamine resin composite which is treated with conventional WPC machinery and achieves the strength of thermoset resin wooden composites (Müller et al., 2005). At the international wood machinery fair Ligna⁺ 2007, Hannover, AMI presented a decking with direct digital printed surface (Fig. 14.8). The wall thickness of this profile is much thinner than that of conventional WPC decking. As a result the

14.8 Hype®wood profile from AMI with direct digital printed surface in design 'Carinthean Beech' (source: Kompetenzzentrum Holz GmbH (Wood K plus), Austria).

(a)

(b)

14.9 (a) Rotomoulded show room chair and (b) stools and table in spruce, beech and oak made from Rotowood® by PHK Polymertechnik, Germany (source: Dr Hans Korte).

profile weighs only 1.5 kg per running metre compared with about 2.5 kg of conventional polyolefine-based material.

Another new polymer for WPC formulation is PET (polyethylene terephthalate) from recycled bottles. Egger, one of the big wood panel producers in Europe, claimed WPC formulations with PET in EP 1664193 B1. Tecnaro, Germany developed Arboform, a lignin-based polymer with natural fibre reinforcement.

Foaming is not yet a viable process for WPC because of high wood contents which impede foaming. But it will become of high interest for those companies working with wood contents lower than 50%. At the AMI WPC conference in Vienna, 2007, Strandex Europe together with Cincinnati Milacron presented foamed WPC parts blown with chemical blowing agents. Not with wood as fibre matrix but with hemp, MöllerTech GmbH, Germany, developed a method called AquaCell® to foam PP with hemp shives in injection moulding by use of water as a physical blowing agent (Salamon and Armsen, 2005).

Rotomoulding with WPC is a unique development of PHK-Polymertechnik GmbH, in co-operation with IPT Institute of Polymer Technologies, Germany (Fig. 14.9). With co-grinding of wood and polymer, PHK succeeds in getting a free-flowing powder of plastic-coated wood particles called Rotowood®. The process and a showroom chair were presented at the First German WPC Congress, 8–9 November 2005, Cologne, Germany. An English version is given at Korte and Hansmann (2006). Besides the fact that rotomoulding is ideal for making big parts it is a pressure-less process in which densities of 0.5 g/cm or even less can easily be realised without foaming ingredients.

14.5.1 Interviews with different experts

Interviews with different experts yielded the following trends: WPC is one of the future trends of the wood material industry! Therefore a lot can be expected from this industry and also from the research sector in the future. In this context it is necessary to utilise the technical advantages of WPC in a much better way and not only to open up markets by means of the price advantage. Many technical advantages and special features have not yet been sufficiently examined and accordingly are also not used for marketing purposes. The hydrophobing of wood fibres yields advantages for WPC with regard to moisture resistance. Directly printable surfaces can offer an enormous potential for surface design, as already described above (Fig. 14.8) by means of a realised product. PVC as matrix offers many technical advantages, and despite certain reservations by end customers, several manufacturers are on the market, while others are also experimenting with the material and will follow.

Experts unanimously expect a large growth from the following products: park benches, façades, cable conduits and carriers for grinding materials (use of bast fibres) are technically mature and should gain more market share with an increasing degree of popularity. However, the development of new products by raw material suppliers (wood fibres) or mechanical engineering companies (extruders), as frequently happened in the past, is now too expensive for these players – besides the wood industry, it is also up to the plastics industry. Bioplastics and recycled granulates will also gain in importance in Europe in the future.

14.6 Conclusions

Wood–plastic composites (WPCs) have demonstrated technical and price requirements to become new mass-produced materials, similar to the use of thermoplastic recycled materials (Table 14.4). Injection moulding, thermoforming and possibly other technologies will gain in importance in Europe beyond the already well-developed decking sector. Here the injection moulding technique has the biggest application potential compared with other techniques. The machines are available and can be used in many ways – both in and beyond the automotive industry.

Bound by durable biopolymers, a 100% natural material can emerge from today's WPC research. Its market chances mainly depend on political frameworks: packaging ordinance, end-of-life vehicle ordinance and many other regulations as well as the globally increasing CO_2 discussion must be mentioned here as potential 'catalysts'. On a political level, natural materials should

Table 14.4 Factors that influence future development

Mineral oil price	With increasing mineral oil prices, WPC will become even more competitive; since the latest oil price increases, many companies have been intensively searching for less mineral oil-dependent materials
Availability of wood	Wood globally is the most important material. Owing to wood shortages on the world market, demand will increase for other renewable resource materials that do not exclusively consist of wood
Production technique	Petrochemical products were the focus of the R&D efforts in the past 50 years. There is considerable technological backlog in the field of the establishment and optimisation of new products and processes of renewable resources
Regulations	Regulations can create markets and promote them respectively, but also restrict them; in many cases even cost-neutral for the legislature (without tax or aid expenditure). Examples: amendment of the packaging ordinance and end-of-life ordinance, ban on the impregnation of wood with CCA salts in the USA

Source: Karus and Müssig, 2006.

Market and future trends for wood–polymer composites in Europe 317

urgently be acknowledged as CO_2 repositories, possibly giving another impetus to WPC.

In view of the comprehensive experiences from North America, technical difficulties with outdoor applications should be solvable along with marketing problems, by targeted practice-oriented promotion by public authorities: optimisation of the properties of WPC through wood/natural fibre mixtures, improvement of the extrusion process, moisture absorption as well as suitable additives, combined with a targeted market introduction of WPC products.

In the current state of global economy, the sales development benefits from the industry's search for less mineral oil-dependent plastics. According to experts, the present handicaps are mainly the many small manufacturers that neither have access to major customers nor can provide the required amounts or quality and support on a long-term basis. Among today's manufacturers, many companies are pursuing their own strategies and goals; only a few of them are organised in associations, such as the 'Working Group of Wood-Polymer-Composites' ('Fachgruppe Holz-Polymer-Werkstoffe') in the Association of the German wood-based panel industries (Holzwerkstoffindustrie eV, VHI) (see Section 14.7.2).

14.7 Wood–polymer composite codes, standards, research and manufacturing in Europe

14.7.1 Standardisation and quality

As with all new materials, at some time quality and standardisation issues arise. In Europe in mid-2003 an *ad-hoc* group was founded by Belgian and British participants to create a working group in the Comité Européen de Normalisation Technical Committee CEN/TC 249 'Plastics' (Grymonprez, 2007). WPCs are now handled under CEN TC 249 WG 13. It soon became clear that WPC standardisation could not be done from a plastic point of view alone. Therefore CEN/TC 112 'Wood-based panels' was invited to participate.

Instead of going for an EN European Standardisation procedure which has to consider a lot of European and national regulations the decision was made to go for a technical specification CEN/TS instead. The advantages are: only one language instead of three, having no public enquiry, a maximum lifetime of six years (after an initial three years, a three-year extension is possible), no standstill and no withdrawal of national standards. But the TS has to be withdrawn if an EN is published on the same subject. In November 2006 a proposal with three parts was sent to the CE secretary for enquiry and formal vote. All three parts were published in June 2007. Part 1: pr CEN/TS 15534-1 covers 'Test methods for the characterization of WPC materials and products' structured into six chapters: mechanical properties, physical properties, durability, thermal properties, burning properties and other. Part 2: pr CEN/TS 15534-2 covers

'Characterization of WPC materials' and Part 3: pr CEN/TS 15534-3 covers 'Characterization of WPC products'.

In 2005 Austria started with standardisation of WPC on a national level regarding WPC more as a special type of wooden composite. The Austrian working group ON-FNA087 AG-Holz-Kunststoff-Verbundwerkstoffe (WPC) has developed the Austrian standard ÖNORM B 3030 – 3032, where 3030 defines terminology and classification, 3031 properties and general testing methods – material properties and 3032 properties and general testing methods – product properties. Publication of the first part started in December 2005 and is ongoing.

The working group WPC within VHI (see Section 14.7.2) started its work with quality issues and developed a quality label which was published in December 2007. It will be awarded by Qualitätsgemeinschaft Holzwerkstoffe eV, Gießen (Quality Association Wooden Composite). The Label 'Qualitätszeichen Holzwerkstoffe' (Quality Label Wooden Composite) will be awarded to products that are harmless to health and technically safe (Fig. 14.10). The quality label is foreseen for decking profiles according to CEN/TS 15534-1-3).

According to the quality label the fibre base of WPC has to be a 100% wood only from certified origin (FSC or PEFC certified) and the polymers have to be a 100% virgin material or materials from industrial processes processed only once (to be proved to the Qualitätsgemeinschaft Holzwerkstoffe eV). Processing waste material of WPC can be recycled in the same process.

There are seven physical properties: bending behaviour at room temperature, bending behaviour after exposure to elevated temperature, bending behaviour after changing conditions from cold water storage to freezing and drying, swelling in cold water, swelling in boiling water, slip resistance and thermal expansion. Companies applying for the quality label will be audited in a first audit by Qualitätsgemeinschaft Holzwerkstoffe eV and later on by third-party auditing companies on a yearly basis. Each decking profile that is allowed to carry the quality label needs to have a technical data sheet that clearly records producer, product name, polymer type and wood content. With publication of CEN/TS 15534-1-3 and the quality label 'Qualitätszeichen Holzwerkstoffe'

14.10 Quality label wooden composite of the VHI for decking profiles (published in December 2007) (source: nova-Institut GmbH).

Market and future trends for wood–polymer composites in Europe 319

European producers are on track to secure the WPC market against collapse due to potential product failures of insufficient product quality.

14.7.2 Influence and work of industry associations

In Germany it is traditional to merge the interests of different companies into associations. Although these companies are competitors on the market, they know that cross-company issues, important particularly during the introduction of a new material, can only be coped with jointly. Standardisation, basic research, a joint seal of approval, technology and knowledge transfer to engineers, comprehensive marketing and information of the decision makers and customers are particularly important here. The largest German WPC manufacturers have allied in a specialised group within the Association of the German wood-based panel industries (VHI), and two other plastics associations are setting up other special sections and working groups respectively.

VHI – working group of wood–polymer composites in Germany

Among the association of the German wood-based panel industries (VHI, Verband der Deutschen Holzwerkstoffindustrie eV, Gießen, www.vhi.de) at the end of 2005, the producers of wood polymer composites founded their own working group (www.vhi.de/wpc.cfm). The major tasks of the WPC working group are:

- to act as a meeting point of all WPC producers;
- to follow up and, where needed, steer the activities in respect to technical and environmental topics which have an impact on the WPC producers;
- to attend the standardisation process of WPC in Europe;
- to initiate R&D activities.

The working group represents seven producers, including two in neighbouring countries. A main topic of the current working programme is the determination of technical standards and the establishment of an own quality scheme on the German market (Peter Sauerwein, personal communication, VHI, May 2007).

Pro-K – industrial association semi-finished products and consumer products made from plastics eV (Industrieverband Halbzeuge und Konsumprodukte aus Kunststoff eV)

The industrial association Pro-K (www.pro-kunststoff.de) was founded in 2006 as 'General Association of the Plastics-Processing Industry' ('Gesamtverband Kunststoffverarbeitende Industrie eV'/GKV) from a fusion of the trade associations 'Plastics consumer goods' and 'Building, furniture and industrial semi-finished products made from plastics'. This association has taken up the cause of 'innovation power and dynamics' for the industry. Pro-K divides into market oriented and product oriented groups, in which the leading companies of the respective subsectors of the plastics industry are involved. The foundation of a specialised group for WPC is imminent.

Federation of Reinforced Plastics eV (Industrievereinigung Verstärkte Kunststoffe eV/AVK)

The AVK (www.avk-tv.de) is a trade organisation for processors of reinforced and filled plastics and thermosets, the raw material suppliers and suppliers of semi-finished products. Members from the mechanical engineering industry, engineering firms, inspection departments and scientific institutes complement this broad spectrum. The type of reinforcement, fibre type and length, filler materials, processing techniques and matrix is irrelevant for membership. Any national or foreign company, scientific institute or organisation that meets these criteria can become a member.

The primary task is to promote the use of reinforced plastics, to advance the materials used and to improve the working processes. Other tasks are: organisation/realisation of the annual congress, education and training programmes, publishing of brochures, handbooks and publications, cooperation with public authorities and other associations at home and abroad (especially the European umbrella organisation EuCIA concerning the drafts for changing existing and new regulations), collaboration in standardisation committees (DIN, ISO, CEN), coordination of state-aided research and development projects, awarding of the approval mark for thermoset moulding materials/parts as well as the selection and awarding of the AVK innovation prizes.

To complete the many tasks, today there are 14 marketing and technical working groups, the results of which contribute greatly to the solution of the industry's central issues. The interests of the natural fibre reinforced plastics sector are represented by the working group 'Natural fibre reinforced polymers'; the foundation of an own working group 'Wood plastic composites' has been adjourned indefinitely for the time being, after a poorly attended foundation meeting in January 2006.

14.7.3 List of the most important German/European producers

Tables 14.5–14.7 are not claimed to be complete and include only those who are actively involved in development, whose products are freely available in series on the market, and who are doing active marketing. In addition to the companies mentioned above, there are numerous other suppliers on the European market; among these a total of more than 25 decking manufacturers alone. The number of all injection moulding companies with WPC experience by far exceeds the

Market and future trends for wood–polymer composites in Europe

Table 14.5 The companies in the specialised group 'Wood polymer materials' of the VHI

Company	Location	Web	Main focus
German members			
Haller Formholz GmbH	Schwäbisch Hall	www.haller-formholz.de	Granulate, international consulting
Kosche Profilummantelung GmbH	Much	www.kosche.de	Extrusion
Möller GmbH & Co. KG	Meschede-Eversberg	www.moeller-profilsysteme.de	Extrusion
Novo-Tech GmbH & Co. KG	Groß-Schierstedt	www.novo-tech.de	Extrusion
Werzalit GmbH & Co. KG	Oberstenfeld	www.werzalit.de	Extrusion, injection moulding
Other members			
Deceuninck NV	Belgium	www.deceuninck.com	Extrusion
Tech-Wood Nederland BV	The Netherlands	www.tech-wood.com	Extrusion

Table 14.6 German manufacturers not organised in the VHI

Company	Location	Web	Main focus
J. Rettenmaier & Söhne GmbH & Co. KG	Rosenberg	www.jrs.de	Granulate
Kunststofftechnik Gebr. Lenz GmbH	Bergneustadt	www.leni.de	Injection moulding
PHK – Polymertechnik GmbH	Wismar	www.phk-polymertechnik.de	Rotomoulding, rotowood
PINUFIN Oberflächentechnik GmbH	Ulm	www.holzverbundwerkstoffe.de	Granulate, extrusion
ProPolyTec GmbH	Redwitz a.d. Rodach	www.propolytec.de	Extrusion, injection moulding
TECHNAMATION Technical Europe GmbH	Geilenkirchen	www.technamation.com	Extrusion, injection moulding
TECNARO GmbH	Ilsfeld-Auenstein	www.tecnaro.de	Sustainable thermoplastic material

Table 14.7 European producers

Company	Location	Web	Main focus
Decodeck, Neofibra NV	Belgium	www.decodeck.com	Extrusion
Eco-Profil nv	Belgium	www.eco-profil.com	Extrusion
Euribix SA	Switzerland	—	Extrusion, international consulting
Inter Primo A/S	Denmark	www.interprimo.dk	Granulate, extrusion
Isosport Verbundbauteile GmbH	Austria & Slovenia	www.isokon.si	Granulate, pressed and extruded sheets
Kareline natural composites OY Ltd	Finland	www.kareline.fi	Granulate, injection moulding
Polyplank AB (publ)	Sweden	www.polyplank.se	Extrusion, injection moulding
Renolit Automotive	Italy	www.renolit.com	Sheets
Scandinavian Wood Fiber AB	Sweden	www.woodfiber.se	Granulate
Silvadec SA	France	www.silvadec.com	Extrusion
WoodN Industries srl	Italy	www.woodn.com	Extrusion
WTL International Ltd	UK	www.wtl-int.com	Granulate

space available here; it is estimated that approximately 70 companies have products on the market or samples that are immediately ready for market entry in case of demand.

14.7.4 List of German universities and institutes, working in the field of wood–polymer composites

At present several universities and private research institutions in their own specific ways are dealing with topics around WPC. These are noted in Table 14.8.

14.8 The nova-Institut and Innovationsberatung Holz & Fasern

The nova-Institut is globally active in market research, industry consulting and political advisory work, project management and online media, uses and creates expert knowledge and innovative technologies for advancing the use of renewable resources for material and energetic recovery. We use and create expert knowledge to actively arrange the raw material and energy shift on the basis of renewable resources.

- *Market research*: For almost 15 years, staff members of nova-Institut have been researching the global markets for fossil and renewable resources and

Table 14.8 German universities and institutes in the field of WPC

Institution	Location	Web	Main focus
Fraunhofer-Institut für Holzforschung Wilhelm-Klauditz-Institut	Braunschweig	www.wki.fraunhofer.de	Wood technology
Fraunhofer-Institut für Werkstoffmechanik	Halle	www.iwm.fraunhofer.de	Composites, polymer technology
Institut für Holztechnologie Dresden GmbH	Dresden	www.ihd-dresden.de	Furniture, surfaces, testing methodologies
Institut für Polymertechnologien eV	Wismar	www.ipt-wismar.de	Injection moulding, extrusion, rheology, development of rotowood technology
Dr Hans Korte – Innovationsberatung Holz & Fasern	Wismar	www.hanskorte.de	Engineering research and developments
nova-Institut GmbH	Huerth/ Cologne	www.nova-institut.de/nr	Market research, project management
Sueddeutsches Kunststoff-Zentrum	Wuerzburg	www.skz.de	Pipe and profile extrusion
University of Applied Sciences Rosenheim	Rosenheim	www.fh-rosenheim.de	Wood technology and industrial engineering

their applications. This comprehensive background knowledge is necessary to derive market and technology trends as well as strategies for the future from the dynamic developments of recent years.

- *Economical analysis*: The economical analyses focus on the price developments of raw materials and work materials as well as on the production processes for renewable resources, from primary production and processing to industrial biotechnology.
- *Feasibility studies and resource management*: Feasibility studies on investment projects combine market and economical analyses with technical feasibility investigations. Our customers are both investors and public institutions. Resource management deals with the question of how to optimally use the 'resource land' in the future for foods, energy and raw materials on a regional and global level.
- *Marketing support and knowledge transfer*: Support of companies for the

marketing of topics, projects and products. The knowledge transfer includes workshops, congresses, lectures, professional articles, books and databases.
- *Industry consulting and political advisory work*: Together with partners from our networks, we conduct comprehensive consulting for politics and industry on all technical, economical and ecological issues in the context of the energy and raw material shift and the use of renewable resources in existing and new industrial process chains. Typical customers are ministries, associations and foundations, small and medium-sized companies as well as major corporations from the automotive, plastics and wood material industry.
- *Project development and management, moderation*: In the past 15 years, nova-Institut has initiated and coordinated numerous projects in the field of applied science and industrial implementation, as well as moderated respective processes: from the project idea, the acquisition of a team, the project proposal, the leadership and management of the project to the elaboration of the reports and presentations. The acquisition of national and international funds for reducing the costs and risks of innovative projects is also important.
- *Expert networks in science and technology*: The nova-Institut is part of a comprehensive cooperative network of scientists and engineers from all over the world who deal with technical and ecological issues. These experts can be used on an individual basis for consultation and development of projects as well as the establishment of strong cooperation.

The nova-Institut is a member of the Federation of Reinforced Plastics eV (Industrievereinigung Verstärkte Kunststoffe eV/AVK), Frankfurt, founding member of the 'Cluster Industrial Biotechnology (CLIB)', Düsseldorf, member of the steering committee of the 'Hard Fibre Group' of FAO, Rome, member of the FAO management group for the 'International Year of Natural Fibre 2009 (IYNF)', as well as the branch office of the 'European Industrial Hemp Association (EIHA)', Hürth. Furthermore, the staff members are involved in numerous working groups of different associations.

Contact
nova-Institut GmbH, Chemiepark Knapsack, 50351 Hürth, Germany
Tel: +49-(0)2233-48-14 40
Fax: +49-(0)2233-48-14 50
E-mail: contact@nova-institut.de
Internet: www.nova-institut.de/nr

14.8.1 Innovationsberatung Holz & Fasern

Dr Hans Korte (Fig. 14.11) is involved with engineering research and developments in the field of wood and fibres. His company specialises in leading edge solutions for the development, introduction and marketing of new products and/

14.11 Innovationsberatung.

or procedures. Dr Korte undertakes developmental activities for his clients to provide an early insight into areas such as copyright prior to full product development and marketing. His company researches potential funding opportunities and identifies a range of solutions that will assist clients with the wider development and sales of their products.

Professional project management ensures that deadlines and targets are achieved to budget. Additional partners can be factored into solutions such as universities and, if appropriate, other businesses. Dr Korte's clients can benefit from constant review of projects and milestone checks to ensure satisfactory progress and adjustments to deadlines where necessary. Typical areas of activity include:

- the development of impregnation processes for solid wood;
- development of thermal insulating materials and the corresponding production processes;
- wood–plastic composites;
- improving the ply bond strength of timber;
- recycling of high-performance fibres from composite materials;
- structural mats;
- development of special plastics;
- quality inspection and standardisation issues are addressed; and
- investigation of industrial and wood damage.

Contact
Innovationsberatung Holz & Fasern
Dr Hans Korte, Lübsche Str. 77, 23966 Wismar, Germany
Tel: +49-(0)3841 224722
Fax: +49-(0)3841 224711
E-mail: Info@HansKorte.de
Internet: www.HansKorte.de

14.9 Examples of wood–polymer composite products

Figures 14.12–14.17 on the following pages show examples of wood–polymer composite products.

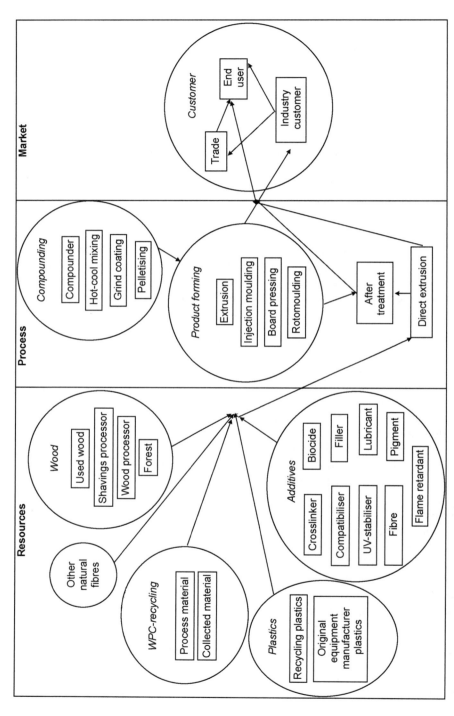

14.12 Interaction of resources, processes and markets for wood–polymer composites influencing the range of products shown in the following figures (source: Dr Hans Korte).

Market and future trends for wood–polymer composites in Europe 327

14.13 Shelving system made from WPC, based on classic aluminium constructions true to original (source: Maschinenbau Kitz GmbH, Troisdorf).

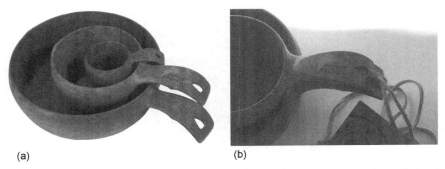

(a) (b)

14.14 Dishwasher safe and food safe: WPC cups and vessels from Finland (source: Christian Gahle).

Wood–polymer composites

14.15 Overview of different deckings (source: Christian Gahle).

14.16 Collection of different WPC board materials (source: Christian Gahle).

Market and future trends for wood–polymer composites in Europe 329

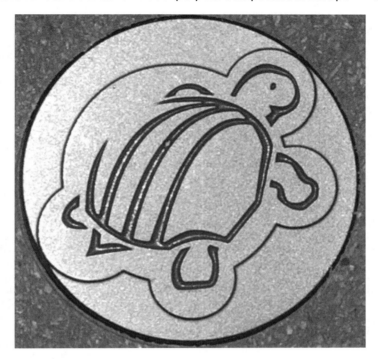

14.17 The turtle – symbol of longevity on the profiles of novo-tech (source: Christian Gahle).

14.10 References

Anon. (1977) *Verfahren zum Herstellen von plattenförmigen Formwerkstück-Rohlingen, DE2743873A*, 29 September 1977, Bison-Werke Bähre und Greten GmbH & Co KG, Springe.

Applied Market Information (AMI) (2003) Wood Plastic Composites on the cusp of take-off in Europe, Press Release of 2003-10-14.

Eder, A. (2003) Holz/Kunststoff-Verbundwerkstoffe- Marktvolumen und Marktchancen. *4th Int. Seminar für die Holzindustrie*, Vienna.

Eder, A. (2007) '3rd Wood Fibre Polymer Composite Symposium, Bordeaux', *WPCs – An Updated Worldwide Market Overview Including a Short Glance at Final Consumers*, Bordeaux, 26–27 March 2007.

Gahle, C. (2006) Neues Design mit WPC im Möbelbau – Beispiele aus Entwicklung und Praxis. In *Fourth N-Fibre-Base Congress*, 27–28 June 2006, Hürth, Germany.

Grymonprez, W. (2007) European standardisation on WPC. In *Wood–Plastic Composites 2007*, 6–8 February, Vienna, Austria.

Hackwell, B. and Pritchard, G. (2005) The market for wood plastic composites in Europe. *2nd Wood Fibre Polymer Composites Symposium - Applications and Perspectives*, 24–25 March, Bordeaux.

Kaczmarek, D. and Wortberg, J. (2003) Holz aus dem Extruder, *Kunststoffe* 2/2003, Munich.

Karus, M. and Müssig, J. (2007) Markt für Bio-Polymerwerkstoffe sowie holz- und naturfaserverstärkte Kunststoffe, *Dokumentation zur Marktanalys Nachwachsende Rohstoffe – Teil II*, Printversion herausgegeben von der Fachagentur Nachwachsende Rohstoffe eV, Gülzow.

Kikuchi, T. (2002) The Development of EinWood™ Composites, *New Wood Fiber Plastic Composites from Japan*, Tokyo.

Kikuchi, T. (2006) WPC: Marketing und Normen in Japan. In *6th Global Wood and Natural Fibre Composites Symposium*, Kassel.

Kirsch, E. and Daniel, M. (2004) Naturstoff Holz – Neue Anwendungsgebiete in der Spritzgieß- und Extrusionstechnik, *Österreichische Kunststoff-Zeitschrift* **35** 1/2.

Korte, H. (2007) Ein Produkt aus Holz + Kunststoff: WPC aus dem Jahre 1977, *Holztechnologie* **4**.

Korte, H. and Hansmann, H. (2006) WPC in rotomoulding. In *Progress in Woodfibre-Plastic Composites 2006*, 1–2 May 2006, Toronto.

Marutzky, R. (2007) Innovationen und Holzwerkstoffe – eine Einführung in das Thema. In *Holzinnovations Workshop*, Köln, 8 May 2007.

Müller, U. et al. (2005) Processing and properties of HipeWood – advanced wood melamine resin composites. In *Erster Deutscher WPC-Kongress*, 8–9 November 2005, Cologne, Germany.

Ortmann, S., Schwill, R., Karus, M. and Müssig, J. (2005) Die kommende Werkstoffgruppe – Naturfaser-Polypropylen-Spritzgießen. In: *Kunststoffe*, **95**, 23–28.

Pritchard, G. and Hackwell, B. (2006) Wood plastics composites in Europe – poised to take off. In *Fourth N-Fibre-Base Congress*, 27–28 June 2006, Hürth, Germany.

Salamon, M. and Armsen, S. (2005) Aquacell – mit Naturfasern geschäumtes Polypropylen: Verfahren, Bauteile, Eigenschaften. In *Third N-Fibre-Base Congress*, 9–10 June 2005, Hürth, Germany.

Schulte, M. (2006) Durability of WPC correlated to wood content, influence on properties of decking profiles. In *6th Global Wood and Natural Fibre Composites Symposium*, 5–6 April 2006, Kassel, Germany.

Vogt, D., Karus, M., Ortmann, S., Schmidt, Chr. and Gahle, Chr. (2006) *Studie 'Wood-Plastic–Composites' – Holz-Kunststoff-Verbundwerkstoffe*, Schriftenreihe 'Nachwachsende Rohstoffe', Band 28, Fachagentur Nachwachsende Rohstoffe eV, Gülzow.

15
Improving wood–polymer composite products: a case study

A A KLYOSOV, MIR International Inc., USA

15.1 Introduction: wood–polymer composite decking

Decking is defined as a platform either attached to a building, or unattached, as in the case of boardwalks, walkways, piers, docks, and marinas. The decking market includes deck boards, railing systems (consisting of a top rail, balusters, bottom rail, and posts), and accessories, such as stairs and built-in benches. According to Principia Partners, US demand for decking (wood and wood–plastic composites) in 2005 reached $5.1 billion, or approximately 1.2 billion board meters, and grew to $5.5 billion and 1.3 billion lineal meters in 2006.[1]

Consider a generic wood–plastic composite (WPC) deck, preferably of a premium quality. What should be done in order to avoid complaints from the deck owner? Which properties of the deck should we consider, in order to extend its lifetime as much as possible, preferably longer than that of a common pressure-treated lumber deck? In other words, what is required to make a material that is both durable enough to meet the warranty guidelines, and at the same time to be cost-efficient so that it is competitive in the marketplace? What can happen to the WPC deck in use, and how can problems be prevented? Which properties of the composite material should we aim at, what should we study in that regard, what should we test and how, what should we optimize in order to make a premium product, or, at least – for a less ambitious manufacturer – to pass the building code? These are questions considered in this chapter.

In a simple case, WPC deck is assembled with boards made of a composite material. The boards can be solid or hollow, or of an 'opened', engineered design, they can be extruded (in a common case) or compression molded. Typically, but not always, WPC boards have width of 14 cm, height (thickness) of 3.18, 2.54, 2.38, or 2.06 cm, and – for standard boards – are 3.7, 4.9, or 6.1 m in length. The board's surface can be smooth (unbrushed), brushed, embossed, or have an 'exotic' pattern, such as streaks, or simulated wood texture. The great majority of WPC boards, manufactured and sold today, are based on polyethylene (PE), polypropylene (PP), and polyvinyl chloride (PVC) and wood flour. The reason is simple: since WPC boards are competing on the market with common lumber, their price should be at the same level. In practical terms, their cost should be no more than two to three times higher than that of wooden boards, and that increase should be justified by, say, aesthetics (good look, and the absence of knots, splinters, warping, and checking), acceptable mechanical properties, good durability, low maintenance, lack of microbial degradation, resistance to termites, and possibly even fire resistance. Many customers would pay a premium price to have such a material on their decks. So far, only three plastics, named above (PE, PP, and PVC) can fit the respective pricing category, at the same time having properties necessary for the WPC material to pass the building code.

Types of wood used for common lumber decking include pressure-treated lumber, redwood, cedar, and other imported wood. Pressure-treated lumber accounted for about 78% of wood demand of decking in 2006. WPC building materials consist of a blend of wood flour or other natural plant-derived fiber and industrial grade polymers, such as PE, PP, and PVC.

15.2 Brands and manufacturers

In 2008, there are 22 companies commercially manufacturing WPC deck boards in the United States and Canada. Seventeen of them make PE-based boards (some of them make railing systems as well), two make PP-based boards, and three make PVC-based boards. These are companies which have registered their products in a form of ICC-ES (International Code Council Evaluation Service) reports approved in the United States. It should be noted here that ICC, by providing respective reports does not certify, approve or otherwise make the product 'accepted'. ICC does not carry any responsibility for the products and, in fact, does not commonly even take a look at them. Actual products are not submitted to ICC-ES for their consideration. Practically speaking, ICC-ES just approves a format in which the report is issued and prints the report. Manufacturers who make and sell their WPC products and obtained ICC-ES reports (with some exceptions noted below) are listed below.

Polyethylene-based products

- Advanced Environmental Recycling Technologies & Weyerhaeuser (ChoiceDek, Dreamworks, LifeCycle, MoistureShield, AERT)
- Epoch Composite Products (Evergrain)
- Integrated Composite Technologies (EverGreen) (there is no valid ICC-ES report)
- Universal Forest Products Ventures II (EverX, Latitudes, Veranda)
- Fiber Composites; LMC (Fiberon, Perfection, Veranda)
- LDI Composites (GeoDeck)
- Brite Manufacturing (Life Long)

- Alcoa Home Exteriors (Oasis)
- Composatron Manufacturing (Premier)
- Master Mark Plastic Products (Rhino Deck)
- SmartDeck Systems (SmartDeck) (there is no valid ICC report)
- HB&G Building Products; Tendura Industries (Tendura) (there is no valid ICC report)
- TimberTech (TimberTech)
- Trex Company (Trex)
- Midwest Manufacturing Extrusion (UltraDeck)
- Carney Timber (XTENDEX, E-Deck)
- Louisiana Pacific (WeatherBest, LP Composite, Veranda)

Polypropylene-based products

- Correct Building Products (CorrectDeck)
- Elk Composite Building Products (Cross Timbers)

PVC-based products

- CertainTeed (Boadwalk)
- Millenium Decking (Millenium)
- Procell (Procell)

15.3 Improving the performance of wood–polymer composite decking

15.3.1 Mechanical performance

The most obvious requirement is that deck should not collapse under a certain, reasonable weight (load). What is a reasonable load, though? The code specifies it as service load, and employs a 'fail' term rather than 'collapse'. The ICC requirement is a uniformly distributed load of 490 kg/m^2 of a deck, that is 4.9 kPa. This roughly correlates to a load that a common hot tub, filled with water and having five adult occupants in it, would uniformly distribute on its support. The typical dimensions are 2.4×2.0 m^2, hence it occupies an area of 4.8 m^2. Total weight of the hot tub consists of its own weight (600 kg), water (1400 kg), and people (5×70 kg $= 350$ kg), for a rather heavy scenario, total 2350 kg. Therefore, the hot tub produces a uniformly distributed load of about $2350/4.8 = 490$ kg/m^2, or 4.9 kPa. However, the ICC code also requires a $\times 2.5$ safety factor, on top of the 4.9 kPa (490 kg/m^2) requirement, that is a deck should hold a live uniform load of 12.25 kPa.

What about WPC deck boards? As a brief example, let's consider two WPC boards – Trex and GeoDeck. Trex has reported that flexural strength of their boards (solid boards of 14 cm width and 3.18 cm thick) is 9.8 MPa. It means that a Trex board placed on two joists at 40.6 cm span, would have a break load derived from the formula

$$S = \frac{PLh}{8I} \qquad 15.1$$

where S = flexural strength (9.8 MPa in this case, that is 98 kg/cm^2), P = break load, or a center point load (kg), L = span (40.6 cm in this case), h = board height/thickness (3.175 cm in this case), I = moment of inertia, equal to $bh^3/12$, in this case of a solid board, with b = board width, 14 cm. This calculation shows that the moment of inertia is equal to 37.3 cm^4 for a standard Trex board. From the above equation, a break load (an ultimate load) for a standard Trex board equals to 227 kg. This would translate to an ultimate uniformly distributed load of 80 kPa. The latter value was calculated using a standard formula for an ultimate uniformly distributed load:

$$W = \frac{16 \times S \times I}{bhL^2} \qquad 15.2$$

where W = uniformly distributed load, in kg/cm^2, b = board width (14 cm), and other factors were defined above. As can be seen, a Trex deck is able to hold 80 kPa, that is more than six times higher load than the ICC required load, including the necessary safety factor.

Similar calculations for the GeoDeck deck show that at flexural strength of the board (hollow boards of 14 cm width and 3.18 cm thick, moment of inertia of 30.5 cm^4) of 19.2 MPa, a break load at 40.6 cm span (center point load) would be 363 kg. This would translate to an ultimate uniformly distributed load of 128 kPa, which is more than 10 times higher than the ICC required load, including the necessary safety factor.

These examples illustrate that flexural strength of composite deck boards is quite satisfactory in terms set by building code organizations. It is several times higher than the respective building code requirements. Indeed, out of hundreds of thousands of composite decks installed in the United States, none is known to have collapsed during service since composite boards appeared.

How strong can a WPC deck board can be? We know that wood is very strong, at least for the same purposes WPCs are intended. The flexural strength of wood can reach 140 MPa. In WPC wood fiber is blended with a much weaker polymer matrix, which for high-density polyethylene (HDPE) has flexural strength of about 10 MPa. In a very simplified case, when, say, 50% plastic–50% wood fiber are *ideally* blended into the WPC, and wood fiber is oriented along the flow, that is longitudinally, the flexural strength would be equal to a symmetrical superposition of the flexural strength of the matrix and the fiber, that is about 75 MPa.

In reality the flexural strength of wood–HDPE composites is of 9.5–30 MPa for commercial deck boards, up to 35 MPa for laboratory WPC, obtained at carefully controlled conditions, and up to 63 MPa, obtained in laboratory conditions and in the presence of coupling agents. At the highest end of this range are wood flour (pine, 61–63%) filled HDPE composites, obtained in finely optimized and carefully controlled conditions, using best available lubricants and having flexural strength of 32.2 ± 0.6 MPa (without coupling agents) and 63 ± 1 MPa (in the presence of 3% Polybond 3029) (Jonas Burke, Ferro Corporation, private communication). Hence, in the last case flexural strength of the WPC reaches 84% of the theoretical maximum of 75 MPa.

If flexural strength is directly related to a break load of a board (in this context) placed on supports, flexural modulus is directly related to a deflection of a board, placed on supports, under a certain load. Unlike the flexural strength of composite boards, typically significantly exceeding building code requirements at commonly accepted spans (such as 40.6 cm on center), flexural modulus of plastic-based composite boards often puts certain restrictions on their installation.

There are two main situations concerning deflection of boards which may not pass the building code requirements: deck boards at a certain span (distance between neighboring joists) and stair tread at a certain span. Let us consider these situations using the same examples, Trex composite deck boards and GeoDeck composite deck boards. These examples would illustrate general shortcomings of plastic-based composite deck boards in terms of their flexibility and deflection.

The building code requires that the maximum load at certain deflection of the test span shall be recorded (ASTM D 7042, Section 5). A common load requirement for measuring deflection of deck boards is uniformly distributed live load of 4.9 kPa (see above). To choose common requirements for flooring, a deflection shall not exceed 1/360 of the span (BOCA® National Building Code/1999, Section 1604.5.4). Deflection under uniformly distributed load is determined by the following formula:

$$D = \frac{5WbL^4}{384EI} \qquad 15.3$$

where W = uniformly distributed load (4.9 kPa in this case), D = deflection, in cm, at the load, W (should not exceed $40.6/360 = 0.11$ cm in this case), b = board width (14 cm in this case), L = support span, in cm, E = flexural modulus, MPa, I = moment of inertia, cm^4.

For Trex boards ($E = 1.2$ GPa as reported by Trex, with $I = 37.3$ cm^4), deflection under uniformly distributed load of 4.9 kPa at 40.6 cm span would be 0.054 cm, that is within the building code requirements. However, at a support span of 61 cm, deflection at the same conditions would be 0.28 cm, which significantly exceeds the allowable limitation ($61/360 = 0.17$ cm).

For GeoDeck boards ($E = 2.6$ GPa, $I = 30.5$ cm^4), deflection under uniformly distributed load of 4.9 kPa at 40.6 cm span would be 0.031 cm, which is within the building code requirements (0.11 cm). Furthermore, at support span of 61 cm, deflection at the same conditions would be 0.16 cm, which is also within the allowable limitation (0.17 cm).

As a result, for Trex boards ($3.18 \times 14 \, cm^2$, solid board) maximum decking span at 40.6 cm at 4.9 kPa is allowed (ICC-ES Report ESR-1190), and for Geodeck (same overall dimensions, hollow board) that at 61 cm is allowed (ICC-ES Report ESR-1369). Only 2 WPC deck boards, GeoDeck and TimberTech (ICC-ES Report ESR-1400), are allowed to employ 61 cm span on decks; 3 more WPC commercial deck boards are allowed to have 48–51 cm span; 14 WPC commercial deck boards on ICC-ES record have 40.6 cm allowable span, and 1 WPC board has only 30 cm span allowed on decks.

These records show that flexural modulus of commercial WPC deck boards (and the respective span on decks) certainly has room for improvement. This in turn will improve quality of WPC boards and save money and material on deck joists. This conclusion is supported by consideration of support spans for stair treads (see below).

The building code requires that the maximum deflection of deck boards used as stair treads under a concentrated load of 136 kg placed at midspan shall be 3.2 mm or 1/180th of the span (AC 174, Section 4.1.1; 2000 International Building Code, Section 1607.1). For a 40.6 cm span, the allowed deflection is either 3.2 mm, or 40.6/180 = 2.3 mm.

At a span of 40.6 cm on center, deflection of stair tread under 136 kg of load will be approximately defined by the following equation:

$$D = \frac{PL^3}{48EI} \qquad 15.4$$

where D = deflection, cm, P = 136 kg, center point load, L = span, 40.6 cm, E = flexural modulus, I = moment of inertia.

For Trex solid board (see above) deflection at a span of 40.6 cm would be equal to 0.42 cm. It is too much for both criteria, that is 0.31 and 0.23 cm allowable deflection (see above). The span would not pass. Even for a span of 30.5 cm, with the allowed deflection of 30.5/180 = 0.17 cm, the deflection for this solid board under concentrated load of 136 kg would be 0.18 cm, that is slightly higher than the allowed one ($L/180$). Indeed, in ICC-ES Report ESR-1190 maximum stair tread span for Trex boards is listed as 26.7 cm.

For hollow GeoDeck, a calculated deflection at a span of 40.6 cm would be equal to 0.24 cm, that is slightly higher than 40.6/180 = 0.23 cm, but within the allowed 0.32 cm. Direct experiments with GeoDeck boards as stair treads showed that the 40.6/180 deflection was reached at an average 137 kg, which was technically satisfactory compared with the designated 136 kg.

Overall, for 12 WPC deck board brands for which allowable stair tread span is on ICC-ES record (published in the respective ICC-ES reports), only two (CorrectDeck and GeoDeck) have an allowable span of 40.6 cm, six have an allowable span of 30.5 cm, and four have an allowable span of 27, 23, or even 20 cm. This again shows that stiffness of commercial WPC deck boards (and the respective span on decks and stair treads) certainly can and should be improved. This in turn will improve the quality of WPC boards and bring them closer in this regard to the stiffness of real wood. This is one of the most challenging tasks for WPC materials.

In a similar manner, as it was discussed in the preceding section, we can ask – how stiff can a WPC deck board possibly be, if not filled with mineral fillers? We know that wood is very stiff, at least in applications WPCs are intended for, since the flexural modulus of wood is about 10 GPa. Polymers are much more flexible, and the flexural modulus for HDPE is at best 1 GPa. Again, in a very simplified case, for 50% HDPE–50% wood fiber composites, in which both principal ingredients are ideally mixed and wood fiber is oriented along the flow, that is longitudinally, the flexural modulus would be equal to a symmetrical superposition of the flexular moduli of the matrix and the fiber, that is about 5.5 GPa.

In reality for industrial WPCs, exemplified again with Trex, it is 1.2 GPa, which is four or five times less. For best laboratory WPCs, flexural modulus is close to 4.8 GPa in the absence of coupling agents (4.8 ± 0.2 and 4.9 ± 0.2 GPa for a wood flour filled HDPE in the presence of two different lubricants) and slightly higher in the presence of coupling agents (5.0 ± 0.2 and 5.3 ± 0.1 GPa, respectively; Jonas Burke, Ferro Corporation, personal communication). Hence, in the last case flexural modulus of the WPC reaches 91–96% of the alleged theoretical maximum of 5.5 GPa. It fits rather well with a similar 84% figure for flexural strength of experimentally available WPCs with respect to the alleged theoretical maximum (see above).

15.3.2 Thermal expansion–contraction

This is a rather unpleasant phenomenon of decks made of WPC boards, hence, a very important area for R&D of WPC. Almost exclusively (except for specially engineered and aerospace-designed materials), all solid materials expand almost linearly (in every direction) with increasing temperature, and contract with decreasing temperature. It is this degree of expansion–contraction that can make the phenomenon an unpleasant one, and at the same time challenging for designers with plastic and composite decking. Would a consumer like it if the ends of deck boards would quite visibly stick out of the deck frame for a few inches on a hot day, and completely disappear under the deck frame on a chilly night?

The coefficient of linear expansion–contraction (CTE, for Coefficient of Thermal Expansion) is a measure of 'how much'. In fact, the coefficient numerically describes a fraction of the board length that would be added to (expansion) or subtracted from (contraction) per one degree temperature. If, for example, a 6 m WPC board is elongated by 1.5 cm when the board surface temperature increased from 20 to 60 °C, the coefficient of linear expansion is 1.5 cm/600 cm/40 deg = 6.25×10^{-5} 1/deg C. This, by the way, is in the neighborhood of a very typical value for expansion-contraction of WPC boards.

At first, though, it might appear that a 60 °C temperature on a deck is too high. However, it is not too high for some situations. Commonly, on a summer afternoon a deck surface temperature is higher than the air temperature. To be more specific, it is about 20 degrees higher in the North, and 25 degrees higher in the South. Hence, if the air temperature increases from 20 degrees in the morning to 35 degrees in the afternoon, by about 2 o'clock a deck surface temperature will be about 55 degrees (North) and 60 degrees (South).

For neat plastics, the CTE is about twice that of WPC boards, which are about 50% filled with non-plastic materials, such as wood fiber and sometimes minerals. Since the coefficients of expansion-contraction of both wood fiber and minerals are about ten times lower than those for WPC materials, the reduction in the coefficient's value for filled WPC. In reality, the picture is somewhat more complicated, since it is the expansion-contraction of wood *along the grain* that is 10 times lower compared with common WPC. Expansion-contraction of wood across the grain is close to that of WPC. That is, an orientation of wood fiber in a WPC material can increase or decrease the coefficient of expansion-contraction.

The longer the fiber (and the higher the fiber aspect ratio) and the more it is oriented longitudinally, along the deck board, the lower is the CTE. Overall, for different commercial WPC deck boards the coefficient is in the range of 4×10^{-5} to 8×10^{-5} 1/deg C. In other words, some commercial WPC boards can expand-contract by 200% more than others. These 'overexpanded' decks are very noticeable, and sometimes cause complaints from deck owners.

There are two principal ways to decrease a magnitude of expansion-contraction of WPC deck boards. First, the formulation of WPC material (less plastic, different fillers, higher fiber aspect ratio) and/or the extrusion regime (the faster the extrusion speed, the more longitudinal orientation of the fiber) can be changed. Second, to restrain boards on the deck better, by employing more powerful nails or screws. The moving forces of expanding-contracting boards can be neutralized or blocked by powerful fasteners. Certainly, in those cases the stress has to go somewhere, and it can be expected that at some point the restraint would manifest itself into torsion damage in the joist substructure underneath, or into damage of the boards themselves. However, for those WPC boards (exemplified with GeoDeck) that were observed to be restrained enough on a real deck not to thermally expand-contract, such damages have not been noticed. However, when fasteners (screws) were being removed, those boards were producing sounds like a guitar string. Hence, they indeed 'held' a good deal of stress. Overall, values of expansion-contraction of WPC boards are largely unpredictable, and represent highly empirical values. To make composite deck boards with truly minimized CTEs is a very challenging task, not resolved as yet in the industry.

15.3.3 Shrinkage

Unlike linear thermal expansion–contraction, which is a completely reversible phenomenon, shrinkage of WPC boards is a one-way, irreversible, though limited process. If contraction of deck boards on a chilly night or during winter seasons opens a gap (sometimes 3–6 mm on long decks), the gap is typically closed on a warm day or during summer seasons. However, when boards shrink, the gap never closes back (Figs 15.1 and 15.2).

Shrinkage happens when a plastic-based board, extruded and pulled from the die, cools too fast. This means that the stretched long polymer molecules, coming from the die, do not get enough time to settle to return to their thermodynamically favorable coiled form. They are 'trapped' in the board solidified matrix in an unsettled, stretched shape.

To be exact, these 'distorted in space' polymeric molecules continue to get rearranged into their energetically minimized shape; but at ambient temperature, rates of this rearrangement are too slow, about 100 million times slower than those at hot melt temperature. If it would take five seconds for a polymer molecule to coil from its stretched shape at hot melt temperature; at ambient temperature it would take about 16 years. However, on a deck on a hot summer day it might take only a few weeks. In the North it might take a year or two. This is an explanation how those temperature-dependent figures for deck shrinkage were obtained.

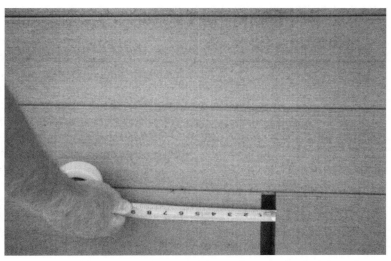

15.1 A 1-inch (25 mm) gap due to shrinkage of composite deck boards on a deck.

15.2 Shrinkage of WPC boards on a deck.

Let's take hot melt temperature (HDPE-based WPC) as 150 °C. The temperature coefficient for polymer molecules conformational rearrangements, that is a change in speed of the process by each 10 °C, approximately equals to 4. This value for so-called cooperative processes is significantly higher compared with common temperature coefficients, typically between 2 and 3. In this case a temperature drop from 150 °C to ambient 20 °C would result in 4^{13} slower rate of the polymer molecules' rearrangements, that is approximately 10^8, or 100 000 000 times. Five hundred million seconds approximately equal to 139 000 h, that is 5800 days, or 16 years. An increase in temperature from 20 to 60 °C on a deck would accelerate the rearrangement of polymer molecules in 4^4 = 256 times, that is from 16 years to 23 days of hot temperature on the deck. At 90 °C deck boards would further accelerate the rearrangement of polymer chains in 4^3 = 64 times faster compared with that at 60 °C (see above), that is in about 9 hours. This is a common annealing time period for WPC deck boards.

In order to eliminate shrinkage, WPC boards are treated by annealing in a chamber between 80 and 90 °C for about a day. It should be noted here that shrinkage is observed, and a respective annealing is required, as a rule, only for profile (hollow) WPC boards. Solid boards, because of their mass, are cooled more slowly than hollow boards, hence, the cooling time for solid boards is typically long enough to have stretched polymer molecules to settle in their coiled form. Therefore, shrinkage often is an issue only for hollow WPC boards.

The amount of post-manufactured shrinkage has several variables, and depends on the WPC formulation (especially the percentage of plastic in

Improving wood–polymer composite products: a case study 341

formulation), the extrusion speed, the cooling regime, the density of the resulting board, and on downstream pooling (and the rate of pool). While still hot, the rate of shrinkage is rapid, so the faster the cooling rate, the higher the post-manufacturing shrinkage. In its worst case, post-manufacturing (in-service) manufacturing shrinkage reaches 0.3–0.5% of the board length, that is 2–3 cm for a 6 m long board. For shorter boards shrinkage is proportionally smaller.

15.3.4 Slip resistance

Slippage on a deck is a very serious matter. A broken limb can financially devastate a good company, particularly if it is not an isolated case. Generally, WPC deck boards are more slippery than wood boards. This is easy to verify using a simple experimental set-up. Take a 1.5 m conditioned (not wet) board, fix it at a certain angle, place on the board a leather-soled shoe with a chunk of a heavy metal in it (to increase the weight of the shoe for its stability on the board), and slowly (or step-wise) incline the board until the shoe starts to slide down. With a wood board (such as pressure-treated lumber) this will happen at an angle of about 29 (at the ratio of an opposite side of the triangle to the adjacent side, that is at the tangent ratio of about 0.55). With WPC boards the same shoe will start sliding down at an angle of about 16–26 for different WPC materials with brushed or unbrushed board surface (the tangent ratio between 0.28 and 0.48). These tangent ratio values in a simplified case are called the coefficient of friction of the board.

The coefficients of friction should be determined in more controlled conditions and using professional equipment,[2] but for illustrative purposes the experiment described above would be good enough. It will show that WPC boards are commonly more slippery than wood boards, that some WPC boards are more slippery than others, and that wet boards, both wood and WPC, are less slippery than dry boards. The last statement sounds counterintuitive; however, thanks to the capillary effect of wood and WPC materials, it is, as a rule, true. For the shoe to slip, wet wood and WPC boards needs to be inclined up to 34–36 (the coefficient of friction of 0.67–0.73). Under more controlled laboratory conditions, the coefficient of friction for dry wood boards is about 0.70–0.90, and that for dry WPC materials is typically between 0.40 and 0.65.

There is a common perception, not supported by building code documents (or supported by some outdated documents), that the coefficient of friction for any materials made for walking surfaces should be not less than 0.50, in order to be safe. Not all WPC deckboards would satisfy this (unofficial) criterion. In order to minimize slippage, some WPC manufacturers texture the surface of their material (typically brushing or deep embossing). It is known that some types of plastic, for example, low-density polyethylene (LDPE), are noticeably less slippery (have higher coefficient of friction) than other plastics (for example, HDPE). However, making WPC boards with predetermined and controlled traction properties is generally not yet among WPC manufacturers' concerns.

15.3.5 Water absorption, swell, buckling

WPC materials will absorb variable amounts of moisture, some more, some less. Why this is so is discussed by Klyosov.[2] When immersed in water, WPCs absorb typically between 0.7% and 3% by weight after 24 h immersion. This can be compared to water absorption by wood, such as pressure-treated lumber, which absorbs about 24% water by weight after 24 h of immersion. When immersed in water for much longer time, commercial WPC materials absorb up to 20–30% of water, wood more than 100% by weight.

Water absorption by WPC materials may lead to a number of unpleasant events, such as board distortions, swelling, and buckling, and mold propagation. Also, saturation of WPC boards with water sometimes decreases the flexural modulus of the boards, hence resulting in a higher deflection under load. Besides, water absorption leads to a faster board deterioration, oxidation (water is a catalyst of plastic and wood oxidation) and other negative consequences.

WPC materials absorb water due to their porosity. The base plastic material of WPC, such as neat HDPE, absorbs little water. However, being filled with cellulose fiber, minerals, pigment additives (which often contain free metals, serving as effective catalysts of plastic oxidation), and during processing at high temperatures, plastic undergoes rather noticeable degradation, depolymerization, which leads to the formation of volatile organic compounds (VOCs) formation. Along with this, moisture in cellulose fiber is converted to steam at hot melt temperatures, and also adds to microbubbling in the hot melt. Steam and VOCs make the material foamed, with non-controlled porosity. This noticeably decreases the density of the final WPC product. For example, Trex's specific gravity (density) theoretically should be 1.10 g/cm^3, whereas in reality it is reportedly 0.91–0.95 g/cm^3 (Trex data). Even the fact that the range of density is listed indicates that this parameter is poorly controllable. These densities indicate that porosity of Trex material is between 16 and 21%. When the material is immersed, water fills this void volume.

Water absorption accelerates mold growth because water is a necessary component for microbial life. Typically, materials that have moisture content of 19% or lower do not support the growth of mold. This amount of moisture can be retained in the very thin upper layer of WPC profiles in humid, moist areas, with inadequate deck ventilation, for an indefinitely long time. Sometimes installation instructions are violated and deck boards are installed too close to the ground, or they are installed high enough, but the deck is 'boxed' and completely isolated underneath, creating a perfect 'greenhouse', which is moist and wet. In these cases moisture content in WPC deckboards can exceed 20–25% and may remain at that level for a time. These are very favorable conditions for mold growth and may create the respective health issues. That is why installation instructions for many composite decks prescribe a deck to be installed at least 30 cm, and preferably 60 cm, from the grade or rooftop, or provide a wider space between boards (such as 4–6 mm). Some installation instructions say that failure to adhere to proper ventilation may void the warranty.

When WPC boards absorb water, they swell. When the boards are in close contact with each other, a very high pressure can develop in the area of contact, reaching over a thousand Newtons. This may lead to boards buckling. Typically, for WPC boards to be buckled they should be in contact with water for a long time, days and weeks. However, the lower the board density, the higher the swell, the more likely boards would buckle after their shorter exposure to water. Buckling typically results from an improper installation of a composite deck – causing a prolonged contact with water (from outside or from inside of deckboards, such as for hollow boards), lack of proper gapping, etc.

In order to minimize water absorption by WPC boards, they should have as high a density as their formulation allows. To achieve this goal, a proper amount of antioxidants should be introduced to the formulation. Antioxidants slow down the plastic degradation under high temperature, attrition, etc., hence, minimize VOC and/or CO_2 formation and the respective decrease of density. Moisture in the ingredients also leads to a decrease of the final material's density, hence cellulose fiber should be dried, if necessary. Finally, vented extruders remove VOCs and steam from hot melt and greatly increase density of the final product.

15.3.6 Microbial degradation

On 28 May 2004, the Superior Court of New Jersey certified a nationwide class action in a case originally filed in 2000 against Trex Company, Inc. and ExxonMobil Corp. The case alleged that the Trex product was defective. A press release which announced the class action on 2 June 2004, said: 'In addition, although the Company claims that the product does not need sealants, after the product exhibits mold, the Company allegedly recommends that consumers apply sealants'. While we are not going to discuss here merits of the class action and history of the case, we just point out the words 'after the product exhibits mold ...' apparently recognized by the manufacturer. Mold of the product was one of main reasons of the class action against Trex. This is how serious it can be.

The issue of where mold is coming from and in which form is covered in detail by Klyosov.[2] Here we just mention that appearance of mold on some WPC decks and stairs can be made more likely by certain types of WPC formulation (and less likely by others), by improper deck installation, and by climatic conditions.

WPC formulations that invite mold are those with a relatively high porosity (typically made using moist wood fiber) and, hence, having lower density, than it might be in the final product. Particularly, this happens if the WPC profile is extruded in the absence of or with too little antioxidant. Typically, these WPC materials absorb more water than other WPC products on the market. Formulations that make mold on the deck are less likely to contain not only antioxidants, but also minerals, which create a natural barrier for microbial degradation of WPC materials. Obviously, biocides and other antimicrobial agents in the formulation help to prevent or slow down mold growth on decks. As an example, a WPC post-sleeve was made with no added antioxidant, unlike regular post-sleeves. As a result, it absorbed water in the amount of 3% per bulk material (after 24 h under water), compared with a regular value of 1%, and after some outdoor exposure developed black mold.

Improper installation of WPC decks is associated typically with lack of ventilation at the bottom of the deck and/or deck level too close to the ground, particularly when the ground is wet. Water in those decks is retained for a long time and that in turn creates more favorable conditions for mold to grow. Naturally, in wet areas rainwater absorbed by decks dries out much more slower than in dry areas, which may lead to mold on decks (see, for example, Fig. 15.3).

Unfortunately, antimicrobial components (often as much as $30–80/kg) are often too expensive to be affordable by WPC deck manufacturers. Biocides for plastics are commonly designed aiming at a quite different, and higher, price structure, e.g. for small plastic-made biomedical devices. If the cost is, for example, $10 for a 50 g device, and $80/kg for a biocide, the latter cost is still

15.3 Mold on composite deck boards.

affordable at 0.1% biocide load. This would increase cost of the device by 0.4 cents, or by 0.04% of the total cost of the product. However, 0.1% of the $80/kg biocide in WPC deck boards that cost otherwise $0.70/kg would increase cost of boards by 8 cents, that is by 11% of the cost. Realistically, at 0.2% of an effective antimicrobial agent in a WPC formulation (that is, at 2 g per kg of the composite) and the allowable price of the formulation to be increased by 2 cents per kg (cost of materials), cost of the biocide for WPC boards should not exceed $10/kg.

15.3.7 Termite resistance

On the list of homeowners' problems, termites rank very highly. According to the *Boston Globe* (July 9, 2000), which in turn refers to Bay Colony Home Inspections, between 20 and 25% of the homes sold in most areas of New England have termites or have had them in the past. Toward the south of the United States, the problem is greater. And, of course, termites do not just live around the house. In many cases termites eat as much as 80% and more of all the structural components of a house, including its deck, if it has one. According to Home Inspection data, in 70% of the above cases the termites have been treated and returned.

There are several main types of termites. Some of them require elevated moisture content, such as dampwood termites. Some live deep inside wood, such as drywood termites. Some live in colonies in the ground and build tunnels, using wood as their food. Generally, WPC materials are very resistant to termites. Despite wood fibers not being completely – as a rule – encapsulated into the plastic matrix, but form a sort of continuing chain across WPC materials (unless the ratio of plastic to fiber is really high, more than 80%), termites cannot get into the plastic matrix. At best, termites can only slightly trim cellulose fiber at the WPC surface.

As a result, weight loss of WPC materials by termites is negligible, if anything. Let us consider GeoDeck composite board as an example. It was subjected to termites collected from a colony of subterranean termites *Reticulitermes flavipes*, according to the procedure given in ASTM D3345-74. Five of $2.5 \times 2.5 \times 0.6\,\text{cm}^3$ blocks of southern yellow pine sapwood and five blocks cut from WPC board were exposed to termites for 8 weeks. With the wood samples, weight loss due to termite action was of $9.1 \pm 0.7\%$. With the WPC samples, two out of five samples were practically untouched (no weight loss), and an overall, average weight loss was $0.2 \pm 0.2\%$.

Here are a few examples of termite resistance rating, showed in the respective company records:

- Trex, rating 9.6.
- GeoDeck – No attack, rating 10.
- Nexwood – No damage, rating 10.

It can be seen that commercial WPC deck boards are dramatically more termite resistant than wood lumber.

15.3.8 Flammability

Polyethylene and polypropylene-based WPC materials are flammable. Flammability of materials is characterized in many different ways, one of which is the flame spread index (FSI). As reference values, the FSI for inorganic reinforced cement board surface is arbitrarily set as 0, and for select grade oak surface as 100 under the specified conditions. The FSI for ordinary wood species is typically between 100 and 200, for some special cases is as low as 60–70. An average FSI for about 30 different wood species is 125 ± 45.

For comparison, wood fiber filled HDPE hollow boards have an FSI around 150; solid boards, about 80–100; WPC hollow boards containing minerals, around 100; PVC-based wood-filled deck boards, typically between 25 and 60. Both ordinary wood species and most WPC deck boards belong to Class C category of flammability in terms of flame spread. There are four basic categories, or classes, for flame spread index: Class A, with FSI between 0 and 25; Class B, with FSI between 26 and 75; Class C, with FSI between 76 and 200; and below Class C, with FSI above 200 (unclassified materials). Classes A, B and C sometimes are called Classes I, II, and III.

Until recently, the flammability of WPC decks was not even a concern. Decks were not supposed to be inflammable. What is the point if the house would burn and the deck would stay? Then it was recognized that brushfires often ignite a house via the deck. Now legislatures of several states, California first, are working on a new law, according to which decks should be fire-proofed to some extent. This poses a new challenge for WPC decks. The new law came into effect January 1, 2008 (the California Code of Regulations, Part 12, Title 24, Decking SFM Standard 12-7A-1).

Technically to make a WPC deck of a low flammability is not difficult. Principally, there are two ways to go – either to load a WPC formulation with flame retardant components, or to employ PVC (or other low-flammable plastics) as a base plastic for WPC.[2] As always, optimization is the name of the game. PVC is not considered to be an environmentally friendly material. When ignited, the resin releases hydrogen chloride (HCl), a toxic and volatile strong acid. If not stabilized properly, PVC can release HCl under direct sunlight, at high temperature of a hot deck surface on a sunny summer day. Some flame retardants, particularly polybrominated diphenyl esters, are also far from benign. Mineral flame retardants, such as aluminum trihydrate and magnesium hydroxide, are required at a high loading level (up to 40–50% w/w) to be effective.

Considering that plastic often takes 40–50% w/w of flame retardants, there is no room for wood filler in WPC, which will not be WPC any more, but rather a mineral-filled plastic. At any rate, replacement of wood fiber of 6–10 cents/kg with mineral flame retardant of 40–60 cents/kg would significantly increase the cost of the resulting material. All these questions pose a great challenge to WPC manufacturers aiming at fire-proof composite deck boards.

15.3.9 Oxidation and crumbling

One of the most unpleasant, damaging and unexpected features of some WPC materials is their elevated vulnerability to oxidation, leading to board crumbling (Figs 15.4 and 15.5). In the process of crumbling, the WPC board shows tiny and then developing cracks, its surface becomes dustier and softer, until one can easily scratch it, leaving deep tracks. Eventually the board can collapse under its own weight.

There are number of factors leading to accelerated WPC oxidation, and lack of antioxidants (in the initial, incoming plastics) and/or insufficient amounts added to the formulation is the most important of them. The addition of antioxidants aims both at preserving the plastic during the processing at high temperatures, and protecting the WPC profile during service on a deck under the damaging effects of sunlight, air oxygen, water, pollutants, and other elements.

Briefly, antioxidants quench free radicals that are formed in the process of plastic degradation by oxygen and initiated by temperature and UV light, and assisted by moisture, stress, and the presence of metals and other catalysts of plastic oxidation. If not intercepted by antioxidants, the polymeric plastic is

15.4 An advanced step of crumbling of WPC deck boards (Arizona).

348 Wood–polymer composites

15.5 An advanced step of crumbling of a WPC deck board (Arizona).

degraded (depolymerized) so much that it loses its integrity and ceases to be a plastic anymore. It is converted to a loose powdery material, mainly a filler.

Other factors that accelerate WPC oxidation, and hence decrease durability and shorten a deck's lifetime, are decreased density (specific gravity) of the board compared with the maximum density for the same board, the presence of metals (in pigments, lubricants, other additives), moisture content, and unsettled stress in boards. Decreased density is the result of an increased porosity of the boards, due to moisture presence in the initial ingredients of the WPC (of wood fibers first of all) and plastic degradation during the processing (due to overheating, excessive shear, and/or lack of antioxidants). An excessive porosity allows oxygen to permeate into the WPC material 'from inside', significantly increasing the accessible surface area, along with the rate of oxidation. Metals, particularly free metals, often are efficient catalysts of plastic oxidation. Moisture is also an effective catalyst of plastic oxidation.

Until recently, the effects of these factors and their quantitative manifestation were practically unknown and not recognized either in the WPC industry or even in academic research in the area. That is why the acute deterioration and crumbling of some WPC boards turned out to be quite unexpected and puzzling, and resulted, in some cases, in an avalanche of warranty claims. These cases are considered in detail by Klyosov.[2] It turned out that the progressive deterioration and crumbling resulted from WPC oxidation, and as soon as it was recognized, measures were taken. The oxidative induction time (OIT) parameter was introduced into characterization of WPC products and evaluation of their lifetime in the real world, on real decks.

Essentially, the OIT value quantitatively describes a lifetime of a composite

Improving wood–polymer composite products: a case study 349

(or actually any organic-based) material during its accelerated oxidation in pure oxygen at an elevated temperature, such as 190 °C. For example, for unstabilized (without added antioxidants) WPC materials the OIT can be as low as 0.3–0.5 min. The lifetime of such WPC boards in the south (Arizona, Texas, Florida) can be as low as only several months. The 'lifetime' in this context is a time period by the end of which the consumer can see there is something wrong with the deck and calls for help. In real terms, the deck owner contacts with the manufacturer and files a warranty claim.

For partially stabilized WPC materials the OIT can be between 1 and 10 min. A number of commercial WPC deck boards being sold on the market fall into this range. Depending on a deck profile (solid or hollow) and the board density, and, of course, on location/geography/climatic conditions, the lifetime of the deck can vary, but there is a risk that these boards would not live long enough to see the end of their warranty time period, particularly in the south. For well-stabilized WPC boards the OIT can be in the range of stabilized plastics (15–100+ min).

Figure 15.6 shows the OIT values for commercial wood-plastic decking boards. Twelve of them have an OIT lower than 10 min. This is a troubling

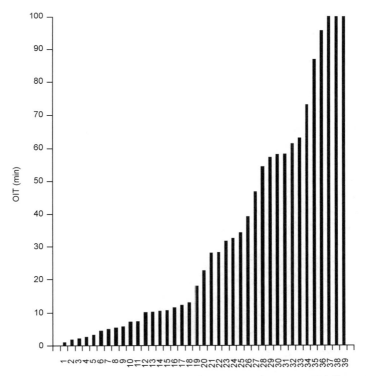

15.6 The oxidative induction time (OIT) values for commercially available WPC deck boards. The manufacturers and board names are numbered in the order of increasing OIT values.

observation, since these boards can be time bombs for the manufacturers. Boards with an OIT above 15–20 min are not going to suffer from deterioration and crumbling due to oxidation, at least caused by hot summer season temperature and UV light. Certainly, these boards can be damaged by other mechanisms (water, mold, bacteria, or algae), or can be broken by force, or burned by fire (which is a very rapid oxidation), but there are means to minimize each of those effects as well. All these aspects are also covered by Klyosov.[2]

15.3.10 Photo-oxidation and fading

Fading is a generally accepted feature of WPCs, probably because people get used to it with common wooden decks, hence, this phenomenon has a kind of grandfather status. However, some composite materials fade less than others, and some much more. Clearly, customers generally prefer to have their deck not fading at all; however, they are either not informed on the prospective fading, or do not know that some WPCs practically do not fade, or accept the fading as given. When the sun irradiation on their deck is uniform throughout the day, it does not create a problem. However, in many cases after just a few months, a difference in color on their deck is noticeable (see Fig. 15.7).

Figure 15.8 shows a difference in fading (in terms of lightness) of 32 commercially available WPC deck boards after 1000 h of the accelerated weathering. A difference between ΔL (on the Hunter Lab color scale) is between 0.4 and 35 L units. In a simplified manner, one unit is the first shade difference that the naked eye can normally detect. That is, a difference in lightness by 0.4 units one cannot detect, while 35 units of fading from, say, the initial $L = 53$ to 88 results in the final lightness of almost a white sheet of paper.

It is rather difficult, if possible at all, to quantitatively translate the fading in the weathering box to the real world. However, some very approximate comparisons can be made. Depending on the material color, one day in the weathering box under 'standard' conditions (340 nm, 0.35 W/m^2, 102:18 cycle, 63 °C black panel temperature) often corresponds to 9 ± 4 days of natural weathering in the US Midwest and New England. This figure is often called the

15.7 Fading of a composite deck. The 'Welcome' mat has just been removed.

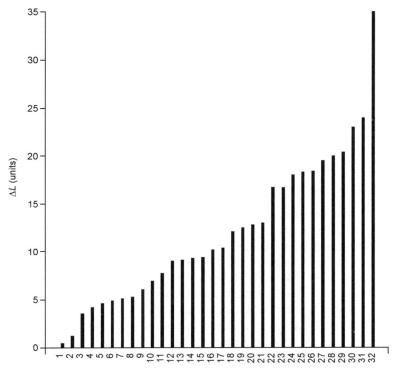

15.8 Fading of commercial WPC boards in terms of lightness (*L* in the Hunter Lab color scale) shift after 1000 h of accelerated weathering (vertical axis) in Q Sun-3000 weathering chamber at 0.35 W/m² at 340 nm, 102:18 min (UV light: UV light + water spray) cycle, 63 °C black panel temperature. ΔL values vary from 0.4 to 35 units/1000 h, that is in almost 100 times, for the boards available on the market. Since some boards can be outdated compared with current manufacturing, they are not named but numbered.

'acceleration factor'. In Arizona and Florida the acceleration factor is about 50% of the above, that is around 3.5.

Hence, a detectable level of lightness change ($\Delta L = 1$) for the WPC board with the lowest degree of fading in Fig. 15.8 (0.4/1000 h) will be reached in the Midwest and New England after about two and a half years, and a level of $\Delta L = 5$ (a surely noticeable change of lightness) will be reached after at least 15 years, taking into account some slowing down the fading process with time. On the other hand, a detectable level of lightness ($\Delta L = 1$) for the WPC board with the highest degree of fading in Fig. 15.8 (35/1000 h) will be reached in these geographical areas after about 10 days, and a level of $\Delta L = 5$ will be reached after about 2 months. This is confirmed by direct observations.

Indeed, if some commercial composite deck boards were placed outside under direct sunlight, their fading would be noticeable after only a couple of weeks. This can be observed for boards in Fig. 15.8 with $\Delta L = 20/1000\,\text{h}$.

Approximate calculations show that the 'theoretical' figure, based on the acceleration factor of 9 ± 4 (see above) would be equal to 19 ± 8 days, that is close to the observed time period.

Fading of composite materials depend on many factors, some of which are related to the WPC composition (wood fiber content, type of cellulosic fiber, amount of UV stabilizers and antioxidants, amount and type of colorants), and some to the outdoor conditions (covered or open deck, amount of moisture on the deck, other climatic conditions). It does not appear that processing of WPC and the profile manufacturing noticeably affect the material's fading.

15.4 Conclusions

At present, the main share of WPCs goes for decking and railing systems (deck boards, stairs, posts and post sleeves, handrails and bottom rails, post caps, balusters, and other small accessories), and similar structures attached to the exterior of dwellings, as well as boardwalks. A relatively small amount of commercially produced WPC goes for siding, fencing, pallets, roofing tiles, and window frame lineals. Other products, such as pilings, railroad ties, marinas, window blinds, and sound barriers, are rather experimental, not commercial as yet, or sold in very small quantities compared with principal WPC products. Automotive products (interior panels, trunk liners, spare tire covers, package trays, etc.) form a separate category of composite products, often use long cellulose fiber, and fall into a quite different price category. They are not considered here.

The public view of WPCs is hard to evaluate objectively. Many have never heard of WPC. Many prefer 'real wood', and they are hard to blame. Wood is an excellent material, far exceeding WPC in many properties, first of all in strength and stiffness, in slip resistance, and – with many types of wood – in fire resistance. Only PVC-based wood composites are generally less flammable than wood. Common wood, however, is an inferior material compared with WPC with respect to water absorption, microbial degradation, and durability. There are exceptional types of wood that satisfy the taste of a sophisticated customer, but those types are too expensive for the general market. Overall, many customers gladly accepted the appearance of WPC on the market, many are doubtful, many rejected it outright. Nevertheless, WPC-based building products have been capturing the market quickly for the last ten years.

What is so attractive about WPC products? This question is fair only with respect to WPC decks, since other products have not attracted enough attention on the market. Well, WPC decks, as a rule, look pretty good. I have one. Anyone can walk on them barefoot without risking getting a splinter in a foot. There are no splinters whatsoever. Then, WPC decks indeed require minimum maintenance. And maintenance with wood decking means – first of all – regular staining and painting. WPC decks do not require them, since they are colored – if colored properly – for life. If WPC boards contain the correct colorants in accurate quantities, along with precise amounts of the right antioxidants and/or UV absorbers, they would not fade. Unfortunately, very few brands of WPC deck boards are made this way. Hence, they commonly fade. Besides, in the south, decks often require treatment with antimicrobial and antitermite chemicals. WPC boards do not require it, since they are much more resistant to biological degradation. It should be noticed here that bioresistance of WPC deck boards diminishes with increase of wood fiber content (above 40%), and increases with mineral content (silica, calcium carbonate, talk, etc.).

WPC decks require, though, normal washing, cleaning and other care, as conventional wood decks do. It is obvious that barbeque on a deck would unavoidably lead to grease and fat stains, that potato salad dropped on a deck made of either wood or WPC, leaves stains that are not easy to remove. In fact, it is much easier to remove grease from a WPC deck than from a wood deck.

Overall, a WPC deck is much more durable than a wooden deck, and requires much less work in the long run. This is certainly attractive for some people. However, it requires a steep payment upfront. This is off-putting for many people. Both features of WPC affect public acceptance, and both are considered to be a practicality issue.

A key issue in public perception regarding a new product in the building industry is an appreciation of the product both by builders and homeowners. It seems that WPCs hit the right spot. Deck installers commonly like WPC deck boards as they are safe to work with due to lack of splinters, easiness to cut, saw, nail, and screw (except PP-based WPC deck boards, which are too tough, but this problem is generally solved with development of special fastening systems). These properties of WPC deck boards result in the ultimate goal of any installer for hire to be accomplished: a good speed of deck installation, hence, faster, and a better pay.

Another important factor in success in the builder's market is the market accessibility for the product. Technically this means the speed from the plant's warehouse to a lumberyard, to distributors, dealers, suppliers, retailers, and to the end user. This is called 'strong channel position to access the market' and 'distribution channels', and leads to a competitive advantage of some manufacturers over others.

15.5 References

1. Principia Partners, *Composite Decking, 2004 vs. 2005*, March 2006.
2. Anatole A. Klyosov, *Wood Plastic Composites*, John Wiley & Sons, 2007, p. 698.

Index

accelerated tests 262–5, 271
  creep 175–6
  weathering 150, 151–2
acceleration factor 350–2
acceptance criteria 260
acetylation 155
acid-base interactions 47
acid scavengers 24, 25, 32
acrylic acid grafted polypropylene
    (AAPP) 48
additives 23–40, 161–2
  biocides 24, 25, 36–7, 161, 251, 344–5
  coupling agents *see* coupling agents
  density reduction additives 24, 25, 36
  fillers *see* fillers
  flame retardants 24, 25, 38–40, 346–7
  foamed WPCs 228, 250–2
  future trends 40
  lubricants 24, 25, 26–9, 195, 196, 250–1
  moisture absorption prevention 156
  product aesthetics additives 24, 25, 37
  protection against weathering 157–60
  rheology control additives 24, 25, 26–9
  stabilisers *see* stabilisers
adhesion, interfacial 41, 42, 68–9
  assumption of perfect adhesion in
      modelling 126
  and creep 181–3
  evaluation of interfacial interactions
      and 48–60
  failure modes 43
  improving 60–6
  mechanisms 45–8
  wetting, dispersion and 43–8
  work of 44–7

aesthetics, additives for 24, 25, 37
agricultural fibres 224
air pollutants
  greenhouse gases 283, 289, 290, 296
  metal emissions 283, 284, 290, 291
Alcan 209
allowable design stresses 260
aluminium trihydrate 39–40
aluminosilicate glasses, amorphous 28–9
American Society for Testing and
    Materials (ASTM) 258
  specifications 259–60, 263–4
AMI (Agrolinz Melamine Industries)
    313–15
amides 27
ammonium phosphate 39
amorphous glass additives 28–9
antimicrobial additives 24, 25, 36–7, 161,
    251, 344–5
antimony trioxide 38
antioxidants 24, 25, 31–2, 156, 343, 347
appearance of WPC products 308–9
AquaCell method 315
Arboform 315
Arrhenius equation 173
ash 12
Asia 301
aspect ratio 120, 125
assembly stage 278–9, 286, 287, 288–9,
    293, 294
attenuated total reflect FTIR (ATR-FTIR)
    49
Austria 318
automotive industry 264–5
  Europe 305, 306–7, 310

life-cycle assessment of car door
    panels 273–99
AVK 320

Bakelite 4–5
barytes (barium sulphate) 35
batch processing 245
belt feeder 77–9
biaxial orientation 209
biocides 24, 25, 36–7, 161, 344–5
biological degradation
    decking 343–5
    wood 18, 143–4
    WPCs 152–5
        protection against 160–1
Bison-Werke 312
blocking/screening 32
blowing agents 36, 235–48
    chemical 36, 239–44
    mechanisms of blowing agent-based
        foaming 235–9
        cell growth control 236, 238–9
        cell nucleation 236, 238
        phase changes 235–7
        polymer/gas solution 236, 237–8
    physical 36, 244–8
board materials, WPC 328
Boltzmann superposition principle 170–1
branching of polymers 4, 5
break load 333–5
bridges 219, 225, 268–70
brown-rot fungi 144, 153–4
buckling
    decking 342–3
    hygroexpansion 132–3
building codes 257–60, 271
    performance measures and 259–60, 271
building industry 257–72, 353
    applications of WPCs 265–70
        product opportunities 266–70
    Europe 305, 306, 310
    properties of WPCs 260–5
bulk materials 72–3
    storage, transportation and conveying
        74–5
Bürgers model 168–70, 171, 177, 178,
    179–80
Burgess Bridge Co. 219
BUWAL 250 database 277

calcium carbonate 220–2
calcium-zinc heat stabilisers 32–3
Canada 258, 332–3
car manufacturing see automotive
    industry
carbon dioxide 283, 289, 290
carbon fibres 19, 20, 118–19
carbonates 35
cell coalescence 238–9
cell collapse 238–9
cell density 230
cell growth control 236, 238–9
cell nucleation 236, 238
cell size 230
cellulose 11, 12, 102
centre line distance 84
centric pelletiser on air 95
chairs 303, 307, 311–12, 314
chemical blowing agents (CBAs) 36,
    239–44
    endothermic and exothermic 239–42
    extrusion foaming of WPCs 242–3
    injection moulding foaming of WPCs
        243–4
chemical bonding 45, 46
chlorothalonil 161
cladding/siding 224, 266, 305
climate change see greenhouse gases
CML method 281, 282, 294
coefficient of friction 341–2
coefficient of thermal expansion (CTE) 9,
    19, 337–9
cold face cutter 90–1, 92
cold plasma treatment 61
colorants 251
colour fading 151–2, 159–60, 350–2
combined fuel economy 279
Comité Européen de Normalisation
    (CEN) standards 317–18
commercial public databases 274–5, 277
compatibilisers see coupling agents
compliance 168, 169–70
    non-linear creep 173–4
composite cylinder assemblage (CCA)
    models 127–9
compounding technologies 79–90,
    228–9
    extruder systems see extrusion
    hot-cold mixers 80–1

compression moulding 225, 229
Conex Wood Extruder 313
confocal microscopy 52–4
construction *see* building industry
consumer goods 305, 308, 310, 327
contact angle 44
continuous double belt press 312
continuous fibre models 127
continuum approach 123
contraction, thermal 337–9
conveying 74–5
conveying elements 84, 86
copolymers 3–4
corotating twin screw extruder 79, 80, 81, 83–90
cost 34
co-stabilisers 33
counter-rotating twin screw extruder 79, 80, 81, 82–3
coupling agents 24, 29–30, 41, 42
 and creep 181–2, 186
 foamed WPCs 231, 251
 improving interface interactions 60–6, 69
 optimisation of mechanical properties 113–15
 prevention of moisture absorption 156
 and rheology 194, 196
creep 150, 157, 166–89
 creep failure and material damage 183–5
  effect of applied load 183–5
  humidity effects 185
 creep-rupture performance 263–4
 future trends 185–6
 interphase modification 67
 non-linear 173–4
 polymers 7, 178–9
 viscoelasticity and 167–76
  experimental methods 174–6
  theoretical background 167–74
 in WPCs 176–83
  filler concentration and temperature effects 177–81
  interfacial adhesion 181–3
creep modulus 168
critical fibre length 103, 136
crosslinks 4
 and creep resistance 67, 182–3, 186

crumbling 347–50
crystalline melting point 5
crystalline silicates 34
crystallinity 247, 248
crystallisation temperature 247, 248
cups 308, 327

DaimlerChrysler 296
damage, creep failure and 183–5
damping factor (tan δ) 58, 59
damping index 64
databases, public 274–5, 277
decay
 wood 144
 WPCs 152–4, 160–1
deck boards 268, 270
decking 186, 266, 331–53
 appearance 308–9
 brands and manufacturers 332–3
 Europe 303, 304, 305, 306, 311, 328
 improving performance 333–52
  flammability 346–7
  mechanical performance 333–7
  microbial degradation 343–5
  oxidation and crumbling 347–50
  photo-oxidation and fading 350–2
  shrinkage 339–41
  slip resistance 341–2
  termite resistance 345–6
  thermal expansion-contraction 337–9
  water absorption, swell and buckling 342–3
 deflection 335–7
degassing 88, 90
degree of polymerisation 3
density
 decreased and oxidation 348
 density reduction additives 24, 25, 36
 extruded WPCs 202–3
 polymers 9
 wood 15–16
dent resistance 107–8
diameter ratio 84, 85
dicumyl peroxide 62, 63
die drawing 210, 215–19
differential scanning calorimetry (DSC) 60
diffusion
 diffusivity of foamed WPCs 247, 248

interdiffusion 46, 47–8
direct extrusion 191, 192
directly printable surfaces 313, 315
dishwasher safe products 308, 327
dispersion 43–8
doors 266, 305
double belt press 99–100
draw ratio 209
  and tensile strength 210, 211
drumsticks 212–13
dry blending 191
dry cutting pelletisers 94–5
dry mixing 191–2
durability 142–65
  biological attack 152–5, 160–1
  changes in WPCs with exposure 145–55
  characteristics of raw materials 142–5
    polymers 144–5
    wood 17–18, 142–4
  future trends 161–2
  methods for protection 155–61
  moisture effects 145–9, 155–6, 161
  thermal changes 149–50, 156–7
  weathering 150–2, 157–60
dynamic mechanical thermal analysis (DMTA) 58–60

EBS (ethylene *bis* stearamide) 27
eccentric pelletiser 94–5
Ecoindicator95 method 281, 282, 294
Ecoindicator99 method (EI99E/E) 280–2, 285–91
economical analysis 323
elastic moduli
  flexural modulus *see* flexural modulus
  kappa number and 130–1
  modelling in-plane Young's moduli 123–5
  prediction of stiffness 128–9
  tensile modulus *see* tensile modulus
  weathering and 151–2, 157, 158, 159–60
elastic properties 119–31
  approach to determine reinforcement efficiency of wood fibres 129–31
  material microstructure 120–3
  micromechanical modelling for stiffness prediction 125–9

electron probe microanalysis (EPMA) 49–50, 51
electron spectroscopy for chemical analysis (ESCA) 48–9
electrostatic bonding 45–7
elongation at break
  polymers 9
  WPCs 111, 112
end of life 280
  EU Directive on end-of-life vehicles (ELVs) 295–6
  impact assessment 286–9, 293, 294
endothermic chemical blowing agents 240, 241
energy use 282
environmental SEM (ESEM) 51–2
epoxidised oils/resins 33
EPS 2000 method 281, 282, 295
Eshelby's equivalent inclusion result 125
esters 27
ethylene/propylene/diene terpolymer (EPDM) 114–15
Europe 300–30
  development of European market 301–4
  examples of WPC products 325–9
  future trends 309–16
    markets 309–10
    processing and materials 311–16
  list of most important producers 320–2
  most significant WPC products 304–9
  standardisation and quality 317–19
European Union Directive on ELVs 295–6
evaluation service reports (ESRs) 260, 332
exothermic chemical blowing agents 240, 241
expanded microspheres 249
expansion, thermal *see* thermal expansion
expert networks 324
extensional viscosity 232, 233
external lubricants 26, 195
extractives 12
extrusion 157, 158, 190–207, 229
  commercial WPCs 197–207
    extrusion melt pressure 198–201
    extrusion parameter influences on composite properties 201–7
    output rate 197–8, 199

current extrusion processing methods 191–3
extruder systems 79–90
　corotating twin screw 79, 80, 81, 83–90
　counter-rotating twin screw 79, 80, 81, 82–3
　single screw 79, 80, 81, 82, 192
foamed WPCs
　chemical blowing agents 242–3
　phase changes 235–6
　physical blowing agents 245–8
　rheology of an extruded WPC 193–6
　　influence of formulation components on 194–6
　　influence on product quality 196

fading 151–2, 159–60, 350–2
failure modes 43
falling ball impact resistance test 108
fasteners 338
　holding strength tests 108–9
feasibility studies 323
Federation of Reinforced Plastics (AVK) 320
feeding systems 75–9, 191–2
fencing systems 266, 267, 305
fibre length
　critical 103, 136
　and strength 103, 135–6
fibre orientation 125
　fibre orientation function 124
　and strength 135–6
　interfacial effects 137
fibre weight fraction 293, 294, 297
fillers 24, 25, 33–5
　concentration and creep 177–8
　effect on mechanical properties 109–12
　oriented WPCs 220–4
　　agricultural fibres 224
　　inorganic fillers 220–2
　　reactive fillers 222–4
　from wood 13–15
fines 35
flame retardants (FRs) 24, 25, 38–9, 346–7
flame spread index (FSI) 346
flammability 346–7
flax 19–20

flexural modulus 105
　decking 335, 337
　drumsticks 213
　extruded WPCs 203–5
　influence of raw materials 110, 111
flexural strength 104–5
　decking 333–5
　drumsticks 213
　extruded WPCs 203–4, 205
　freeze-thaw cycling and 264
　influence of raw materials 110
　moisture effects 148, 149
flexural testing 58
　creep 174–5, 176
foam density 229, 230
foamed WPCs 36, 227–56, 315
　additives 228, 250–2
　applications 228
　blowing agent-based foaming mechanisms 235–9
　chemical blowing agents 36, 239–44
　critical issues in production 231–5
　future trends 252
　heat expandable microspheres 249
　materials 228
　physical blowing agents 36, 244–8
　processing methods 228–9
　significance 227–8
　structure characterisation 229–30
　void formation using stretching technology 250
Folkes equation 45
food safe products 308, 327
Ford Motor Co. 296
formulation design
　influence on rheology 194–6
　output rate 197–8, 199
fossil fuels 289, 290, 296
four-point bending test 174–5, 176
Fourier transform infra-red spectroscopy (FTIR) 49, 50, 150–1
fracture energy 45
France 304
free radical scavengers 157–9
freeze-thaw cycling 148, 149, 264
friction 341–2
fuel consumption 279
fuel economy 279
functional unit 275, 276–7

fungal attack 143–4
    decay 144, 152–4, 160–1
    mould 18, 144, 154, 161, 342–5
furniture 305, 307, 310, 311–12, 327

gain-in-weight feeders 191–2
gas injection system 246
gas/polymer solution 236, 237–8
GeoDeck 334, 335–6, 345
Germany 300–30
    development of WPC market 301–4
    industry associations 317, 319–20, 321
    list of most important producers 320–2
    Quality Label Wooden Composite 318
    universities and research institutes 322–5
glass fibres 19, 20, 118–19
    glass fibre-reinforced polypropylene car door panels 273–99
    impact of producing 278, 285–6
glass transition temperature 5
grades, polymer 10
gravimetric feeding 75, 76
greenhouse gases 283, 289, 290, 296
gypsum 222, 223

halogenated flame retardants 38–40
Halpin-Tsai equations 126
hardness testing 107–8
hardwoods 10, 11
    flooring 213–14
    and mechanical properties of WPC 111, 112
hazard triangle 74
heat stabilisers 32–3
heavy metals 34, 284, 290, 291
hemicellulose 11, 12, 102
high-density polyethylene (HDPE) 4, 5
hindered amine light stabilisers (HALS) 32, 158–9
hollow profile decking 311
Hooke's law 167–8
hot-cold mixers 80–1
hot face pelletising 90, 91–5
hot pressing preform 120–2
hydroforming 225
hygroexpansion 131–4
    determination for wood fibres 133–4

free deformation due to change in moisture content 132–3
Hype®wood 313–15

Ikea 303, 307, 311
impact assessment 276, 285–91
    assembly of panels 286, 287, 288–9, 293, 294
    end of life 286–9, 293, 294
    life-cycle comparisons 288–9
    methods 280–2
        sensitivity analysis 293–5
    production of fibre 285–6
    substances' contribution to impact categories 289–91
impact energy 107
impact modifiers 66, 114–15
impact strength 107
    effect of raw materials 111, 112
impact tests 58, 106–7
in-line compounding 98–9, 100
    and injection moulding 97–8
incineration 280, 292–3
indentation tests 107–8
industry associations 317, 319–20, 321
industry consulting 324
infra-red (IR) spectroscopy 49
injection moulding 95–8, 100, 229, 279, 303, 316
    foamed WPCs 243–4
    in-line compounding 97–8
    oriented WPCs 225
    standard injection with pre-compounded WPC 96–7
    weathering performance 157, 158
Innovationsberatung Holz & Fasern 324–5
inorganic fillers 220–2
inorganic lubricants 28–9
insects 37, 144, 154–5
    termite resistance of decking 345–6
interdiffusion 46, 47–8
interface 41–71
    adhesion and creep 181–3
    coupling agents and 30, 41, 42, 60–6, 69
    effects on strength 136–7
    evaluation of interfacial interactions 48–60

improving interface interactions 60–6
interphase and 42–3
modification and mechanical properties
    113–15
wetting, adhesion and dispersion 43–8
interfacial shear strength (IFSS) 55–8
and fibre length 103, 135–6
intermeshing counter-rotating twin screw
    79, 80, 82–3
internal lubricants 26, 195
International Building Code 257–8
International Code Council (ICC) 257–8,
    332
    Evaluation Service (ICC-ES) 258, 259,
        260, 332
interphase 41–71
    effects on other properties 66–8
    interface and 42–3
    wetting, adhesion and dispersion 43–8
intumescent flame retardants 39
inventory 276, 282–4
inverse gas chromatography (IGC) 50–1
isocyanates 62, 63
isothiazole 161
Italy 304
Izod impact test 106–7

Janka test 107

kappa number 129–31
Kelvin-Voigt element 168–9
kinetic mixers (K-mixers) 192–3
kneading blocks 86
knowledge transfer 323–4
Korte, Hans 324–5

lamellae 6
laminate analogy 123–33
landfill 280
landscape architecture 224
laser confocal scanning microscopy 52–4
laser Raman spectroscopy 57–8
life-cycle assessment (LCA) 273–99
    allocation 280
    assembly 278–9, 286, 287, 288–9, 293,
        294
    end of life 280, 286–9, 293, 294
    functional unit and system boundaries
        275–7

goals 275, 276
impact assessment 276, 280–2, 285–91
inventory 276, 282–4
life-cycle comparisons 288–9
possible effect of EU Directive on end-
    of-life vehicles 295–6
process 274–6
    data collection 274–5
    modelling 275–6
sensitivity analysis 291–5
sources of data 277
use phase 279, 288–9, 293, 294
Lifshitz-van der Waals (LW) forces 47
lightness 151–2, 159–60, 350–2
lignin 11, 12, 13, 102
    kappa number 129–31
linear low-density polyethylene (LLDPE)
    4, 5
linear viscoelasticity 167–70
long-term creep 179–81
loss-in-weight feeders 75, 76–7, 78,
    191–2
loss modulus 58
low-density polyethylene (LDPE) 4, 5
low-density WPCs 215–19
lubricants 24, 25, 26–9, 195, 196, 250–1
lumens 11
    lumen filling 102, 122

m-isopropoenyl-dimethyisocyanate
    (m-ID) 65
magnesium hydroxide 39–40
maintenance of decking 352–3
maleated ethylene/propylene/diene
    terpolymer (EPDM-MA) 114–15
maleated polyethylene (MAPE) 47, 63
maleated polyolefins 30, 45, 63–5, 69
maleated styrene-ethylene/butylene-
    styrene triblockcopolymer
    (SEBS-MA) 66, 114–15
maleic anhydridepolypropylene
    copolymer (MAPP) 4, 48, 63–5,
    113–15, 181–2, 244
manufacturers
    decking 332–3
    European 320–2
manufacturing technologies 72–100, 225
    compounding technologies 79–90,
        228–9

foamed WPCs 228–9
future trends 100
  in Europe 311–16
  injection moulding see injection moulding
  orientation of polymers 209–10
  pelletising systems 90–5, 192–3
  profile extrusion 95, 96, 192–3
  protection against moisture absorption 155
  protection against weathering 157, 158
  raw material handling 72–9
  sheet extrusion 98–100
  wood fibre requirements 72
marine borers 144, 155
market research 322–3
marketing support 323–4
mean field methods 125
mechanical bonding 46, 47
mechanical conveying systems 75
mechanical properties 101–17, 261
  anisotropic nature of wood fibres and role of polymer 101–2
  critical parameters affecting 109–15
    effect of filler and polymer 109–12
    optimisation by interfacial modification 113–15
  foamed WPCs 230
  improving performance of decking 333–7
  interfacial shear strength and fibre length 103
  micromechanical modelling see micromechanical modelling
  polymers 7, 8, 9
  test methods 58, 59, 104–9
  wood 19–20
  see also under individual properties
melt blenders 192–3
melt pressure 198–201
metal airborne emissions 283, 284, 290, 291
metal hydroxides 39–40
methacrylic acid (MAA) 61
methane 283, 289, 290
microbial degradation see biological degradation; mould
microbond test 54–6
microcellular foams 230

microfibril helix angle 12–13
micromechanical modelling 118–41
  elastic properties 119–31
    determination of reinforcement efficiency of wood fibres 129–31
    in-plane Young's moduli 123–5
    material microstructure 120–3
    stiffness prediction 125–9
  hygroexpansion 131–4
  strength 134–7
microscopy 51–4
microspheres 36, 249
modelling
  LCA 275–6
  micromechanical see micromechanical modelling
  viscoelasticity and creep 167–74
modulus of elasticity see elastic moduli
modulus of rupture (MOR) see flexural strength
moisture see water/moisture absorption
molecular grafts 61
molecular mobility 58–60
molecular weight 3
Mooney equation 193
Mori-Tanaka approach 125–6, 127
mould 18
  decking 342–5
  wood 144
  WPCs 154, 161

nail withdraw resistance 109
nanotechnology 162
  nanocellular foams 230
  nano-clay 252
National Fire Protection Association (NFBA) (US) code 258
natural polymers 2
Newtonian liquids 167–8
nitrogen 240
nitrous oxide 283, 289, 290
non-linear creep 173–4
non-Newtonian fluids 193
nova-Institut 322–4
Novo-Tech 329
number-average molecular weight 3
nylon-wood composites 156–7

off-gassing 201–2

oil price 316
oleo-derived lubricants 26
optical microscopy 52
orientation, fibre *see* fibre orientation
oriented wood fibre mat 120, 121
oriented WPCs 208–26
  applications 212–19, 224–5
    drumsticks 212–13
    hardwood flooring 213–14
    low-density WPCs 215–19
  current developments 219–25
    fillers 220–4
    future trends 225
  orientation of polymers 208–12
    materials 210–12
    processes 209–10
  stretching technology 250, 251
outdoor durability *see* durability
output rate 197–8, 199
oxidation 347–50
oxidation induction time (OIT) 348–50

Palltrusion process 192–3, 312
panel production 279, 312
paraffin wax 62
parallel-to-face tensile test 105
particle size
  distribution and fillers 35
  and moisture absorption 146, 147
  wood flours 14–15
pelletising systems 90–5, 192–3
  hot face pelletising 90, 91–5
  strand pelletiser 90–1, 92
performance measurement 257–72
  and building codes 259–60, 271
  building construction applications 265–70
  properties of WPCs 260–5
perpendicular-to-face orientation test 106
petroleum-derived lubricants 26
phase changes 235–7
phosphate flame retardants 39
photodegradation *see* ultraviolet (UV) radiation
photo-oxidation 350–2
photostabilisers 157–60
physical blowing agents (PBAs) 36, 244–8
  batch process foaming 245

extrusion foaming 245–8
physical property tests 104
pigments 24, 25, 32, 37, 158–60
Pinatubo eruption 264–5
planed surfaces 157, 158
planetary roller extruder 313
plasticisers 28
pneumatic conveying systems 74, 75
polarity mismatch 219
political consulting work 324
polyaminoamide-epichlorohydrin/stearic anhydride 65–6
polydiphenylmethane diisocyanate/stearic anhydride 66
polydispersity index 3
polyethylene (PE) 2, 7–8, 9, 144–5, 210, 302
  additives for PE-based composites 24
  LCA 292, 293, 297
polyethylene terephthalate (PET) 2, 212
polyethylene wax 62
polymer/gas solution 236, 237–8
polymers 1, 2–10
  creep 7
    effects of temperature 178–9
  effect on mechanical properties of WPCs 109–12
  molecular organisation 4–7
  molecular structure 2–4, 5
  orientation of 208–12
  outdoor durability 144–5
  properties 7–10
  role of 101–2
  storage, transportation and conveying 74–5
polypropylene (PP) 2, 7–8, 9, 210–12, 278, 302
  additives for PP-based composites 25
  degradation 144–5
polystyrene (PS) 2
polyvinyl chloride (PVC) 2, 7, 9, 10, 145, 302, 313–15
  additives for PVC-based composites 25
  LCA 292, 293, 296–7
  lubricants 27–8
  PVC heat stabilisers 32–3
porosity 348
Portland cement 222–4

power-law model 170, 177, 178, 179
pre-compounded extrusion 191, 192
pre-compounded WPCs 96–7
price 302–3
primary creep 167
Pro-K 319–20
processing technologies *see*
    manufacturing technologies
product aesthetics additives 24, 25, 37
profile extrusion 95, 96, 192–3
project management 324, 325
Prony series 170
PSAC LLC 219
public databases 274–5, 277
public perception of WPCs 352–3
pull-out test 54–6

quality 317–19
  influence of rheology on product
    quality 196
Quality Label Wooden Composite 318

radial compression 210
ram extrusion 209, 215
random polypropylene copolymer 4
rapid-pressure-drop-rate die 238
Rattlesnake Creek bridge, Montana 268, 270
raw materials
  car doors 278–9
  handling 72–9
    feeding systems 75–9
    storage, transportation and
      conveying 74–5
  see also polymers; wood
reactive fillers 222–4
recovery, creep 170–1
recycling 280, 292–3, 295–6, 297
refilling of feeders 77, 78
regulations 316
  EU Directive on ELVs 295–6
Reimelt-Henschel MischSysteme 312
reinforcements, wood 13–15
research institutes 322–5
residence time, control of 234–5
resin transfer moulding (RTM) 120–2
resource management 323
respiratory inorganics 283, 284, 291
rheology 193–6

influence of formulation components
    on 194–6
  influence on product quality 196
  rheology control additives 24, 25, 26–9
rice husks 224
Rockwell hardness tester 107, 108
roll stack 98–9
roofing 266
rotomoulding 314, 315
Rotowood 314, 315
roughness 47

scanning electron microscope EPMA
    (SEM-EPMA) 50, 51
scanning electron microscopy (SEM) 41,
    42, 51–2, 53, 216–18
scope of LCA 275, 276–7
screw feeder 77
screw holding strength 108–9
SEBS-MA 66, 114–15
secondary creep 167
self-consistent scheme 126
semicrystalline polymers 5, 6–7
sensitivity analysis 291–5
shape 34
shavings, from wood planing 312
shear lag models 126–7
shear rate 193
shear viscosity 194–5
  WPC foams 232, 233
sheet extrusion 98–100
shelving 327
shift factors 171–3
shrinkage 339–41
siding/cladding 224, 266, 305
silanes 45, 62–3, 69
  crosslinking and creep 67, 182–3
sill plate 267–8
Simapro method selector 280–2
Simplex model 260–1
single fibre fragmentation (SFF) test 56–7
single fibre microbond test 54–6
single fibre pull-out test 54–6
single screw extruders 79, 80, 81, 82, 192
single screw feeders 77
slip resistance 341–2
smoke suppressants 24, 25, 38–40
soft-rot fungi 144
softwoods 10, 11

and mechanical properties of WPCs 111, 112
solubility 247, 248
specific torque 84
spherulites 6
stabilisers 31–3
   antioxidants 24, 25, 31–2, 156, 343, 347
   PVC heat stabilisers 32–3
   UV stabilisers 24, 25, 32, 157–60, 251
stair treads 336–7
standard injection moulding 96–7
standards/standardisation 258, 317–19
stearates 27, 30
stearic acid 62
storage of bulk materials 74
storage modulus 58, 59
strand pelletising 90–1, 92
strength
   flexural *see* flexural strength
   impact strength 107, 111, 112
   interfacial shear strength 55–8, 103, 135–6
   micromechanical modelling 134–7
      empirical models 135–6
      interfacial effects 136–7
   tensile *see* tensile strength
   weathering 151–2, 157, 158, 159–60
stress
   allowable design stresses 260
   creep
      effect on creep damage 183–5
      progressive deformation 166–7
   stress-strain curves 64–5, 129, 130, 215–16, 261
stretching technology 250, 251
substitution potential 309, 310
surface analysis 48–51
surface modifications 60, 61, 69
surfaces
   changes during weathering 150–1
   textured 341–2
swelling 148
   decking 342–3
   see also hygroexpansion
synergists 38–9
synthetic polymers 2
synthetic zeolite 33
system boundaries 275–7

tacticity 4
talc 35, 251–2
tan δ (damping factor) 58, 59
tandem extrusion system 245–6
Technopartner Samtronic (TPS) 312
temperature
   critical processing temperatures for WPC foams 234–5
   crystalline melting point 5
   crystalline temperature 247, 248
   effect on creep 178–9
   glass transition temperature 5
   profiles and extrusion output rate 197–8, 199
   properties of polymers and 7, 8
   tensile strength 210, 211
tensile modulus (stiffness)
   effect of MAPP 113, 114
      plus impact modifiers 115
   effect of raw materials 109–10, 111, 112
   polymers 9
   wood 19–20
tensile strength 109
   moisture uptake and MAPP 67–8
   optimisation of mechanical properties
      with MAPP 113, 114
      with MAPP and impact modifiers 115
   polymers 9
   and draw ratio 210, 211
   and temperature 210, 211
   tests 58, 105–6
   wood 19–20
termite resistance 345–6
   see also insects
tertiary creep 167
textured surfaces 341–2
thermal conductivity 8, 9, 104
thermal degradation 149–50
   extruded WPCs 201–2
   foamed WPCs 231
   protection against 156–7
   wood 18
thermal expansion 149, 157
   coefficients of 9, 19, 337–9
   and contraction of decking 337–9
thermal-oxidative degradation 150, 156
thermoforming 225

thermoplastic microspheres 249
thermoplastics 5, 119
thermosets 4–5, 119
thickness swell 148
three-point bending test 104–5, 174–5, 176
time-temperature superposition (TTS) principle 171–3, 186
  long-term creep and 179–81
tin 32–3
top-cut 35
transmission electron microscopy (TEM) 52, 53
transportation of raw materials 74–5, 279
transverse bulk modulus 128
transverse shear modulus 128–9
Trex 333–4, 335, 336, 337, 342, 343
twin screw extruders 79, 80, 81, 82–90, 191
twin screw feeders 77

ultra-high molecular weight polyethylene (UHMWPE) 210
ultraviolet (UV) radiation
  degradation of polymers 144–5
  degradation of wood 17–18, 143
  UV stabilisers 24, 25, 32, 157–60, 251
underwater pelletisers 92–3
unexpanded microspheres 249
uniaxial orientation 208–9
United Kingdom (UK) 304
United States (US) 301
  ASTM specifications 259–60, 263–4
  building codes 257–8
  decking manufacturers 332–3
  demand for decking 331
  standard mesh sizes 14–15
universities 322, 323
use phase 279, 288–9, 293, 294

venting 88–90
venting port 90
Verband der Deutschen Holzwerkstoffindustrie (VHI) 318, 319, 321
vibration feeders 77
vinyltriethoxysilane 65
viscoelasticity 7
  and creep 167–76

  experimental methods 174–6
  theoretical background 167–74
  linear viscoelasticity 167–70
viscosity 193
  wood content and shear viscosity 194–5
  WPC foams 232–4
void formation, by stretching 250, 251
void fraction 229
  effect of CBA types 240, 242
volatile emissions 231–2
  processing temperature and residence time 234–5
volume expansion ratio 229
volumetric feeding 75

wall slip 193, 195, 196
water/moisture absorption 102, 261–2
  decking 342–3
  effect on creep 185
  extruded WPCs 205–7
  foamed WPCs 231–2
  hardwood flooring 214
  hygroexpansion 131–4
  interface/interphase and 67–8
  outdoor durability 145–9
    protection 155–6, 161
  polymers 9
  water soak test 104
  wood 16–17
water ring pelletisers 93–4
waterfront facilities 265–6
waxes 24, 25, 26–9
weathering
  wood 144–5
  WPCs 150–2
    protection 157–60
weight-average molecular weight 3
weight belt feeders 77–9
wetting 43–8
white-rot fungi 144, 153–4
Williams–Landel–Ferry (WLF) equation 173
windows 266, 305
wood 1, 10–20
  availability 316
  chemical constituents 11–13
  content
    and creep 177–8

and rheology of WPCs 194–5, 196
effect of wood filler on mechanical properties of WPCs 109–12
properties 15–20
  density 15–16
  durability 17–18, 142–4
  mechanical 19–20
  moisture absorption 16–17
  thermal 18–19
sources and production of fillers and reinforcements from 13–15
structural applications compared with WPCs 268, 269
wood anatomy 10–11
wood chips 301
wood fibres 11, 13, 19, 72–3, 119, 301
  acetylation to protect against moisture absorption 155
  anisotropic nature of 101–2
  bulk density 16, 73
  concentration
    and creep 177–8
  and die drawing orientation 216
  determining reinforcement efficiency of 129–31
  dispersion in WPC foams 231
  hygroexpansion 133–4
  impact of producing 278–9, 285–6
  requirements for manufacturing system 72
  storage, transportation and conveying of bulk material 74, 75
wood flour 13, 14–15, 17, 18, 20, 301
  content and moisture absorption 146, 147
  density 15, 16
work of adhesion 44–7

Young's equation 44
Young's modulus *see* elastic moduli

zinc borate 37, 160–1
ZSK corotating twin screw extruder 83–90